AF879377

Sandeep Singh, Bharat Bhushan, Sandeep Malik, Aliakbar Montazer Haghighi, Ömer Kişi (Eds.)

Advanced Computational Techniques in Mathematics

Also of Interest

Inverse Problems on Large Scales
Mathematical Modelling and Computational Methods
Edited by Bochra Mejri, Ronny Ramlau, Otmar Scherzer, 2025
ISBN 978-3-11-135600-6, e-ISBN (PDF) 978-3-11-135727-0

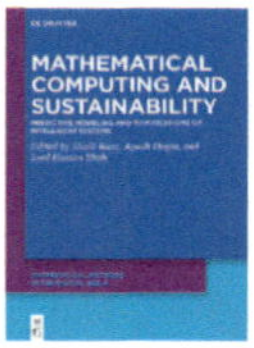

Mathematical Computing and Sustainability
Predictive Modeling and Ramifications of Intelligent Systems
Edited by Shalli Rani, Syed Hassan Shah, Ayush Dogra, 2025
ISBN 978-3-11-161111-2, e-ISBN (PDF) 978-3-11-161203-4

Evolutionary Dynamic Equations
Stabilizability, Observability, and Controllability
Akram Ben Aissa, Khaled Zennir, 2025
ISBN 978-3-11-914539-8, e-ISBN (PDF) 978-3-11-221435-0

Nonlinear Analysis and Computational Techniques
Proceedings of the ICNACT-2024 Conference Held During 8-10 August, 2024
Edited by Hemant Kumar Nashine, Ranis Ibragimov, Hemanta Kalita, 2025
ISBN 978-3-11-172455-3, e-ISBN (PDF) 978-3-11-172463-8

Observability and Mathematics Modeling
Hilbert, Euclid, Gauss-Bolyai-Lobachevsky, and Riemann Geometries
Nikolai Khots, Boris Khots, 2025
ISBN 978-3-11-167946-4, e-ISBN (PDF) 978-3-11-168048-4

Fixed Point Theory and Functional Analysis
Metric Spaces, Banach Spaces, Hilbert Spaces
Edited by Pradip Debnath, Hari Mohan Srivastava, Yeol Je Cho, 2026
ISBN 978-3-11-914503-9, e-ISBN (PDF) 978-3-11-221576-0

Advanced Computational Techniques in Mathematics

Conference Proceedings ICEPAM-2024

Edited by
Sandeep Singh, Bharat Bhushan, Sandeep Malik,
Aliakbar Montazer Haghighi, Ömer Kişi

DE GRUYTER

Mathematics Subject Classification 2020
Primary: 34-01, 35-01; Secondary: 35C08, 65L20

Editors

Sandeep Singh
Department of Mathematics
Akal University, Talwandi Sabo
Bathinda, Punjab 151302
India
sandeepinsan86@gmail.com

Bharat Bhushan
Department of Mathematics
Akal University, Talwandi Sabo
Bathinda, Punjab 151302
India
bgoyal.math@gmail.com

Sandeep Malik
Department of Mathematics
Akal University, Talwandi Sabo
Bathinda, Punjab 151302
India
sndp796@gmail.com

Aliakbar Montazer Haghighi
Department of Mathematics
Prairie View A&M University
Prairie View, Texas 77446-0519
USA
amhaghighi@pvamu.edu

Ömer Kişi
Department of Mathematics
Bartin University
Bartin 74100
Turkey
okisi@bartin.edu.tr

ISBN 978-3-11-223451-8
ISBN 978-3-11-223452-5 (PDF)
ISBN 978-3-11-223453-2 (E-PUB)
DOI https://doi.org/10.1515/9783112234525

Library of Congress Control Number: 2026933918

Bibliographic information published by the Deutsche Nationalbibliothek
The Deutsche Nationalbibliothek lists this publication in the Deutsche Nationalbibliografie; detailed bibliographic data are available on the Internet at https://dnb.dnb.de.

© 2026 Walter de Gruyter GmbH, Genthiner Straße 13, 10785 Berlin, Germany

All rights reserved. No part of this publication may be reproduced, translated, stored in a retrieval system, or transmitted in any form or by any means, electronic, mechanical, photocopying, recording or otherwise, without prior written permission from the publisher. Text and datamining for commercial purposes requires the publisher's permission.

De Gruyter and Walter de Gruyter GmbH are part of De Gruyter Brill.
www.degruyterbrill.com

Questions about General Product Safety Regulation: productsafety@degruyterbrill.com

Cover illustration: gremlin / E+ / Getty Images
Printing: CPI books GmbH, Leck

MIX
Papier | Fördert gute Waldnutzung
FSC® C083411

Dedicated to God

Contents

Preface

The present volume brings together a diverse yet interconnected collection of contemporary research works that reflect the breadth, depth, and evolving nature of mathematical, computational, and applied sciences. The chapters assembled here are unified by a common objective: to employ rigorous mathematical modeling, analytical techniques, and computational methods to address complex problems arising in science, engineering, biology, and social systems.

Several contributions in this volume explore mathematical models motivated by real-world phenomena in health, biology, and ecology. These studies demonstrate how mathematical frameworks ranging from differential equations and fractional calculus to stability and bifurcation analysis can provide meaningful insights into the dynamics of infectious diseases, population interactions, ecological toxicity, and predator–prey systems. Such models not only deepen theoretical understanding but also support decision-making related to disease control, environmental protection, and resource management.

Another significant theme of this book lies in nonlinear science and fractional-order systems. Several chapters are devoted to the analysis of nonlinear partial differential equations that arise in fluid dynamics, wave propagation, plasma physics, and optical systems. By employing modern analytical tools such as Lie symmetry analysis, expansion methods, homotopy-based techniques, and spline-based numerical schemes, the authors derive exact solutions, investigate stability, and validate results through numerical simulations. These contributions highlight the power and versatility of fractional and nonlinear modeling in capturing memory effects, dispersion, and complex system behavior.

The volume also reflects the growing importance of optimization, computation, and intelligent algorithms in contemporary research. Several chapters present advanced optimization techniques, bioinspired algorithms, and scheduling models, emphasizing efficiency, convergence, and comparative performance. These studies are particularly relevant in an era where large-scale and complex optimization problems are increasingly common in engineering, logistics, and data-driven applications.

Interdisciplinary perspectives are further strengthened by contributions in graph theory, social network analysis, fuzzy systems, algebraic structures, and geometric analysis. These chapters illustrate how abstract mathematical concepts such as graph invariants, fuzzy ideals, neutrosophic logic, and geometric connections can be developed systematically and applied to model uncertainty, structure, and interactions in both theoretical and applied contexts.

In addition, the volume addresses emerging challenges in data analysis, pattern recognition, and security applications. Research on image processing, face recognition, and feature extraction demonstrates how mathematical theory, combined with computational intelligence, can enhance accuracy, robustness, and efficiency in real-world systems.

https://doi.org/10.1515/9783112234525-204

Taken as a whole, this book serves as a platform for researchers, academicians, and graduate students to engage with current advances across multiple domains of applied mathematics and computational science. The diversity of topics reflects the interdisciplinary nature of modern research, while the underlying mathematical rigor ensures coherence and scholarly depth. We hope that this volume will stimulate further research, encourage crossdisciplinary collaboration, and contribute meaningfully to the advancement of both theory and application.

The Editors

About the editors

Dr. Sandeep Singh

is an Associate Professor in the Department of Mathematics at Akal University, Talwandi Sabo, Bathinda, Punjab, India. He earned his PhD in Group Theory from Thapar University, Patiala, and was a postdoctoral fellow at IIT Roorkee under the SERB-NPDF program. Dr. Singh's research focuses on Group Theory, Automorphism Groups, Number Theory, Sum-Set Problems, and Optimization Techniques. He has published over 30 research articles in reputed international journals and edited four books. A recipient of CSIR Junior and Senior Research Fellowships, he has presented his work at numerous international conferences with travel support from DST, India. With over a decade of teaching and research experience, he has supervised three PhD students and is currently guiding four more, along with mentoring several Master's students. E-mail: chahal.maths@gmail.com.

Dr. Bharat Bhushan

is currently working as an Assistant Professor in the Department of Mathematics at Akal University, Talwandi Sabo, Punjab, India. Dr. Bhushan completed his PhD in Mathematics from Punjabi University, Patiala under Dr. Deepak Kumar. His research interests include Algebra, Derivations in Rings, commutative algebra with endomorphisms, and differential equations. An experienced educator, he has taught courses such as Linear Algebra, Modern Algebra, Real Analysis, and Mathematics for Artificial Intelligence, and has multiple publications in SCI and Scopus-indexed journals. E-mail: bgoyal.math@gmail.com.

Dr. Sandeep Malik

is an Assistant Professor in the Department of Mathematics at Akal University, Talwandi Sabo, Bathinda, Punjab, India. He obtained his PhD in Applied Mathematics from the Central University of Punjab, Bathinda, Punjab. Dr. Malik has actively participated in numerous international conferences and workshops. He has authored more than 50 research articles published in peer-reviewed journals, with his work receiving over 1,100 citations and an H-index of 20 (Scopus). Dr. Malik's research interests include investigating the integrability, phase portraits, bifurcation analysis, and derivation of the soliton solutions of nonlinear partial differential equations. E-mail: sndp796@gmail.com.

Prof. Aliakbar Montazer Haghighi

is a Professor of Mathematics at Prairie View A&M University, Texas, USA. He is the cofounder and Editor-in-Chief of Applications and Applied Mathematics: An International Journal (AAM). He earned his PhD from Case Western Reserve University and has held academic positions at Prairie View A&M University, Benedict College, and the National University of Iran. Prof. Haghighi served as Head of the Mathematics Department (2002–2021) and contributed to various academic committees. He has received multiple awards, including the Excellence in Research Award and the Life-Time Honor Award from Marquis Who's Who. He has authored several books with Springer and CRC Press and is a life member of the American Mathematical Society. His research is supported by various grants. E-mail: amhaghighi@pvamu.edu.

https://doi.org/10.1515/9783112234525-205

Prof. Ömer Kişi

is a distinguished mathematician affiliated with Bartın University, Turkey. His research interests lie in operator algebras, functional analysis, and sequence spaces, with a particular focus on statistical convergence, fuzzy normed spaces, and neutrosophic analysis. He has made significant contributions to the field through numerous publications, including studies on arithmetic statistically convergent double sequences in intuitionistic fuzzy normed spaces and neutrosophic I-statistically convergent sequences. His work has played a crucial role in advancing the mathematical understanding of these complex topics. E-mail: Okisi@bartin.edu.tr.

Sonia Goel, Khushboo Malhotra, Deepak Gupta, and A. K. Tripathi

Comparative analysis of optimization techniques for a two-stage flow shop scheduling model

Abstract: This study examines the optimization of a two-stage flow shop scheduling model consisting of identical parallel machines in the first stage and a single machine in the second stage. The proposed Branch and Bound method is employed to solve the model, and its performance is compared with that of the Cuckoo Search algorithm and Johnson's algorithm.

Keywords: Cuckoo Search algorithm, Branch and Bound algorithm, makespan, identical parallel machines

MSC 2020: 90B35, 90C27, 90C59, 68T20, 90C10

1 Introduction

Flow shop scheduling is a fundamental problem in manufacturing and production systems that involves sequencing jobs through a series of machines or processes to optimize performance criteria such as makespan. The two-stage flow shop scheduling problem, although simpler than multistage variants, remains significant due to its NP-hard nature. Classical approaches, such as Johnson's algorithm, provide optimal solutions for specific problem settings, whereas modern heuristic methods, including the Cuckoo Search algorithm (CS), are capable of generating near-optimal solutions with reduced computational effort. The Branch and Bound (BB) method, a well-established exact optimization technique, guarantees optimal solutions but often requires substantial computational resources. This study presents a comparative analysis of these methods to evaluate their effectiveness, efficiency, and practical limitations. The novelty of this study is reflected in application of the proposed branch and bound method for large-scale scheduling problems. Computational results have been produced using a MATLAB tool.

Acknowledgement: The authors are highly thankful to Mr. Chhering Dorje for his diligent efforts in technical assistance.

Sonia Goel, Deepak Gupta, A. K. Tripathi, Department of Mathematics, MMEC, MM (Deemed to be a University), Mullana, Ambala, Haryana, India
Khushboo Malhotra, Department of Mathematics, D. A. V. College (Lahore), Ambala City, Haryana, India, e-mail: khushi19nov88@gmail.com

https://doi.org/10.1515/9783112234525-001

2 Practical significance

Healthcare facilities: Hospitals and healthcare institutions face complex scheduling challenges involving the allocation of resources such as operating rooms, medical equipment, and personnel to effectively meet patient demands. By applying the optimization techniques proposed in this study, healthcare providers can enhance scheduling efficiency, reduce patient waiting times, and improve the utilization of medical resources.

Construction projects: Construction projects require the coordination of multiple activities, resources, and interdependencies to ensure timely completion within budgetary constraints. The application of the proposed scheduling approaches can assist construction firms in improving project planning, reducing delays, and optimizing resource utilization, thereby achieving cost savings and increased profitability.

3 Literature review

The proposed scheduling model has attracted considerable attention in the fields of operations research and manufacturing optimization due to its practical applicability in a wide range of industrial settings. The primary objective of such scheduling problems is generally to minimize the makespan. To address this optimization challenge, numerous algorithms have been proposed and investigated, including classical methods such as Johnson's algorithm, exact techniques like the Branch and Bound (B&B) method, and metaheuristic approaches such as the Cuckoo Search (CS) algorithm, Ant Colony Optimization (ACO), the Genetic Algorithm (GA), and Particle Swarm Optimization (PSO). Johnson's algorithm, introduced by S. M. Johnson in 1954, is a classical approach for solving two-machine flow shop scheduling problems with machines arranged in series. The algorithm is simple and computationally efficient, operating in polynomial time, which makes it particularly suitable for small- to medium-sized problems. However, its effectiveness decreases in more complex environments involving parallel machines or additional constraints [9]. The Branch and Bound method is an exact optimization technique that systematically explores the solution space by dividing it into subproblems and applying bounds to eliminate nonoptimal solutions. While this method guarantees an optimal solution by exhaustively evaluating all possible job sequences, its computational complexity increases exponentially with the number of jobs, limiting its applicability to small and moderately sized problems. Several researchers have applied the B&B approach to single-machine scheduling problems [8, 12]. Extensions of this technique to two-stage flow shop scheduling models with parallel machines have also been investigated by various authors [4–6, 14, 17]. We acknowledge that the Branch and Bound (BB) technique has been previously applied to flow shop scheduling problems, including the studies reported in [5, 8]. However, the motivation and contribution of the present work differ from the existing BB-based approaches in several important aspects. First, the present study focuses on a two-stage flow shop scheduling model with identical parallel machines in the first stage and a single machine in the second stage, which introduces additional combinatorial complexity compared to the models addressed in [8] and [5]. While earlier works primarily concentrate on the application of BB as an exact solution technique, the current work emphasizes the comparative performance evaluation of the BB method against both a classical algorithm (Johnson's algorithm) and a modern metaheuristic approach (Cuckoo Search algorithm) under a unified problem setting. Second, the proposed BB framework in this study is employed not only to obtain optimal schedules for small- and medium-sized problem instances but also to establish benchmark optimal solutions that facilitate a meaningful comparison with heuristic and metaheuristic methods. This comparative perspective is not the primary focus of the earlier studies. Third, the present work provides an extensive computational analysis for a wide range of job sizes, including large-scale instances, to examine the scalability, efficiency, and practical limitations of the BB method relative to heuristic approaches. This comprehensive comparison highlights the tradeoff between optimality and computational effort, thereby offering additional insights into the practical applicability of BB-based solutions in real-world scheduling environments. Finally, the study contributes to the literature by consolidating classical, exact, and metaheuristic methods within a single analytical framework, which helps practitioners and researchers better understand the relative strengths and limitations of each approach for two-stage flow shop scheduling problems with parallel machines. Furthermore, studies addressing three-stage flow shop scheduling models using the B&B method have been reported in the literature [3, 15, 16]. Metaheuristic algorithms, particularly the Cuckoo Search algorithm, have emerged as effective tools for solving complex and large-scale scheduling problems. The CS algorithm combines random search mechanisms with Lévy flight behavior to efficiently explore the solution space, enabling it to escape local optima and converge toward global solutions more effectively than traditional heuristics. Due to its simplicity and robustness, CS has been successfully applied to a variety of optimization problems, including scheduling [10, 19–21, 23]. In addition, several other heuristic and metaheuristic techniques, such as ACO, PSO, and GA, have been widely employed in scheduling theory to solve different problem formulations. Relevant studies on these approaches can be found in [1, 2, 7, 11, 18, 22].

4 Problem construction

Consider a scheduling environment consisting of n jobs to be processed on two types of machines, denoted as V and W. Machine type V comprises mmm identical parallel processors, whereas machine type W consists of a single processor. The processing times of job iii on machine types V and W are represented by vi and wi, respectively. Each job is not required to be processed on all parallel processors of type V; instead, it may be partially processed on one or more processors at the first stage. Upon completion of processing at the first stage, the job proceeds to the second-stage machine W. The available processing time for each identical processor, denoted by $n1j$, is assumed to be known. The objective of the scheduling problem is to determine an optimal job sequence that minimizes the total project completion time (makespan). The mathematical formulation of the model is presented in matrix form in Table 1.

4.1 Mathematical form of the model

The model of the problem depicted in matrix form is given in Table 1.

Table 1: Proposed mathematical model.

Jobs	**Machine V**		**Machine W**
I	$V_1\ V_2\ V_3\ V_4\ V_5 \ldots V_m$	**Utilization time on machine V in h (v_i)**	**Utilization time on machine W in h (w_i)**
1	$v_{11}\ v_{12}\ v_{13}\ v_{14}\ v_{15} \ldots v_{1m}$	v_1	w_1
2	$v_{21}\ v_{22}\ v_{23}\ v_{24}\ v_{25} \ldots v_{2m}$	v_2	w_2
3	$v_{31}\ v_{32}\ v_{33}\ v_{34}\ v_{35} \ldots v_{3m}$	v_3	w_3
$\vdots$	$\vdots$	$\vdots$	$\vdots$
n	$v_{n1}\ v_{n2}\ v_{n3}\ v_{n4}\ v_{n5} \ldots v_{nm}$	v_n	w_n
Available time	$n_{11}\ n_{12}\ n_{13}\ n_{14}\ n_{15} \ldots n_{1m}$		

5 Solution approach

5.1 Branch and bound method

We solve this problem as follows:

Step 1:

Check the condition

$$\sum_{j=1}^{m} n_{1j} = \sum_{i=1}^{n} v_i \ldots \tag{1}$$

and apply the Modified Distribution method, otherwise make arrangements to satisfy condition (1). The Modified Distribution Method is employed to allocate the processing time of each job among the identical parallel machines at the first stage in such a way that the feasibility condition of the scheduling model is satisfied. Since each job is not required to be processed on all parallel machines, this method distributes the processing load across the available processors while respecting the given utilization constraints. The method ensures that the total assigned processing time on the parallel machines does not exceed the available processing capacity of each machine. Once the feasibility condition is achieved, the resulting distributed processing times serve as input for the Branch and Bound procedure to determine the optimal job sequence.

Step 2:

Apply the formulas

a)

$$m' = \max_{i \in J_r} \left(\sum_{i=1}^{n} v_{ij} \right) + \min_{i \in J_{r'}} (w_i) \tag{2}$$

$$m'' = \max_{i \in J_r} (v_{ij}) + \sum_{i=1}^{n} (w_i). \tag{3}$$

b) Find $m = \max\{m', m''\}$.
c) Find m for all permutation of tasks.
d) Select the minimum value from all the available mmm values. This will establish the basis for creating the separating tree.
e) Construct the output table corresponding to the desired task sequence.

5.1.1 Flow chart of the branch and bound method

The flow chart showing the Branch and Bound Method is given in Figure 1.

5.2 Cuckoo search algorithm

The CS algorithm is a population-based heuristic approach for finding the near-optimal solution of a performance measure.

The flow chart of the Cuckoo Search Algorithm is shown in Figure 2.

5.3 Johnson's algorithm

Johnson's algorithm is a classical sequencing rule developed for two-machine flow shop scheduling problems with the objective of minimizing the makespan. The algorithm operates by partitioning jobs into two sets based on their processing times on the first and second machines. Jobs with shorter processing times on the first machine are scheduled earlier in ascending order, whereas jobs with shorter processing times on the second machine are scheduled later in descending order. Owing to its simplicity and optimality for two-machine flow shop problems, Johnson's algorithm is employed in this study as

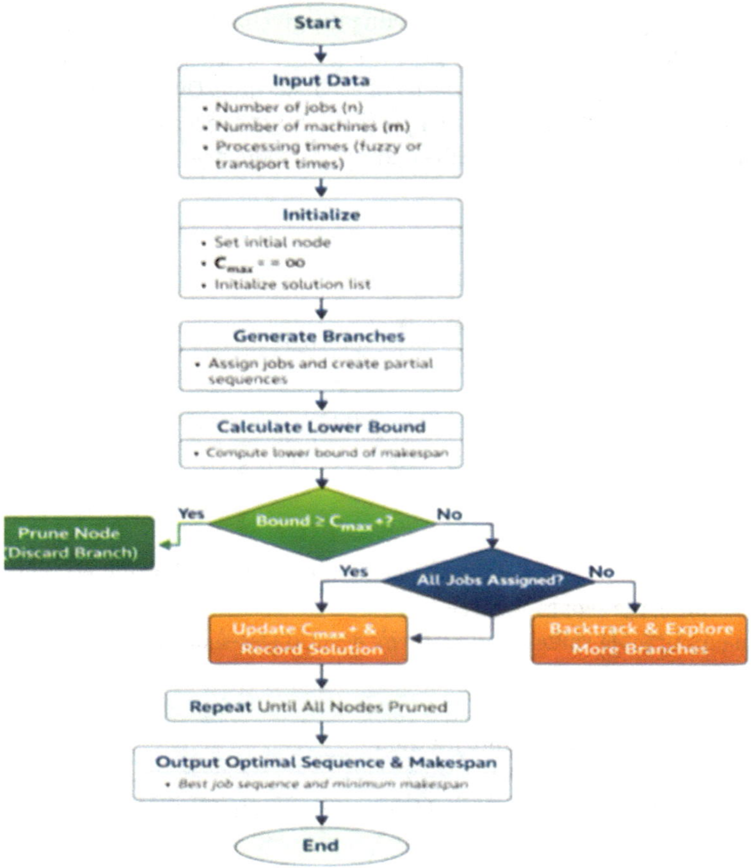

Figure 1: Branch and bound methodology.

a benchmark to evaluate the performance of the proposed Branch and Bound method and the Cuckoo Search algorithm.

5.4 Numerical approach

The illustration for supporting the proposed model is given in Table 2.

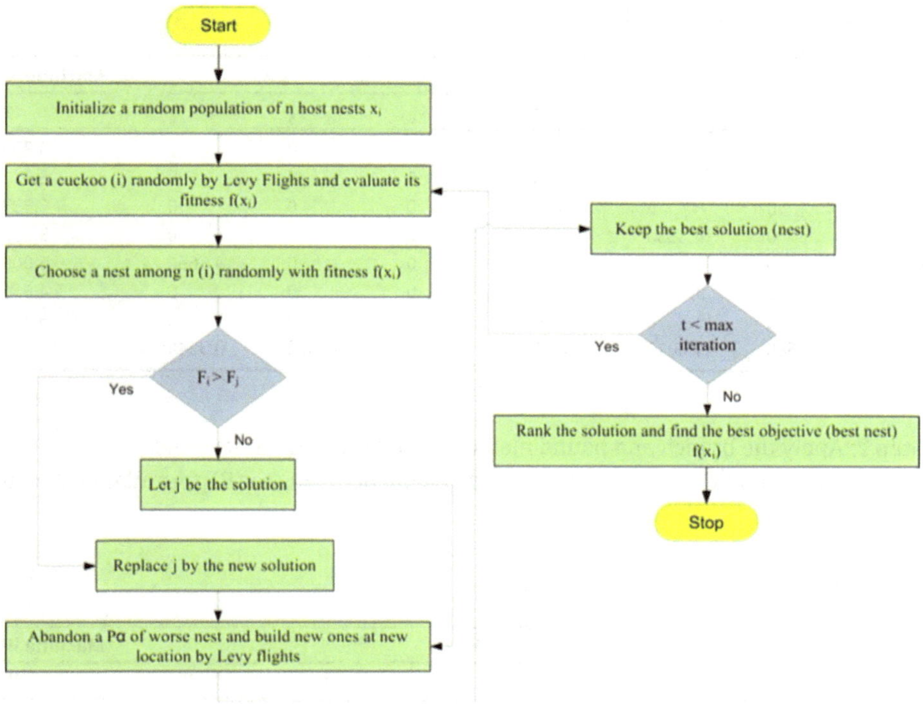

Figure 2: Flow chart showing the cuckoo search algorithm [13].

Table 2: Illustration showing the proposed model.

Jobs	**Machine *V***							**Machine *W***
I	V_1	V_2	V_3	V_4	V_5	V_6	**Processing time on machine *V* in h (v_i)**	**Processing time on machine *W* in h (w_i)**
1	1	6	6	6	7	7	3.666	3.333
2	5	6	7	8	9	9	2.666	6
3	2	2	2	3	3	2	3	5.333
4	5	6	7	8	9	9	4.333	5.333
5	2	2	3	3	3	2	4.666	6
6	5	6	7	8	9	9	3.666	4.666
7	5	6	7	8	9	9	4.333	5.333
n_{1j}	9	9	8	7	6	5		

5.4.1 Solution by the BB algorithm

We solve the problem in two steps:

Step 1: See the Modified Distribution method in Table 3:

Table 3: Modified distribution method.

Jobs	Machine *V*						Machine *W*
I	V_1	V_2	V_3	V_4	V_5	V_6	*W*
1	3.6660	0	0	0	0	0	3.333
2	2.6660	0	0	0	0	0	6.000
3	0	0	3.0000	0	0	0	5.333
4	0	4.3330	0	0	0	0	5.333
5	0	0	0	0	0	4.6660	6.000
6	0	1.6640	1.0010	0	0	0	4.666
7	0	0	4.3330	0	0	0	5.333
8	2.6680	1.0010	0.6670	7.0000	6.0000	0.3340	0

Step 2: Apply the branch and bound method to find the optimal sequence.

Find the optimal schedule of the required sequence using the BB formulas as shown in Table 4.

Table 4: In–out table for best possible sequence.

Jobs	Machine *V*						Machine *W*
I	V_1	V_2	V_3	V_4	V_5	V_6	*W*
6	–	0–1.6640	0–1.0010	–	–	–	1.6660–6.3300
2	0–2.6660	–	–	–	–	–	6.3300–12.3300
8	2.666–5.334	1.664–2.665	1.0010–1.6680	0–7	0–6	0–0.3340	12.3300–12.3300
7	–	–	1.6680–6.0010	–	–	–	12.3300–17.6630
5	–	–	–	–	–	0.3340–5.000	17.6630–23.6630
4	–	2.6650–6.9980	–	–	–	–	23.630–28.6630
3	–	–	6.0010–9.0010	–	–	–	28.6630–34.3290
1	5.3340–9.0000	–	–	–	–	–	34.3290–37.6620

Hence, the required elapsed time = 37.6620 h.

Thus, smaller problems, where either the number of jobs or the number of identical parallel machines is limited, can be solved manually with relative ease. However, for larger-scale problems, we have developed a MATLAB program using the R2018a version to determine the makespan. The results, displayed in Table 5 show scenarios with an increased number of parallel machines at the first level and a greater number of jobs.

5.4.2 Computational results by the cuckoo search algorithm

The output results by the cuckoo search algorithm run in MATLAB are presented in Table 6.

Table 5: Computational results by the proposed BB algorithm.

Tasks	Processors at level 1	Optimal sequence	Optimum makespan in h
4	6	4, 3, 5, 2, 1	26.6650
5	6	2, 5, 6, 4, 3, 1	36.6650
7	6	6, 2, 8, 7, 5, 4, 3, 1	37.6620
10	6	1, 2, 11, 10, 9, 8, 6, 5, 3, 4	51.6630
15	6	9, 10, 2, 13, 12, 15, 1, 14, 7, 3, 5, 6, 8, 11, 4	115.9997
20	6	10, 1, 11, 2, 20, 15, 17, 9, 14, 16, 4, 5, 8, 12, 19, 7, 10, 6, 13, 3	137.6583
30	6	26, 24, 1, 11, 18, 17, 29, 28, 20, 12, 15, 9, 16, 2, 4, 5, 8, 19, 23, 27, 30, 7, 10, 22, 25, 14, 6, 21, 13, 3	195.6600
50	6	46, 49, 1, 45, 42, 48, 50, 9, 47, 44, 43, 11, 16, 24, 26, 31, 39, 41, 2, 4, 5, 8, 12, 15, 17, 19, 20, 23, 27, 30, 32, 34, 35, 38, 7, 10, 22, 25, 37, 40, 14, 29, 6, 21, 36, 3, 13, 18, 33, 28	328.6550
80	6	9, 14, 1, 12, 11, 13, 80, 10, 15, 2, 79, 16, 24, 26, 31, 39, 41, 46, 54, 56, 62, 67, 75, 77, 4, 5, 8, 17, 19, 20, 23, 27, 30, 32, 34, 35, 38, 42, 45, 47, 49, 50, 53, 57, 60, 63, 66, 68, 70, 71, 74, 78, 7, 22, 25, 37, 40, 52, 55, 61, 73, 76, 29, 44, 59, 65, 6, 21, 36, 51, 72, 3, 18, 28, 33, 43, 48, 58, 69, 64	522.6420

Table 6: Computational results by the cuckoo search algorithm.

Tasks	Processors at level 1	Optimal sequence	Optimum makespan in h
4	6	2, 4, 1, 3	29.3310
5	6	2, 1, 4, 3, 5	36.6666
7	6	2, 7, 6, 5, 4, 3, 1	38.6640
10	6	1, 6, 5, 7, 10, 4, 3, 8, 2, 9	51.6633
15	6	2, 8, 4, 13, 14, 5, 7, 12, 9, 1, 10, 11, 15, 3, 6	116.9997
20	6	7, 15, 17, 18, 2, 3, 11, 13, 5, 19, 8, 1, 10, 16, 14, 12, 20, 6, 4, 9	151.3330
30	6	26, 1, 16, 12, 7, 10, 13, 23, 21, 8, 19, 27, 5, 11, 3, 17, 6, 18, 2, 29, 30, 4, 20, 24, 9, 25, 14, 15, 28, 22	198.9920
50	6	39, 20, 17, 6, 30, 24, 14, 19, 10, 38, 33, 4, 29, 2, 27, 34, 9, 43, 31, 40, 46, 21, 35, 11, 44, 42, 16, 12, 25, 13, 36, 32, 26, 3, 7, 37, 50, 45, 41, 22, 23, 5, 18, 47, 15, 49, 1, 48, 28, 8	328.9870
80	6	39, 59, 1, 16, 23, 46, 78, 73, 32, 68, 55, 41, 30, 58, 34, 74, 53, 72, 27, 80, 18, 13, 36, 43, 28, 40, 24, 26, 47, 75, 21, 52, 8, 38, 67, 71, 6, 64, 19, 10, 3, 4, 69, 65, 6, 9, 31, 50, 5, 22, 48, 42, 7, 45, 14, 11, 17, 33, 49, 76, 15, 63, 12, 35, 25, 44, 61, 51, 77, 37, 57, 2, 54, 20, 62, 66, 56, 79, 29, 70	524.3130

5.4.3 Computational results by Johnson's algorithm

The output results by Johnson's algorithm run in MATLAB are presented in Table 7. Pictorial presentations related to Johnson's Algorithm for scheduling different numbers of jobs are shown in Figure 3.

Table 7: Computational results by Johnson's algorithm.

Tasks	Processors at level 1	Optimal sequence	Optimum makespan in h
4	6	5, 3, 4, 2, 1	34.1120
5	6	6, 1, 4, 2, 3, 5	37.6666
7	6	8, 1, 6, 3, 4, 7, 2, 5	39.6670
10	6	1, 4, 3, 9, 2, 7, 10, 5, 8	51.6630
15	6	1, 6, 10, 11, 15, 2, 5, 7, 9, 3, 4, 8, 14, 12, 13	116.9997
20	6	1, 16, 6, 10, 11, 17, 18, 15, 2, 5, 7, 9, 19, 20, 3, 4, 8, 14, 12, 13	151.3330
30	6	1, 11, 16, 26, 4, 9, 14, 19, 24, 29, 7, 10, 13, 22, 25, 28, 2, 12, 17, 27, 3, 18, 5, 8, 15, 20, 23, 30, 6, 21	195.3260
50	6	1, 11, 16, 26, 31, 41, 46, 4, 9, 14, 19, 24, 29, 34, 39, 44, 49, 7, 10, 13, 22, 25, 28, 37, 40, 43, 2, 12, 17, 27, 32, 42, 47, 3, 18, 33 , 48, 5, 8, 15, 20, 23, 30, 35, 38, 45, 50, 6, 21, 36	325.3210
80	6	1, 11, 16, 26, 31, 41, 46, 56, 61, 71, 76, 4, 9, 14, 19, 24, 29, 34, 39, 44, 49, 54, 59, 64, 69, 74, 79, 7, 10, 13, 22, 25, 28, 37, 40, 43, 52, 55, 58, 67, 70, 73, 2, 12, 17, 27, 32, 42, 47, 57, 62, 72, 77, 3, 18, 33, 48, 63, 78, 5, 8, 15, 20, 23, 3035, 38, 45, 50, 53, 60, 65, 68, 75, 80, 6, 21, 36, 51, 66	524.3130

6 Results and discussion

Table 8 presents the makespan (Cmax) obtained by the proposed Branch and Bound (BB) algorithm, Cuckoo Search (CS), and Johnson's algorithm for the test instances. The BB method consistently achieves the minimum makespan, confirming its effectiveness in generating optimal schedules. The CS algorithm provides near-optimal solutions with slightly higher makespan, reflecting a balance between solution quality and computational effort. Johnson's algorithm, designed for classical two-machine flow shops, produces comparatively larger makespan values due to its heuristic nature.

Figure 4 illustrates these results, showing that the BB method outperforms both CS and Johnson's algorithms, particularly as the problem size increases. This demonstrates the robustness and superiority of the proposed BB approach in minimizing the total completion time in flow shop scheduling with parallel machines.

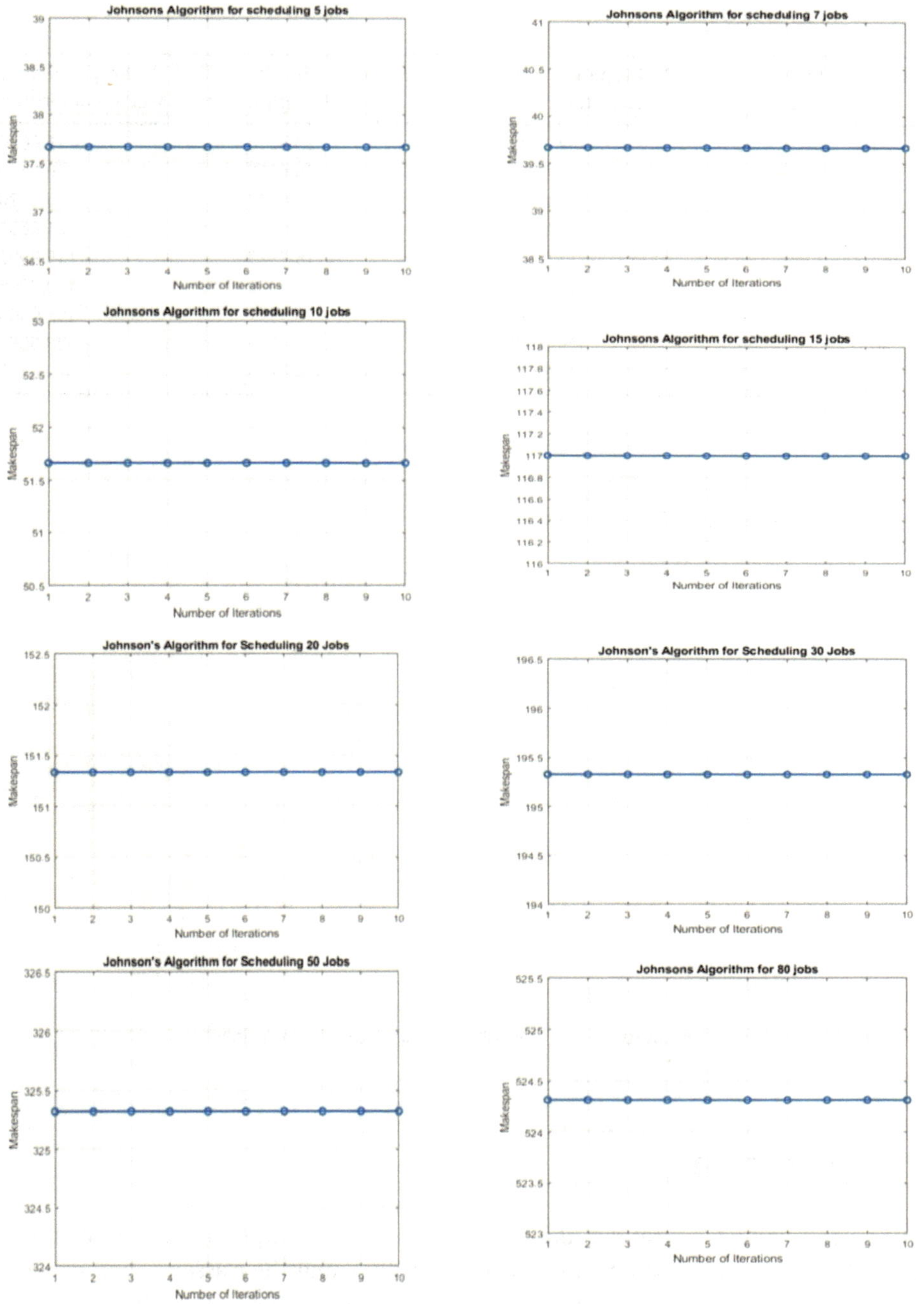

Figure 3: Matlab graphs showing makespan by Johnson's algorithm for different numbers of jobs.

Table 8: Output table by Johnson's algorithm.

Tasks	Processors at level 1	Makespan in h by Johnson's algorithm	Makespan in h by cuckoo search algorithm	Makespan in h by proposed BB algorithm
4	6	34.1120	29.3310	26.6650
5	6	37.6666	36.6666	36.6650
7	6	39.6670	38.6640	37.6620
10	6	51.6630	51.6633	51.6630
15	6	116.9997	116.9997	115.9997
20	6	151.3330	151.3330	150.9996
30	6	195.3260	198.9920	195.6600
50	6	325.3210	328.9870	328.6550
80	6	524.3130	524.3130	522.6420

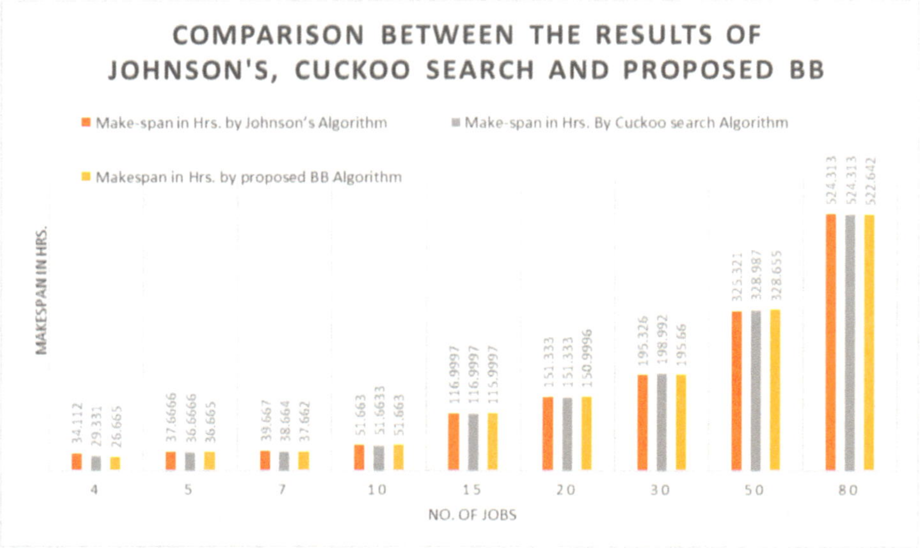

Figure 4: Comparison of BB, cuckoo search, and Johnson's algorithm for large jobs.

7 Conclusion

The study's findings reveal that the proposed BB method outperforms both the metaheuristic approach, represented by the CS algorithm, and the classical approach embodied by Johnson's algorithm in the context of two-stage flow shop scheduling with identical parallel machines in the first stage and a single machine in the second stage. The proposed BB method emerges as the preferred choice for makespan minimization in two-stage flow shop scheduling, demonstrating its superiority over both classical and metaheuristic approaches. Its precision and scalability make it a robust tool for tackling large-scale scheduling optimization challenges, offering significant advancements in efficiency and solution quality.

8 Future scope

In future research, we can extend our model upto the third and fourth stages. In the extended models, different heuristics and metaheuristic approaches can be applied.

Bibliography

[1] Duan, P., & Yong, A. I. (2016). Research on an improved ant colony optimization algorithm and its application. *International Journal of Hybrid Information Technology*, *9*(4), 223–234. https://doi.org/10.14257/ijhit.2016.9.4.20.

[2] Gad, A. G. (2022). Particle swarm optimization algorithm and its applications: a systematic review. *Archives of Computational Methods in Engineering*, *29*(5), 2531–2561. https://doi.org/10.1007/s11831-021-09694-4.

[3] Gupta, D., & Goel, S. (2018). Three stage flow shop scheduling model with m-equipotential machines. *International Journal on Future Revolution in Computer Science & Communication Engineering*, *4*(3), 269–274.

[4] Gupta, D., & Goel, S. (2020). $N \times 2$ flow shop scheduling problem with parallel machines at every stage, processing time associated with probabilities. *Advances in Mathematics: Scientific Journal*, *9*(3), 1061–1069. https://doi.org/10.37418/amsj.9.3.31.

[5] Gupta, D., & Goel, S. (2022). Branch and bound technique for two stage flow shop scheduling model with equipotential machines at every stage. *International Journal of Operational Research*, *44*(4), 462–472. https://doi.org/10.1504/IJOR.2022.125132.

[6] Gupta, D., Goel, S., & Mangla, N. (2022). Optimization of production scheduling in two stage flow shop scheduling problem with m equipotential machines at first stage. *International Journal of System Assurance Engineering and Management*, *13*, 1162–1169. https://doi.org/10.1007/s13198-021-01411-5.

[7] Haldurai, L., Madhubala, T., & Rajalakshmi, R. (2016). A study on genetic algorithm and its applications. *International Journal of Computer Sciences and Engineering*, *4*(10), 139.

[8] Ignall, E., & Schrage, L. (1965). Application of the branch and bound technique to some flow-shop scheduling problems. *Operations Research*, *13*(3), 400–412. https://doi.org/10.1287/opre.13.3.400.

[9] Johnson, S. M. (1954). Optimal two and three stage production schedules with setup times included. *Naval Research Logistics Quarterly*, *1*(1), 61–68. https://doi.org/10.1002/nav.3800010110.

[10] Joshi, A. S., Kulkarni, O., Kakandikar, G. M., & Nandedkar, V. M. (2017). Cuckoo search optimization: a review. *Materials Today: Proceedings*, *4*(8), 7262–7269. https://doi.org/10.1016/j.matpr.2017.07.055.

[11] Katoch, S., Chauhan, S. S., & Kumar, V. (2021). A review on genetic algorithm: past, present, and future. *Multimedia Tools and Applications*, *80*, 8091–8126. https://doi.org/10.1007/s11042-020-10139-6.

[12] Little, J. D., Murty, K. G., Sweeney, D. W., & Karel, C. (1963). An algorithm for the traveling salesman problem. *Operations Research*, *11*(6), 972–989. https://doi.org/10.1287/opre.11.6.972.

[13] Madni, S. H. H., Abd Latiff, M. S., & Abdulhamid, S. I. M. (2017). Optimal resource scheduling for IaaS cloud computing using cuckoo search algorithm. https://doi.org/10.11113/sh.v9n1-3.1145.

[14] Malhotra, K., Gupta, D., Goel, S., & Tripathi, A. K. (2022). Bi-objective flow shop scheduling with equipotential parallel machines. *Malaysian Journal of Mathematical Sciences*, *16*(3), 451–470. https://doi.org/10.47836/mjms.16.3.04.

[15] Malhotra, K., Gupta, D., Goel, S., & Tripathi, A. K. (2023). Comparison of branch and bound with meta-heuristic approach in optimization of three stage flow shop scheduling with multiple processors. https://doi.org/10.21203/rs.3.rs-2822556/v1.

[16] Malhotra, K., Gupta, D., Goel, S., & Tripathi, A. K. (2023). Scheduling involving equipotential processors with constraint as fuzzy utilization time of jobs. In *International Conference on Data Analytics and Insights* (pp. 491–503). Springer Nature Singapore. https://doi.org/10.1007/978-981-99-3878-0_42.

[17] Malhotra, K., Gupta, D., Goel, S., & Tripathi, A. K. (2024). Optimum solution on two stage flow shop scheduling model with like parallel machines. *Indian Journal of Natural Sciences*, *15*(83), 1031–1036.

[18] Mohan, B. C., & Baskaran, R. (2012). Ant colony optimization based recent research and implementation on several engineering domains. *Expert Systems with Applications*, *39*(4), 4618–4627. https://doi.org/10.1016/j.eswa.2011.09.076.

[19] Safdar, K., Abdul Rani, K. N., Rahim, H. A., Rosli, S. J., & Jamlos, M. A. (2023). A review on research trends in using cuckoo search algorithm: applications and open research challenges. *Przegląd Elektrotechniczny*, *99*(5). https://doi.org/10.15199/48.2023.05.04.

[20] Sharma, A. (2022). A review paper on cuckoo search algorithm (CSA). *International Journal of Innovative Research in Computer Science & Technology*, *10*(2), 126–130.

[21] Shehab, M., Khader, A. T., & Al-Betar, M. A. (2017). A survey on applications and variants of the cuckoo search algorithm. *Applied Soft Computing*, *61*, 1041–1059. https://doi.org/10.1016/j.asoc.2017.02.034.

[22] Song, M.-P., & Gu, G.-C. (2004). Research on particle swarm optimization: a review. In *Proceedings of the 2004 International Conference on Machine Learning and Cybernetics* (pp. 2236–2241). IEEE. https://doi.org/10.1109/ICMLC.2004.1382171.

[23] Xiong, Y., Zou, Z., & Cheng, J. (2023). Cuckoo search algorithm based on cloud model and its application. *Scientific Reports*, *13*(1), 10098. https://doi.org/10.1038/s41598-023-37326-3.

Ajay Kumar, Parthkumar P. Sartanpara, and Ramakanta Meher

Comparative analysis of two novel analytical techniques for solving fractional Navier–Stokes equations

Abstract: This study presents a comparative study of two analytical simulations, namely the q-homotopy analysis generalized integral transform technique and a generalized integral transform Adomian decomposition method, applied to the fractional Navier–Stokes equations. The offered test example demonstrates the usefulness and efficiency of the recommended approaches. A graphical depiction of the solution has been created. The series solution can be effectively constrained within a wide acceptable range utilizing the provided approaches. Numerical and graphical data have been provided to make it easier to compare solutions. The results indicate that the convergence parameter in the q-homotopy analysis generalized integral transform technique produces a more optimal solution. It provides an easy mechanism to alter the convergence region of the solution.

Keywords: Generalized integral transform, fractional Navier–Stokes system, homotopy analysis transform method, generalized integral transform Adomian decomposition method, Caputo fractional derivative

MSC 2020: 26A33, 35R11, 76D05, 65M70, 65H20, 34A08

1 Introduction

In 1822, Navier developed the Navier–Stokes (NS) equation that illustrates the motion of viscous fluid flow. Fluid motion was shown by a wide range of physical phenomena, including the flow of blood, ocean currents, liquids in pipes, and airflow around the wings of an airplane. In the case of an incompressible fluid flow with constant density ρ and kinematic viscosity $\nu = \eta/\rho$, the following is the model of the NS equation in two dimensions [1, 2]:

Acknowledgement: The first author acknowledges CSIR, New Delhi, India, for fellowship support (File No. 09/1007(0013)/2021-EMR-I).

Ajay Kumar, Parthkumar P. Sartanpara, Ramakanta Meher, Department of Mathematics, Sardar Vallabhbhai National Institute of Technology, Surat-395007, Gujarat, India, e-mail: ajaykhator123@gmail.com

https://doi.org/10.1515/9783112234525-002

$$\begin{cases} \frac{\partial U}{\partial t} + (U\cdot\nabla)U = b_0\nabla^2 U - \frac{1}{\rho}\nabla p, & \text{on } \mathcal{D}\times(0,T) \\ \nabla\cdot U = 0, & \text{on } \mathcal{D}\times(0,T) \\ U = 0, & \text{on } \partial\mathcal{D}\times(0,T), \end{cases} \tag{1}$$

where $U = (u, v)$ represents the fluid vector, t – time, and p – pressure. The spatial components (ξ_1, ξ_2) in $\mathcal{D}$ and the boundary of $\mathcal{D}$ is $\partial\mathcal{D}$, where η and ρ are dynamic viscosity and density, respectively. The ratio of η and ρ is b_0, the kinematic viscosity. The aforementioned equation is transformed into the following system when expressed in Cartesian coordinates:

$$\begin{cases} \frac{\partial u}{\partial t} + u\frac{\partial u}{\partial \xi_1} + v\frac{\partial u}{\partial \xi_2} = b_0(\frac{\partial^2 u}{\partial {\xi_1}^2} + \frac{\partial^2 u}{\partial {\xi_2}^2}) - \frac{1}{\rho}\frac{\partial p}{\partial \xi_1}, \\ \frac{\partial v}{\partial t} + u\frac{\partial v}{\partial \xi_1} + v\frac{\partial v}{\partial \xi_2} = b_0(\frac{\partial^2 v}{\partial {\xi_1}^2} + \frac{\partial^2 v}{\partial {\xi_2}^2}) - \frac{1}{\rho}\frac{\partial p}{\partial \xi_2}, \end{cases} \tag{2}$$

where, $g_1 = -\frac{1}{\rho}\frac{\partial p}{\partial \xi_1}$, $g_2 = -\frac{1}{\rho}\frac{\partial p}{\partial \xi_2}$, can be determined, if p is known. In this work, we consider Navier–Stokes equations that can be explained with a nonlinear system with $g_1 = -g_2 = g$ as follows [3]:

$$\begin{aligned} \frac{\partial u}{\partial t} + u\frac{\partial u}{\partial \xi_1} + v\frac{\partial u}{\partial \xi_2} &= b_0\left(\frac{\partial^2 u}{\partial {\xi_1}^2} + \frac{\partial^2 u}{\partial {\xi_2}^2}\right) + g, \\ \frac{\partial v}{\partial t} + u\frac{\partial v}{\partial \xi_1} + v\frac{\partial v}{\partial \xi_2} &= b_0\left(\frac{\partial^2 v}{\partial {\xi_1}^2} + \frac{\partial^2 v}{\partial {\xi_2}^2}\right) - g, \end{aligned} \tag{3}$$

with initial conditions (ICs)

$$\begin{aligned} u(0, \xi_1, \xi_2) &= -\sin(\xi_1 + \xi_2), \\ v(0, \xi_1, \xi_2) &= \sin(\xi_1 + \xi_2). \end{aligned} \tag{4}$$

The exact solution [4] of the given problem at $g = 0$ is

$$\begin{aligned} u(t, \xi_1, \xi_2) &= -\sin(\xi_1 + \xi_2)e^{-2b_0 t}, \\ v(t, \xi_1, \xi_2) &= \sin(\xi_1 + \xi_2)e^{-2b_0 t}. \end{aligned} \tag{5}$$

El-Shahed and Salem [5] extended the conventional Navier–Stokes equation by converting the integer-order derivative to the arbitrary-order derivative $\alpha \in (0, 1]$. Fractional derivatives have found extensive applications in fluid dynamics to represent both memory and hereditary effects in time-fractional Navier–Stokes equations and internal waves. They are also used in heat and mass transfer, porous and viscoelastic media, and magnetohydrodynamic systems to effectively describe anomalous diffusion and nonlocal effects. Here, we can write equation (3) into corresponding fractional-order differential equations as

$$\begin{aligned} D_t^{\alpha} u + u\frac{\partial u}{\partial \xi_1} + v\frac{\partial u}{\partial \xi_2} &= b_0\left(\frac{\partial^2 u}{\partial {\xi_1}^2} + \frac{\partial^2 u}{\partial {\xi_2}^2}\right) + g, \\ D_t^{\alpha} v + u\frac{\partial v}{\partial \xi_1} + v\frac{\partial v}{\partial \xi_2} &= b_0\left(\frac{\partial^2 v}{\partial {\xi_1}^2} + \frac{\partial^2 v}{\partial {\xi_2}^2}\right) - g, \end{aligned} \tag{6}$$

where, $\alpha \in (0,1]$ and D_t^{α} is the αth-order Caputo derivative operator with respect to t.

Nonlinearity is not easy to understand and much more complex when it comes to differential equations. Such nonlinear differential equations have always been the center of attention among researchers. Solving nonlinear equations is a very complex task. In many cases, the exact solutions cannot be obtained for nonlinear equations, which give rise to many numerical methods. Numerical methods gave discrete approximate numerical solutions, and it was quite tough to understand the nature of the solutions correctly because discrete points were not able to explain the complete curve.

To overcome the disadvantage, analytical methods are starting to be developed. Analytical methods gave the series solution that were very useful to find the solution. However, they faced complexity when nonlinearity was involved. The issue of nonlinearity was solved when George Adomian introduced the analytical method called the Adomian Decomposition method (ADM) [6]. ADM used an Adomian polynomial to compete the complexity of nonlinearity. In 1992, using a topological concept named Homotopy, Shijun Liao introduced the Homotopy analysis method (HAM) in his PhD thesis [7]. The homotopy analysis method used the homotopy concept and decreased the complexity of nonlinear problems. Additionally, the controlling parameter h was very useful to check the convergence of the obtained series solution. Later, in 2013, El-Tawil and Huseen [8] modified HAM, which helped to provide a larger range of the auxiliary parameter h and this method was named q-HAM.

ADM [9], HAM, and q-HAM were very useful to provide solutions of nonlinear differential equations. In their study, Kumar and Meher [10] solved nonlinear partial differential equations using HAM, hence proving its usefulness in producing approximative solutions. Verma and Meher [11] demonstrated HAM's adaptability in managing mathematical model uncertainty by solving fuzzy differential equations using HAM. Investigating the effect of fuzzy volume fractions on hybrid-type nanofluid flow and thermal transmission in converging/diverging channels, Verma et al. [12] found that volume fraction influences the thermal performance. Using the homotopy analytic approach, Kumar and Meher [13] investigated micropolar nanofluid flow under heat radiation over porous channel walls. Verma and Meher [14] utilized the Atangana–Baleanu–Caputo fractional derivative to investigate a human liver model, hence emphasizing the use of fractional calculus in the modeling of intricate biological systems. Finally, Sartanpara and Meher [15] performed a computational investigation on the Klein–Gordon equation utilizing HAM, thereby demonstrating the effectiveness of the technique in addressing nonlinear equations arising in physics. While solving the fractional-order differential equation, these methods faced a difficulty with the fractional order. To avoid this difficulty, investigators started using different transforms with methods such as the Sumudu transform and the Elzaki transform, the Aboodh transform, the Shehu Transform, etc., with analytical methods.

Using a formable transform and a homotopy analysis technique, Prajapati and Meher [16] solved the time-fractional Rosenau–Hyman model. Singh et al. [17] solved the Poisson equation with local fractional derivative in fractal media using the q-homotopy analysis transform technique (q-HATM). Using the Elzaki–Adomian decomposition technique, Varsoliwala and Singh [18] solved the foam drainage model. Jena and Chakraverty [19] used the homotopy perturbation Elzaki transform for time-fractional Navier–Stokes equations. With a fractional approach, Patel et al. [20] conducted an analytical study to examine atmospheric internal waves. Using the Aboodh transform homotopy perturbation approach, Jani and Singh [21] solved the time-fractional Swift–Hohenberg problem. Using a new fractional calculus technique, namely q-HATM, Veeresha and Kumar [22] investigated the Ivancevic option pricing model. Utilizing the Sumudu transform and Adomian decomposition approach, Patel and Meher [23] focused on temperature-dependent thermal conductivity to investigate convective radial fins. The fractional-order three-dimensional Wu–Zhang equation was analytically solved by Patel and Patel [24]. For the fractional-order Newell–Whitehead–Segel equation, Jani and Singh [25] employed the Aboodh transform homotopy perturbation approach. Prakash et al. [26] presented a homotopy technique combined with the Laplace transform to solve fractional-order multidimensional telegraph equations. To solve fractional Fokker–Planck equations, Prakash and Kaur [27] applied the q-HATM. Analyzing Jeffery–Hamel nanofluid flow with undetermined volume fractions, Verma and Meher [28] used the Double Parametric Fuzzy Homotopy Analysis Method. Using the Shehu transform, Sartanpara and Meher [29] provided a strong computational method in recent advancements for solving the Zakharov–Kuznetsov equations in ion-acoustic waves within magnetized plasmas. Jhangeer et al. [30] studied a generalized KdV–Zakharov–Kuznetsov model in fluid and plasma physics, where the dynamical system techniques were applied to study the ion-acoustic wave propagation, bifurcation behavior, chaotic dynamics, and soliton structures under the impact of a magnetic field. Talafha et al. [31] investigated a higher-dimensional Davey–Stewartson–Kadomtsev–Petviashvili equation to describe the interactions of internal waves, using Lie symmetry analysis, auxiliary equation methods, and the dynamical system viewpoint to examine exact solutions, bifurcation behavior, chaos, and sensitivity to initial conditions. Schurz and Talafha [32] conducted analysis of Fourier series for a stochastic sine-Gordon equation with a modification, and presented the well-posedness, stability, and energy control of approximate solutions of such equations by truncation of eigenfunctions and Lyapunov-type functional methods. Jhangeer et al. [33] examined an extended (2+1)-dimensional time-dependent dissipative mechanism of a Bogoyavlensky–Konopelchenko equation using both analytical methods of solution and dynamical system theory to study the behavior of internal waves, symmetry, and complicated phenomena such as periodicity, quasiperiodicity and chaos. Kumar and Meher [34] developed an efficient approach by integrating the Shehu transform with the q-HAM, offering an efficient solution methodology for fractional-order partial differential equations. Building on this, Sartanpara and Meher [35] expanded their work by modeling atmospheric internal waves using a fuzzy-fractional technique, therefore providing fresh understanding of fractional dynamics in atmospheric events. Using an analytic approach, Sartanpara and Meher [36] solved the generalized time-fractional Fornberg–Whitham equation, hence extending wave propagation theory in fractional-order systems. Furthermore, helping the field of nonlinear wave equations, Rashid et al. [37] built an approximate analytical solution for the fractional Zakharov–Kuznetsov equation in multidimensional space using the Aboodh–Adomian decomposition approach. These sophisticated techniques help to investigate fractional-order models in many different scientific and technical domains.

In 2021, a generalized integral transform was introduced by Jafari [38]. Later, Researchers started using a generalized integral transform to solve differential equations with analytical methods. Using an efficient multiparametric approach combined with a generalized transform, Sartanpara and Meher [39] analytically solved the fractional Kaup–Kupershmidt equation, therefore offering a suitable technique of solution. Incorporating both singular and nonsingular kernels to describe complicated behavior in the system, Prajapati and Meher [40] used a strong analytical methodology to solve the generalized Burgers–Fisher equation containing derivatives of fractional order.

In this chapter, we employ two analytical methods named the q-Homotopy analysis generalized integral transform method (q-HAGITM) and the generalized integral transform Adomian decomposition method (GITADM) in solving the system of nonlinear Navier–Stokes equations with fractional-order derivative. This study offers q-HAGITM, a unique methodology to solve time-fractional coupled Navier–Stokes equations, enabling increased convergence control through the parameter h. By proving that GITADM is a specific instance of q-HAGITM, the work proposes a generalized and computationally efficient technique for nonlinear fractional PDEs, considerably lowering computing effort while assuring excellent accuracy. The results will be discussed analytically, numerically, and graphically in the results and discussion section. Finally, a conclusion is given.

2 Basic definitions

Definition 1. The αth-order Caputo fractional derivative of the function $g(t)$ [29] is defined as

$$D_t^{\alpha} g(t) = \begin{cases} \frac{1}{\Gamma(n-\alpha)} \int_0^t (t-\mu)^{n-\alpha-1} g^n(\mu) d\mu, & n-1 < \alpha < n, n \in \mathbb{N} \\ \mathfrak{I}^{n-\alpha} D^n g(t) & \alpha = n \in \mathbb{N}. \end{cases} \tag{7}$$

Definition 2. The function $g(t)$ is transformed using the generalized integral transform [29], which is defined as

$$G(s,a)=\mathbb{G}[g(t)]=\int_0^\infty \exp\left(\frac{-st}{a}\right)g(t)dt, \tag{8}$$

where s and a are positive numbers.

One can obtain the following results by applying equation (8):

1. $\mathbb{G}[b]=b\frac{a}{s}$, where b is an arbitrary constant.
2. $\mathbb{G}[t]=(\frac{a}{s})^2$.
3. $\mathbb{G}[t^b]=\Gamma(b+1)(\frac{a}{s})^{b+1}$, $b>-1$.
4. Linearity condition: $\mathbb{G}[b_1g_1(t)\pm b_2g_2(t)]=b_1\mathbb{G}[g_1(t)]\pm b_2\mathbb{G}[g_2(t)]$.

Definition 3. If $\mathbb{G}[g(t)]=G(s,a)$, then the inverse generalized integral transform [29] can be represented as

$$g(t)=\mathbb{G}^{-1}[G(s,a)]=\lim_{\delta\to\infty}\frac{1}{2\pi i}\int_{\gamma+i\delta}^{\gamma-i\delta}\frac{1}{a}\exp\left(\frac{st}{a}\right)G(s,a)ds. \tag{9}$$

Definition 4. Given the generalized integral transform $\mathbb{G}[g(t)]=G(s,a)$, then the αth-order fractional derivative in the Caputo sense under this transform is defined as [29]

$$\mathbb{G}[D_t^\alpha g(t)]=\left(\frac{s}{a}\right)^\alpha G(s,a)-\sum_{i=0}^{m-1}\left(\frac{s}{a}\right)^{\alpha-i-1}g^{(i)}(0^+), \tag{10}$$

where, $\alpha>0$ and $m-1<\alpha\le m$.

3 Application to Navier–Stokes equations

3.1 Generalized integral transform Adomian decomposition method

To employ the generalized integral transform Adomian Decomposition Method (GITADM), let us consider fractional Navier–Stokes equations taken from equation (6) for $0<\alpha\le 1$

$$\begin{aligned}D_t^\alpha u+u\frac{\partial u}{\partial\xi_1}+v\frac{\partial u}{\partial\xi_2}&=b_0\left(\frac{\partial^2u}{\partial{\xi_1}^2}+\frac{\partial^2u}{\partial{\xi_2}^2}\right)+g,\\ D_t^\alpha v+u\frac{\partial v}{\partial\xi_1}+v\frac{\partial v}{\partial\xi_2}&=b_0\left(\frac{\partial^2v}{\partial{\xi_1}^2}+\frac{\partial^2v}{\partial{\xi_2}^2}\right)-g,\end{aligned} \tag{11}$$

with ICs

$$u(0,\xi_1,\xi_2) = -\sin(\xi_1+\xi_2),$$
$$v(0,\xi_1,\xi_2) = \sin(\xi_1+\xi_2). \tag{12}$$

Applying the generalized integral transform on equation (11)

$$\mathbb{G}[u(t,\xi_1,\xi_2)] + \frac{a}{s}\sin(\xi_1+\xi_2) + \left(\frac{a}{s}\right)^{\alpha}\mathbb{G}\left\{u\frac{\partial u}{\partial\xi_1} + v\frac{\partial u}{\partial\xi_2} - b_0\left(\frac{\partial^2 u}{\partial{\xi_1}^2} + \frac{\partial^2 u}{\partial{\xi_2}^2}\right) - g\right\} = 0$$
$$\mathbb{G}[v(t,\xi_1,\xi_2)] - \frac{a}{s}\sin(\xi_1+\xi_2) + \left(\frac{a}{s}\right)^{\alpha}\mathbb{G}\left\{u\frac{\partial v}{\partial\xi_1} + v\frac{\partial v}{\partial\xi_2} - b_0\left(\frac{\partial^2 v}{\partial{\xi_1}^2} + \frac{\partial^2 v}{\partial{\xi_2}^2}\right) + g\right\} = 0. \tag{13}$$

Taking the formulation of the inverse generalized integral transformation, we obtain

$$\begin{aligned} u(t,\xi_1,\xi_2) &= u(0,\xi_1,\xi_2) + \mathbb{G}^{-1}\left[\left(\frac{a}{s}\right)\mathrm{ST}\{g\}\right] \\ &\quad + \mathbb{G}^{-1}\left[\left(\frac{a}{s}\right)^{\alpha}\mathbb{G}\left[-\left(u\frac{\partial u}{\partial\xi_1} + v\frac{\partial u}{\partial\xi_2}\right) + b_0\left\{\frac{\partial^2 u}{\partial\xi_1^2} + \frac{\partial^2 u}{\partial\xi_2^2}\right\}\right]\right], \\ v(t,\xi_1,\xi_2) &= v(0,\xi_1,\xi_2) - \mathbb{G}^{-1}\left[\left(\frac{a}{s}\right)\mathrm{ST}\{g\}\right] \\ &\quad + \mathbb{G}^{-1}\left[\left(\frac{a}{s}\right)^{\alpha}\mathbb{G}\left[-\left(u\frac{\partial v}{\partial\xi_1} + v\frac{\partial v}{\partial\xi_2}\right) + b_0\left\{\frac{\partial^2 v}{\partial\xi_1^2} + \frac{\partial^2 v}{\partial\xi_2^2}\right\}\right]\right]. \end{aligned} \tag{14}$$

Assuming the solutions of $u(t,\xi_1,\xi_2)$ and $v(t,\xi_1,\xi_2)$ in the following series form:

$$u(t,\xi_1,\xi_2) = \sum_{m=0}^{\infty} u_m(t,\xi_1,\xi_2), \quad v(t,\xi_1,\xi_2) = \sum_{m=0}^{\infty} v_m(t,\xi_1,\xi_2), \tag{15}$$

equation (14) can be written as follows using equation (15):

$$\begin{aligned} \sum_{m=0}^{\infty} u_m(t,\xi_1,\xi_2) &= -\sin(\xi_1+\xi_2) + \frac{gt^{\alpha}}{\Gamma(\alpha+1)} + \mathbb{G}^{-1}\left[\left(\frac{a}{s}\right)^{\alpha}\mathrm{ST}\left[-\left(\sum_{m=0}^{\infty}A_m + \sum_{m=0}^{\infty}B_m\right)\right]\right] \\ &\quad + \mathbb{G}^{-1}\left[\left(\frac{a}{s}\right)^{\alpha}\mathbb{G}\left[b_0\left\{\sum_{m=0}^{\infty}\frac{\partial^2 u_m}{\partial\xi_1^2} + \sum_{m=0}^{\infty}\frac{\partial^2 u_m}{\partial\xi_2^2}\right\}\right]\right], \\ \sum_{m=0}^{\infty} v_m(t,\xi_1,\xi_2) &= \sin(\xi_1+\xi_2) - \frac{qt^{\alpha}}{\Gamma(\alpha+1)}\mathbb{G}^{-1}\left[\left(\frac{a}{s}\right)^{\alpha}\mathbb{G}\left[-\left(\sum_{m=0}^{\infty}C_m + \sum_{m=0}^{\infty}D_m\right)\right]\right] \\ &\quad + \mathbb{G}^{-1}\left[\left(\frac{a}{s}\right)^{\alpha}\mathbb{G}\left[b_0\left\{\sum_{m=0}^{\infty}\frac{\partial^2 v_m}{\partial\xi_1^2} + \sum_{m=0}^{\infty}\frac{\partial^2 v_m}{\partial\xi_2^2}\right\}\right]\right], \end{aligned} \tag{16}$$

where A_m, B_m, C_m, and D_m are Adomian polynomials defined as

$$uu_{\xi_1} = \sum_{m=0}^{\infty} A_m, \quad vu_{\xi_2} = \sum_{m=0}^{\infty} B_m, \quad uv_{\xi_1} = \sum_{m=0}^{\infty} C_m, \quad vv_{\xi_2} = \sum_{m=0}^{\infty} D_m.$$

The nonlinearity can be presented in the following terms:

$$\begin{aligned}
A_0 &= u_0\left(\frac{\partial u_0}{\partial \xi_1}\right), & A_1 &= u_0\left(\frac{\partial u_1}{\partial \xi_1}\right) + u_1\left(\frac{\partial u_0}{\partial \xi_1}\right), & \cdots\\
B_0 &= v_0\left(\frac{\partial u_0}{\partial \xi_2}\right), & B_1 &= v_0\left(\frac{\partial u_1}{\partial \xi_2}\right) + v_1\left(\frac{\partial u_0}{\partial \xi_2}\right), & \cdots\\
C_0 &= u_0\left(\frac{\partial v_0}{\partial \xi_1}\right), & C_1 &= u_0\left(\frac{\partial v_1}{\partial \xi_1}\right) + u_1\left(\frac{\partial v_0}{\partial \xi_1}\right), & \cdots\\
D_0 &= v_0\left(\frac{\partial v_0}{\partial \xi_1}\right), & D_1 &= v_0\left(\frac{\partial v_1}{\partial \xi_1}\right) + v_1\left(\frac{\partial v_0}{\partial \xi_1}\right), & \cdots.
\end{aligned}$$

For the solution, initial assumptions can be taken as

$$u_0(t, \xi_1, \xi_2) = -\sin(\xi_1 + \xi_2), \tag{17}$$

$$v_0(t, \xi_1, \xi_2) = \sin(\xi_1 + \xi_2). \tag{18}$$

Using equations (15), (16), (17), and (18), we obtain series solutions in the form

$$u(t, \xi_1, \xi_2) = -\sin(\xi_1 + \xi_2) + \frac{gt^\alpha}{\Gamma(\alpha+1)} + \frac{2b_0 \sin(\xi_1 + \xi_2)t^\alpha}{\Gamma[\alpha+1]} - \frac{4b_0^2 \sin(\xi_1 + \xi_2)t^{2\alpha}}{\Gamma[2\alpha+1]} + \cdots \tag{19}$$

$$v(t, \xi_1, \xi_2) = \sin(\xi_1 + \xi_2) - \frac{gt^\alpha}{\Gamma(\alpha+1)} - \frac{2b_0 \sin(\xi_1 + \xi_2)t^\alpha}{\Gamma[\alpha+1]} + \frac{4b_0^2 \sin(\xi_1 + \xi_2)t^{2\alpha}}{\Gamma[2\alpha+1]} + \cdots. \tag{20}$$

3.2 q-Homotopy analysis generalized integral transform method [29]

To apply q-HAGITM, let us consider fractional Navier–Stokes equations with $0 < \alpha \le 1$ as follows:

$$\begin{aligned}
D_t^\alpha u + u\frac{\partial u}{\partial \xi_1} + v\frac{\partial u}{\partial \xi_2} &= b_0\left(\frac{\partial^2 u}{\partial {\xi_1}^2} + \frac{\partial^2 u}{\partial {\xi_2}^2}\right) + g,\\
D_t^\alpha v + u\frac{\partial v}{\partial \xi_1} + v\frac{\partial v}{\partial \xi_2} &= b_0\left(\frac{\partial^2 v}{\partial {\xi_1}^2} + \frac{\partial^2 v}{\partial {\xi_2}^2}\right) - g,
\end{aligned} \tag{21}$$

with ICs

$$\begin{aligned}
u(0, \xi_1, \xi_2) &= -\sin(\xi_1 + \xi_2),\\
v(0, \xi_1, \xi_2) &= \sin(\xi_1 + \xi_2).
\end{aligned} \tag{22}$$

Applying the formulation of the generalized integral transform to equation (21)

$$\begin{aligned}
&\mathbb{G}[u(t,\xi_1,\xi_2)] + \frac{a}{s}\sin(\xi_1+\xi_2) + \left(\frac{a}{s}\right)^{\alpha}\mathbb{G}\left\{u\frac{\partial u}{\partial \xi_1} + v\frac{\partial u}{\partial \xi_2} - b_0\left(\frac{\partial^2 u}{\partial {\xi_1}^2} + \frac{\partial^2 u}{\partial {\xi_2}^2}\right) - g\right\} = 0 \\
&\mathbb{G}[v(t,\xi_1,\xi_2)] - \frac{a}{s}\sin(\xi_1+\xi_2) + \left(\frac{a}{s}\right)^{\alpha}\mathbb{G}\left\{u\frac{\partial v}{\partial \xi_1} + v\frac{\partial v}{\partial \xi_2} - b_0\left(\frac{\partial^2 v}{\partial {\xi_1}^2} + \frac{\partial^2 v}{\partial {\xi_2}^2}\right) + g\right\} = 0.
\end{aligned} \tag{23}$$

Now, the nonlinear operator can be taken as

$$\begin{aligned}
&N^1[\psi_1(t,\xi_1,\xi_2;q),\psi_2(t,\xi_1,\xi_2;q)] \\
&\quad = \mathbb{G}[\psi_1(t,\xi_1,\xi_2;q)] + \frac{a}{s}\sin(\xi_1+\xi_2) \\
&\quad + \left(\frac{a}{s}\right)^{\alpha}\mathbb{G}\left\{\psi_1(t,\xi_1,\xi_2;q)\frac{\partial\psi_1(t,\xi_1,\xi_2;q)}{\partial\xi_1} - b_0\frac{\partial^2\psi_1(t,\xi_1,\xi_2;q)}{\partial{\xi_1}^2}\right. \\
&\quad \left. + \psi_2(t,\xi_1,\xi_2;q)\frac{\partial\psi_1(t,\xi_1,\xi_2;q)}{\partial\xi_2} - b_0\frac{\partial^2\psi_1(t,\xi_1,\xi_2;q)}{\partial{\xi_2}^2} - g\right\}, \\
&N^2[\psi_1(t,\xi_1,\xi_2;q),\psi_2(t,\xi_1,\xi_2;q)] \\
&\quad = \mathbb{G}[\psi_2(t,\xi_1,\xi_2;q)] - \frac{a}{s}\sin(\xi_1+\xi_2) \\
&\quad + \left(\frac{a}{s}\right)^{\alpha}\mathbb{G}\left\{\psi_1(t,\xi_1,\xi_2;q)\frac{\partial\psi_2(t,\xi_1,\xi_2;q)}{\partial\xi_1} - b_0\frac{\partial^2\psi_2(t,\xi_1,\xi_2;q)}{\partial{\xi_1}^2}\right. \\
&\quad \left. + \psi_2(t,\xi_1,\xi_2;q)\frac{\partial\psi_2(t,\xi_1,\xi_2;q)}{\partial\xi_2} - b_0\frac{\partial^2\psi_2(t,\xi_1,\xi_2;q)}{\partial{\xi_2}^2} + g\right\}.
\end{aligned}$$

Here, $0 \le q \le 1/n$ is the homotopy parameter, $n \le 1$, and $\psi_1(t,\xi_1,\xi_2;q)$ and $\psi_2(t,\xi_1,\xi_2;q)$ are unknown functions.

Now, let us construct homotopy as

$$\begin{aligned}
(1-nq)\mathbb{G}[\psi_1(t,\xi_1,\xi_2;q) - u_0(t,\xi_1,\xi_2)] &= hqN_1[\psi_1(t,\xi_1,\xi_2;q),\psi_2(t,\xi_1,\xi_2;q)], \\
(1-nq)\mathbb{G}[\psi_2(t,\xi_1,\xi_2;q) - v_0(t,\xi_1,\xi_2)] &= hqN_2[\psi_1(t,\xi_1,\xi_2;q),\psi_2(t,\xi_1,\xi_2;q)],
\end{aligned}$$

where, $h(\neq 0)$ is an auxiliary parameter, u_0 and v_0 are the initial guesses.

Now, for values of q of 0 and $1/n$, we have

$$\begin{aligned}
\psi_1(t,\xi_1,\xi_2;0) &= u_0(t,\xi_1,\xi_2), \quad \psi_1(t,\xi_1,\xi_2;1/n) = u(t,\xi_1,\xi_2) \\
\psi_2(t,\xi_1,\xi_2;0) &= v_0(t,\xi_1,\xi_2), \quad \psi_2(t,\xi_1,\xi_2;1/n) = v(t,\xi_1,\xi_2).
\end{aligned}$$

As q moves from 0 to $1/n$, the solution moves from the initial conditions to the exact solution.

Evaluating ψ_1 and ψ_2 with the help of Taylor's theorem in terms of q, we have

$$\psi_1(t,\xi_1,\xi_2;q) = u_0 + \sum_{m=1}^{\infty} u_m(t,\xi_1,\xi_2)q^m,$$

$$\psi_2(t,\xi_1,\xi_2;q) = v_0 + \sum_{m=1}^{\infty} v_m(t,\xi_1,\xi_2)q^m,$$

where

$$u_m = \frac{1}{m!}\frac{\partial^m \psi_1(t,\xi_1,\xi_2;q)}{\partial q^m}\bigg|_{q=0}, \quad v_m = \frac{1}{m!}\frac{\partial^m \psi_2(t,\xi_1,\xi_2;q)}{\partial q^m}\bigg|_{q=0}. \tag{24}$$

The solutions can be written in series form as

$$\begin{aligned} u(t,\xi_1,\xi_2) &= u_0 + \sum_{m=1}^{\infty} u_m(t,\xi_1,\xi_2)\left(\frac{1}{n}\right)^m \\ v(t,\xi_1,\xi_2) &= v_0 + \sum_{m=1}^{\infty} v_m(t,\xi_1,\xi_2)\left(\frac{1}{n}\right)^m. \end{aligned} \tag{25}$$

Now, by taking both sides with m times differentiation in the constructed homotopy equations with respect to q, multiplying by $\frac{1}{m!}$, and putting $q = 0$, we obtain

$$\begin{aligned} \mathbb{G}[u_m(t,\xi_1,\xi_2) - L_m u_{m-1}(t,\xi_1,\xi_2)] &= hR_{1,m}[\vec{u}_{m-1},\vec{v}_{m-1}], \\ \mathbb{G}[v_m(t,\xi_1,\xi_2) - L_m v_{m-1}(t,\xi_1,\xi_2)] &= hR_{2,m}[\vec{u}_{m-1},\vec{v}_{m-1}], \end{aligned} \tag{26}$$

where $R_{1,m}[\vec{u}_{m-1},\vec{v}_{m-1}]$ and $R_{2,m}[\vec{u}_{m-1},\vec{v}_{m-1}]$ expressed as

$$\begin{aligned} R_{1,m}[\vec{u}_{m-1},\vec{v}_{m-1}] &= \mathbb{G}[u_{m-1}(t,\xi_1,\xi_2)] + \left(1-\frac{L_m}{n}\right)\sin(\xi_1+\xi_2)\frac{a}{s} \\ &\quad + \left(\frac{a}{s}\right)^{\alpha}\mathbb{G}\left\{\sum_{i=0}^{m-1} u_i\frac{\partial u_{m-1-i}}{\partial \xi_1} + \sum_{i=0}^{m-1} v_i\frac{\partial u_{m-1-i}}{\partial \xi_2}\right. \\ &\quad \left. - b_0\left(\frac{\partial^2 u_{m-1}}{\partial {\xi_1}^2} + \frac{\partial^2 u_{m-1}}{\partial {\xi_2}^2}\right) - g\left(1-\frac{L_m}{n}\right)\right\}, \\ R_{2,m}[\vec{u}_{m-1},\vec{v}_{m-1}] &= \mathbb{G}[v_{m-1}(t,\xi_1,\xi_2)] - \left(1-\frac{L_m}{n}\right)\sin(\xi_1+\xi_2)\frac{a}{s} \\ &\quad + \left(\frac{a}{s}\right)^{\alpha}\mathbb{G}\left\{\sum_{i=0}^{m-1} u_i\frac{\partial v_{m-1-i}}{\partial \xi_1} + \sum_{i=0}^{m-1} v_i\frac{\partial v_{m-1-i}}{\partial \xi_2}\right. \\ &\quad \left. - b_0\left(\frac{\partial^2 v_{m-1}}{\partial {\xi_1}^2} + \frac{\partial^2 v_{m-1}}{\partial {\xi_2}^2}\right) + g\left(1-\frac{L_m}{n}\right)\right\} \end{aligned} \tag{27}$$

and

$$L_m = \begin{cases} 0, & m \le 1, \\ n, & m > 1. \end{cases} \tag{28}$$

Using the inverse formulation of the generalized integral transformation on equation (26), we obtain

$$
\begin{aligned}
u_m(t,\xi_1,\xi_2) &= L_m u_{m-1} + h\mathbb{G}^{-1}\{R_{1,m}[\vec{u}_{m-1},\vec{v}_{m-1}]\},\\
v_m(t,\xi_1,\xi_2) &= L_m v_{m-1} + h\mathbb{G}^{-1}\{R_{2,m}[\vec{u}_{m-1},\vec{v}_{m-1}]\}.
\end{aligned}
\tag{29}
$$

Now, our initial assumptions can be taken as

$$
\begin{aligned}
u_0(t,\xi_1,\xi_2) &= -\sin(\xi_1+\xi_2),\\
v_0(t,\xi_1,\xi_2) &= \sin(\xi_1+\xi_2).
\end{aligned}
\tag{30}
$$

The solutions in series form can be expressed as

$$
u(t,\xi_1,\xi_2) = -\sin(\xi_1+\xi_2) + \frac{1}{n}\left(\frac{h(-2b_0\sin(\xi_1+\xi_2)-g)t^\alpha}{\Gamma[\alpha+1]}\right) \tag{31}
$$

$$
+\frac{1}{n^2}\left(\frac{h(-2b_0\sin(\xi_1+\xi_2)-g)t^\alpha}{\Gamma[\alpha+1]} - \frac{4b_0^2h^2\sin(\xi_1+\xi_2)t^{2\alpha}}{\Gamma[2\alpha+1]}\right)+\cdots \tag{32}
$$

$$
v(t,\xi_1,\xi_2) = \sin(\xi_1+\xi_2) + \frac{1}{n}\left(\frac{h(2b_0\sin(\xi_1+\xi_2)+g)t^\alpha}{\Gamma[\alpha+1]}\right) \tag{33}
$$

$$
+\frac{1}{n^2}\left(\frac{(n+h)h(-2b_0\sin(\xi_1+\xi_2)+g)t^\alpha}{\Gamma[\alpha+1]} + \frac{4b_0^2h^2\sin(\xi_1+\xi_2)t^{2\alpha}}{\Gamma[2\alpha+1]}\right)+\cdots. \tag{34}
$$

As $n \to \infty$, the solution converges to the exact solution. Taking $n = 1$ and $h = -1$, we obtain

$$
u(t,\xi_1,\xi_2) = -\sin(\xi_1+\xi_2) + \frac{(2b_0\sin(\xi_1+\xi_2)+g)t^\alpha}{\Gamma[\alpha+1]} - \frac{4b_0^2h^2\sin(\xi_1+\xi_2)t^{2\alpha}}{\Gamma[2\alpha+1]}+\cdots, \tag{35}
$$

$$
v(t,\xi_1,\xi_2) = \sin(\xi_1+\xi_2) - \frac{(2b_0\sin(\xi_1+\xi_2)+g)t^\alpha}{\Gamma[\alpha+1]} + \frac{4b_0^2\sin(\xi_1+\xi_2)t^{2\alpha}}{\Gamma[2\alpha+1]}+\cdots. \tag{36}
$$

The solution given in equations (35) and (36) is exactly the same as the solution obtained by the generalized integral transform Adomian decomposition method (GITADM). Now, taking $\alpha = 1$ in equations (21), (35) and (36), we obtain

$$
\begin{aligned}
u(t,\xi_1,\xi_2) &= -\sin(\xi_1+\xi_2) + \frac{(2b_0\sin(\xi_1+\xi_2)+g)t}{1!} - \frac{4b_0^2h^2\sin(\xi_1+\xi_2)t^2}{2!}+\cdots,\\
v(t,\xi_1,\xi_2) &= \sin(\xi_1+\xi_2) - \frac{(2b_0\sin(\xi_1+\xi_2)+g)t}{1!} + \frac{4b_0^2\sin(\xi_1+\xi_2)t}{2!}+\cdots,
\end{aligned}
\tag{37}
$$

which can be written as

$$
\begin{aligned}
u(t,\xi_1,\xi_2) &= -\sin(\xi_1+\xi_2)\left(1-2b_0t+\frac{(2b_0t)^2}{2!}-\cdots\right) = -\sin(\xi_1+\xi_2)e^{-2b_0t},\\
v(t,\xi_1,\xi_2) &= \sin(\xi_1+\xi_2)\left(1-2b_0t+\frac{(2b_0t)^2}{2!}-\cdots\right) = \sin(\xi_1+\xi_2)e^{-2b_0t},
\end{aligned}
\tag{38}
$$

which is the exact solution of the system of Navier–Stokes equations.

4 Results and discussion

In this section, we will discuss the results obtained using both novel methods named q-HAGITM and GITADM using Maple software. Table 1 provides numerical data of the absolute error of $u(t, \xi_1, \xi_2)$ obtained by both methods (upto three terms approximations) for $\alpha = 1$. Here, the q-HAGITM different values of h provides different solutions and sometimes faster convergence, too. Similarly, Table 2 shows the absolute error comparison for $v(t, \xi_1, \xi_2)$ via both methods (upto three terms approximations) for $\alpha = 1$. q-HAGITM and GITADM provide the same error for $h = -1$ and $n = 1$, such as ADM is one in the case of HAM. Hence, HAM had an advantage over ADM. Figure 1 presents the graphical representations of the exact solutions, the q-HAGITM solution, and the GITADM solution for $u(t, \xi_1, \xi_2)$ upto three terms approximations for $\alpha = 1$. Similarly, Figure 2 presents the comparison of surfaces of $v(t, \xi_1, \xi_2)$ upto three terms approximations for $\alpha = 1$. From Figures 2 and 3, we can see that both methods are very accurate and effective for finding solutions. Figure 4 represents the surfaces of $u(t, \xi_1, \xi_2)$ for $\alpha = 0.5$

Table 1: Absolute error of $u(t, \xi_1, \xi_2)$ at $\xi_2 = 0.5$, $b_0 = 0.1$, $g = 0$ and $n = 1$.

ξ_1	t	q-HAGITM			GITADM
		$h = -0.95$	$h = -1$	$h = -1.05$	
0.1	0.1	0.0000179708	7.491×10^{-7}	0.0000405565	7.491×10^{-7}
	0.5	0.0000423018	0.0000918008	0.0005223406	0.0000918008
	0.9	0.0001127654	0.0005249982	0.0017166761	0.0005249982
0.5	0.1	0.0000267813	0.0000011164	0.0000604401	0.0000011164
	0.5	0.0000630414	0.0001368080	0.0007784297	0.0001368080
	0.9	0.0001680514	0.0007823900	0.0025583146	0.0007823900
0.9	0.1	0.0000313637	0.0000013074	0.0000707817	0.0000013074
	0.5	0.0000738280	0.0001602163	0.0009116218	0.0001602163
	0.9	0.0001968055	0.0009162599	0.0029960516	0.0009162599

Table 2: Absolute error of $v(t, \xi_1, \xi_2)$ at $\xi_2 = 0.4$, $b_0 = 0.1$, $g = 0$ and $n = 1$.

ξ_1	t	q-HAGITM			GITADM
		$h = -0.95$	$h = -1$	$h = -1.05$	
0.2	0.2	0.0000209763	0.0000068034	0.0001240510	0.0000068034
	0.6	0.0000788759	0.0001800995	0.0008487975	0.0001800995
	1	0.0001164443	0.0008176713	0.0024604264	0.0008176713
0.6	0.2	0.0000290184	0.0000094119	0.0001716117	0.0000094119
	0.6	0.0001091166	0.0002491489	0.0011742221	0.0002491489
	1	0.0001610885	0.0011311622	0.0034037410	0.0011311622
1	0.2	0.0000324792	0.0000105343	0.0001920785	0.0000105343
	0.6	0.0001221300	0.0002788630	0.0013142628	0.0002788630
	1	0.0001803004	0.0012660674	0.0038096797	0.0012660674

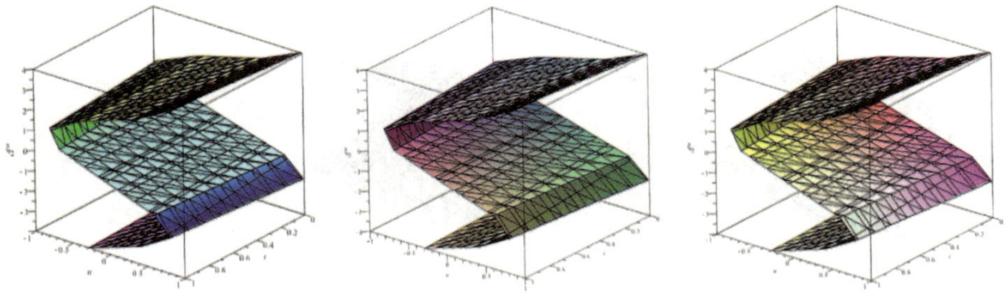

(a) Exact Solution **(b)** q-HAGITM solution **(c)** GITADM solution

Figure 1: Surface of $u(t, \xi_1, \xi_2)$.

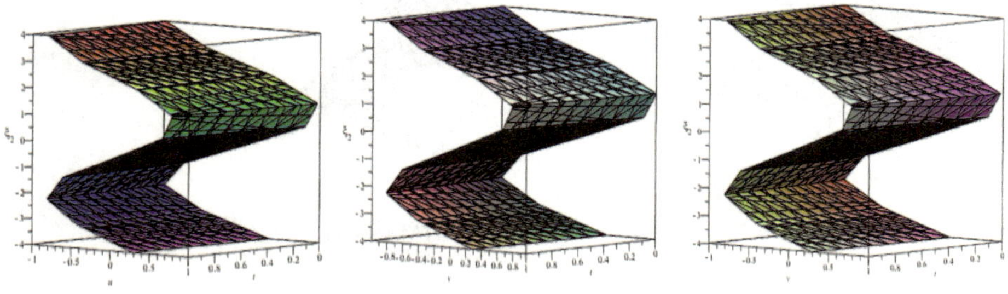

(a) Exact Solution **(b)** q-HAGITM solution **(c)** GITADM solution

Figure 2: Surface of $v(t, \xi_1, \xi_2)$.

(Figures 3(a) and 3(b)) and for $\alpha = 0.75$ (Figures 3(c) and 3(d)). Similarly, Figures 4(a) and 4(b) are the presentation of $v(t, \xi_1, \xi_2)$ at $\alpha = 0.5$ using GITADM and q-HAGITM, respectively, and Figures 4(c) and 4(d) are the presentation of $v(t, \xi_1, \xi_2)$ at $\alpha = 0.75$ using GITADM and q-HAGITM, respectively. Figures 3 and 4 show that both methods are very good at solving fractional-order differential equations.

5 Conclusion

The numerical solution to the coupled Navier–Stokes equations of time-fractional order is found in this chapter utilizing q-HAGITM and GITADM. The results support the reliability and accessibility of these methods. In order to modify the convergence point of a series of solutions, the utilized algorithm q-HAGITM enables us to vary the value of the parameter h. For $h = -1$ and $n = 1$, the solutions of GITADM and q-HAGITM coincide, therefore GITADM can be presented as a special case of q-HAGITM. The ability of q-HAGITM to minimize computational effort while preserving excellent accuracy shows even more its efficiency. q-HAGITM offers a good approach to solve coupled nonlinear fractional partial differential equations, hence lowering computing complexity without

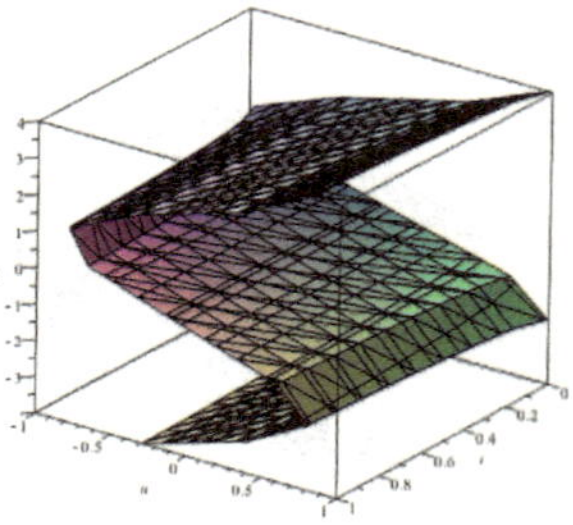

(a) GITADM Solution $\alpha = 0.5$

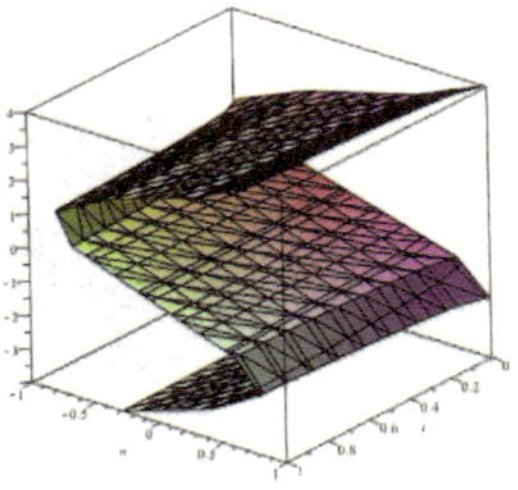

(b) q-HAGITM solution $\alpha = 0.5$

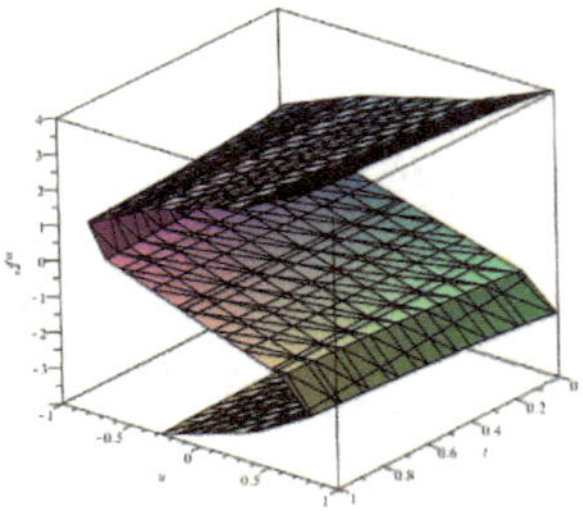

(c) GITADM Solution $\alpha = 0.75$

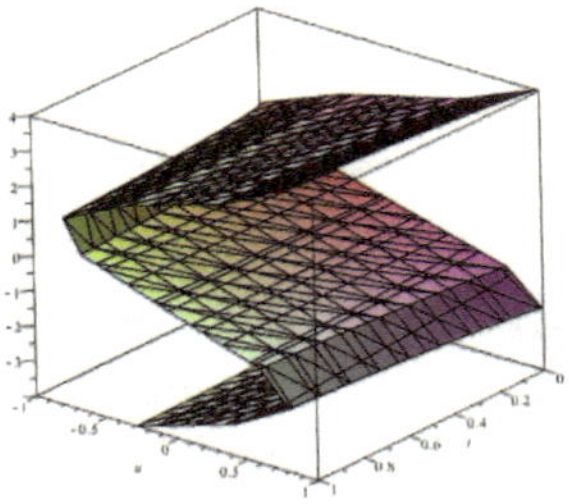

(d) q-HAGITM solution $\alpha = 0.75$

Figure 3: Surface of $u(t, \xi_1, \xi_2)$ at different values of α.

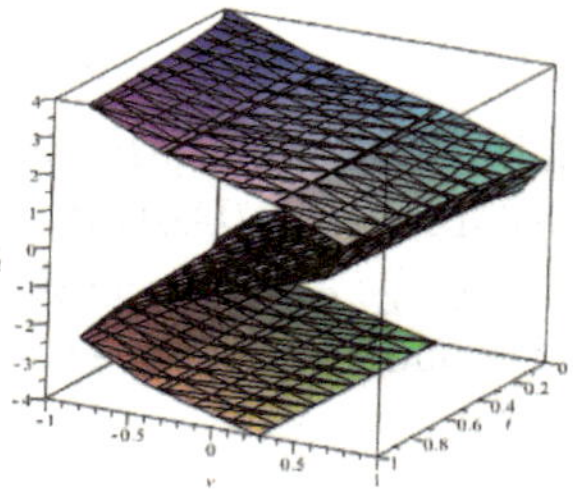

(a) GITADM Solution $\alpha = 0.5$

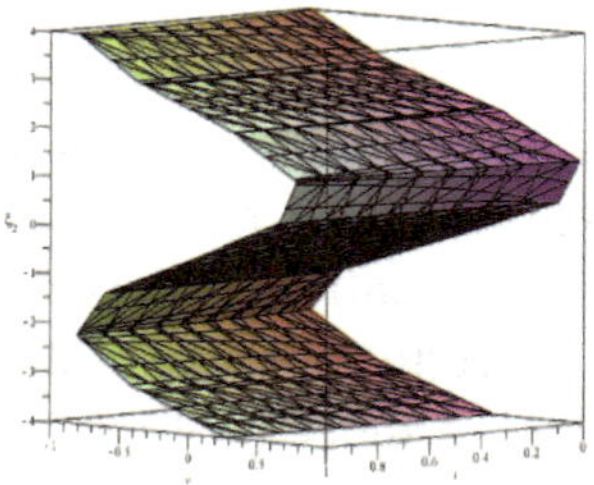

(b) q-HAGITM solution $\alpha = 0.5$

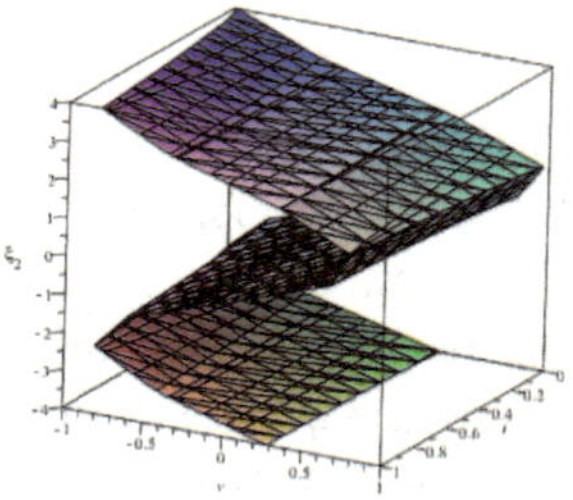

(c) GITADM Solution $\alpha = 0.75$

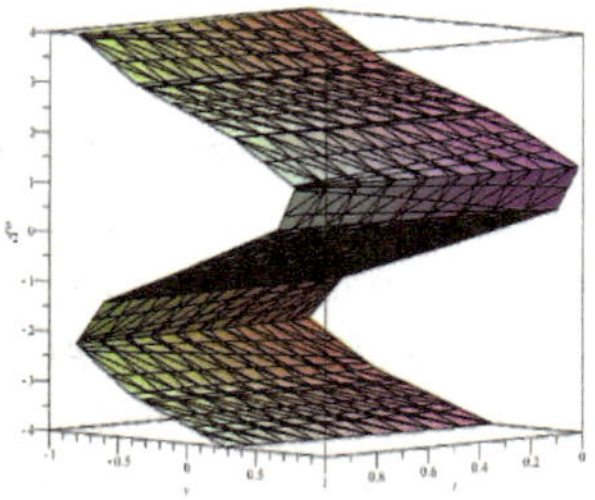

(d) q-HAGITM solution $\alpha = 0.75$

Figure 4: Surface of $v(t, \xi_1, \xi_2)$ at different values of α.

sacrificing accuracy from conventional methods. Furthermore, the adaptability of the approach is improved by the flexibility brought about by the parameter h, so the method is a strong instrument for handling a wide spectrum of fractional-order problems. These results confirm q-HAGITM as a useful method for addressing challenging fluid dynamics models, therefore opening the path for future applications in various engineering and scientific fields. Further work will focus on extending the proposed q-HAGITM to more complex fractional fluid equations, including multidimensional and variable-coefficient systems. The method will also be examined for problems involving nonsingular kernels and coupled thermofluid or magnetohydrodynamic effects.

Bibliography

[1] Birajdar, G. A. (2014). Numerical solution of time fractional Navier–Stokes equation by discrete Adomian decomposition method. *Nonlinear Engineering*, *3*(1), 21–26.

[2] Navier, C. L. M. H. (1823). Mémoire sur les lois du mouvement des fluides. *Mémoires de l'Académie Des Sciences de L'Institut de France*, *6*(1823), 389–440.

[3] Singh, B. K. & Kumar, P. (2018). FRDTM for numerical simulation of multi-dimensional, time-fractional model of Navier–Stokes equation. *Ain Shams Engineering Journal*, *9*(4), 827–834.

[4] Prakash, A., Veeresha, P., Prakasha, D. G. & Goyal, M. (2019). A new efficient technique for solving fractional coupled Navier–Stokes equations using q-homotopy analysis transform method. *Pramāna*, *93*(1), 6.

[5] El-Shahed, M. & Salem, A. (2004). On the generalized Navier–Stokes equations. *Applied Mathematics and Computation*, *156*(1), 287–293.

[6] Adomian, G. (1988). A review of the decomposition method in applied mathematics. *Journal of Mathematical Analysis and Applications*, *135*(2), 501–544.

[7] Liao, S. J. (1992). On the proposed homotopy analysis technique for nonlinear problems and its applications. *Shanghai Jiao Tong University*.

[8] El-Tawil, M. A. & Huseen, S. N. (2012). The q-homotopy analysis method (q-HAM). *International Journal of Applied Mathematics and Mechanics*, *8*(15), 51–75.

[9] Patel, H. S. & Meher, R. (2018). Analytical investigation of Jeffery–Hamel flow by modified Adomian decomposition method. *Ain Shams Engineering Journal*, *9*(4), 599–606.

[10] Kumar, A. & Meher, R. (2022). Solving non-linear partial differential equations using homotopy analysis method (HAM). In *Nonlinear Dynamics and Applications: Proceedings of the ICNDA 2022* (pp. 955–964). Cham: Springer.

[11] Verma, L., Meher, R., Hammouch, Z. & Baskonus, H. M. (2022). Effect of heat transfer on hybrid nanofluid flow in converging/diverging channel using fuzzy volume fraction. *Scientific Reports*, *12*(1), 20845.

[12] Verma, L. & Meher, R. (2022, October). Solution of fuzzy differential equation using homotopy analysis method. In *AIP Conference Proceedings* (Vol. 2451, No. 1, p. 020031). AIP Publishing LLC.

[13] Kumar, A. & Meher, R. (2025). Study on micropolar nanofluid flow with thermal radiation allowance across a resistive porous material between the channel walls. *Archive of Applied Mechanics*, *95*(3), 70.

[14] Verma, L. & Meher, R. (2022). Study on generalized fuzzy fractional human liver model with Atangana–Baleanu–Caputo fractional derivative. *The European Physical Journal Plus*, *137*(11), 1–20.

[15] Sartanpara, P. & Meher, R. (2022, October). Computational study of Klein–Gordan equation using homotopy analysis method. In *AIP Conference Proceedings* (Vol. 2451, No. 1, p. 020030). AIP Publishing LLC.

[16] Prajapati, V. J. & Meher, R. (2022). Solution of time-fractional Rosenau–Hyman model using a robust homotopy approach via formable transform. *Iranian Journal of Science and Technology, Transactions A: Science*, *46*(5), 1431–1444.

[17] Singh, J., Ahmadian, A., Rathore, S., Kumar, D., Baleanu, D., Salimi, M. & Salahshour, S. (2021). An efficient computational approach for local fractional Poisson equation in fractal media. *Numerical Methods for Partial Differential Equations*, *37*(2), 1439–1448.

[18] Varsoliwala, A. C. & Singh, T. R. (2022). Solution of foam drainage model by Elzaki Adomian decomposition method. *Mathematics in Engineering, Science and Aerospace*, *13*(2).

[19] Jena, R. M. & Chakraverty, S. (2019). Solving time-fractional Navier–Stokes equations using homotopy perturbation Elzaki transform. *SN Applied Sciences*, *1*(1), 16.

[20] Patel, T., Patel, H. & Meher, R. (2024). Analytical study of atmospheric internal waves model with fractional approach. *Journal of Ocean Engineering and Science*, *9*(4), 326–337.

[21] Jani, H. P. & Singh, T. R. (2023). Solution of time fractional Swift–Hohenberg equation by Aboodh transform homotopy perturbation method. *International Journal of Nonlinear Analysis and Applications*, *14*(1), 1005–1013.

[22] Veeresha, P. & Kumar, D. (2025). Analysis and dynamics of the Ivancevic option pricing model with a novel fractional calculus approach. *Waves Random Complex Media*, *35*(3), 5376–5393.

[23] Patel, T. & Meher, R. (2019). A study on convective-radial fins with temperature-dependent thermal conductivity and internal heat generation. *Nonlinear Engineering*, *8*(1), 145–156.

[24] Patel, T. & Patel, H. (2023). An analytical approach to solve the fractional-order (2 + 1)-dimensional Wu–Zhang equation. *Mathematical Methods in the Applied Sciences*, *46*(1), 479–489.

[25] Jani, H. P. & Singh, T. R. (2024). Aboodh transform homotopy perturbation method for solving fractional-order Newell–Whitehead–Segel equation. *Mathematical Methods in the Applied Sciences*, *47*(15), 12028–12043.

[26] Prakash, A., Veeresha, P., Prakasha, D. G. & Goyal, M. (2019). A homotopy technique for a fractional order multi-dimensional telegraph equation via the Laplace transform. *The European Physical Journal Plus*, *134*(1). Article 19.

[27] Prakash, A. & Kaur, H. (2017). Numerical solution for fractional model of Fokker–Planck equation by using q-HATM. *Chaos, Solitons and Fractals*, *105*, 99–110.

[28] Verma, L. & Meher, R. (2022). Effect of heat transfer on Jeffery–Hamel Cu/Ag–water nanofluid flow with uncertain volume fraction using the double parametric fuzzy homotopy analysis method. *The European Physical Journal Plus*, *137*(3), Article 312.

[29] Sartanpara, P. P. & Meher, R. (2021). A robust computational approach for Zakharov–Kuznetsov equations of ion-acoustic waves in a magnetized plasma via the Shehu transform. *Journal of Ocean Engineering and Science*. https://doi.org/10.1016/j.joes.2021.11.006.

[30] Jhangeer, A., Jamal, T., Talafha, A. M. & Riaz, M. B. (2024). Exploring travelling wave solutions, bifurcation, chaos, and sensitivity analysis in the (3+1)-dimensional gKdV–ZK model: a comprehensive study using Lie symmetry methodology. *Results in Engineering*, *22*, Article 102194.

[31] Talafha, A. M., Jhangeer, A. & Kazmi, S. S. (2023). Dynamical analysis of (4 + 1)-dimensional Davey–Stewartson–Kadomtsev–Petviashvili equation by employing Lie symmetry approach. *Ain Shams Engineering Journal*, *14*(11), Article 102537.

[32] Schurz, H. & Talafha, A. M. (2021). Existence, uniqueness, and energy of approximate Fourier solutions of modified stochastic sine–Gordon equation with power-law nonlinearity in 1D. *International Journal of Applied and Computational Mathematics*, *7*(4), Article 126.

[33] Jhangeer, A., Talafha, A. M. & Ansari, A. R. (2024). Analytical solutions and dynamical behaviors of the extended Bogoyavlensky–Konopelchenko equation in deep water dynamics. *Physica Scripta*, *100*(1), Article 015279.

[34] Kumar, A. & Meher, R. (2024). Solving time fractional partial differential equations using Shehu transform with q-homotopy analysis approach. In *AIP Conference Proceedings* (Vol. 3242, No. 1). AIP Publishing.

[35] Sartanpara, P. P. & Meher, R. (2022). A robust fuzzy-fractional approach for the atmospheric internal wave model. *Journal of Ocean Engineering and Science*. https://doi.org/10.1016/j.joes.2022.02.001.

[36] Sartanpara, P. P., Meher, R. & Meher, S. K. (2022). The generalized time-fractional Fornberg–Whitham equation: an analytic approach. *Partial Differential Equations in Applied Mathematics*, *5*, Article 100350.

[37] Rashid, S., Kubra, K. T. & Guirao, J. L. (2021). Construction of an approximate analytical solution for multi-dimensional fractional Zakharov–Kuznetsov equation via Aboodh Adomian decomposition method. *Symmetry*, *13*(8), Article 1542.

[38] Jafari, H. (2021). A new general integral transform for solving integral equations. *Journal of Advanced Research*, *32*, 133–138.

[39] Sartanpara, P. P. & Meher, R. (2023). Solution of generalised fuzzy fractional Kaup–Kupershmidt equation using a robust multiparametric approach and a novel transform. *Mathematics and Computers in Simulation*, *205*, 939–969.

[40] Prajapati, V. J. & Meher, R. (2022). A robust analytical approach to the generalized Burgers–Fisher equation with fractional derivatives including singular and non-singular kernels. *Journal of Ocean Engineering and Science*. https://doi.org/10.1016/j.joes.2022.06.035.

Bharat Bhushan, Harjot Singh, Manoj Gaur, and Sourav Kumawat

Exact soliton solutions of time-fractional nonlinear evolution equations using the $(\frac{G'}{G^2})$-expansion method

Abstract: The fractional-order modified equal-width (MEW) equation does a good job of simulating the one-dimensional long water wave propagation in a nonlinear medium, including the dispersion process. The TFFE (Time-Fractional Fisher Equation) describes a wide range of substantial physical phenomena, it is an FPDE generated from the conventional Fisher equation by ousting the time derivative with an FDE of order $1 \geq \psi > 0$. The well-known Burger equation and Fisher equation are combined in the time-fractional generalized Burger–Fisher equation (TF-GBFE), a fractional partial differential equation. Exact solutions for TF-GBFE have been obtained by using the $(\frac{G'}{G^2})$-expansion approach. Soliton waves, which are particularly relevant in nonlinear wave phenomena and are distinguished by localized waveforms that maintain their shape during propagation, are one of the solutions that have been found. Overall, this work emphasizes the potential of the $(\frac{G'}{G^2})$-expansion approach for the study of additional conformable fractional partial differential equations while demonstrating how efficient, useful, and accurate it is in solving the modified equal-width equation, the time-fractional Fisher equation and the time-fractional generalized Burger–Fisher equation.

Keywords: $(\frac{G'}{G})$-Expansion method, exact solutions, conformable derivative, modified equal-width equation, time-fractional Fisher equation, time-fractional generalized Burger–Fisher equation

MSC 2020: 35R11, 35C07, 35C08, 35K57, 26A33

1 Introduction

The field of fractional calculus is an advanced version of regular math equations that we use in many different situations. Even though some might think that this is a new idea, it has been around for a long time. In 1695 Leibniz initially introduced fractional

Acknowledgement: The last author sincerely acknowledges the financial support provided by Akal University, Talwandi Sabo, which made this research work possible.

Bharat Bhushan, Harjot Singh, Sourav Kumawat, Department of Mathematics, Akal University, Talwandi Sabo, Bathinda, Punjab, India, e-mail: bgoyal.math@gmail.com
Manoj Gaur, Apex Institute of Technology (CSE), Chandigarh University, Gharuan, Mohali, India

https://doi.org/10.1515/9783112234525-003

calculus. Since then, investigations into fractional phenomena have taken place in a variety of scientific and engineering disciplines. For example, the mathematical models of the human liver and HIV are given in [8, 30]. In addition, the existence of singular solutions for each of these models is investigated using the Picard–Lindeland fixed-point theory method. Using the well-known fixed-point theorems developed by Dage, Baleanu et al. [6, 7, 9] developed solutions to fractional hybrid equations with hybrid boundary conditions that include variations for the thermostat model, in addition to dealing with fractional integral-differential equation modeling.

Several analytical and numerical techniques have been published recently in an effort to investigate the precise and approximate solutions of NLFDEs, such as the sine-cosine method [2], the Homotopy perturbation method [1, 16], a new version of the fractional F-expansion method [27], the modified Adomian decomposition method [18], the fractional subequation method [5], the tanh method [37], symmetry methods [35, 36], the first integral method [12, 15, 23, 34], the modified Kudryashov method [22], Sumudu transforms [10, 29], the extended tanh method [38], the q-homotopy analysis transform method [28], the Kumar–Malik method [24, 31], the $(\frac{G'}{G})$-Expansion method [14], and so on, but there is no unified method that can be used to deal with all types of nonlinear evolution equations. There are expansion methods that are commonly used to find the soliton solution of partial differential equations, however, these methods are not very commonly used in fractional partial differential equations.

It is important to note that there are various definitions for the fractional derivative existing in the literature. Each fractional derivative has a unique set of benefits as well as drawbacks [25, 32].

However, it is worth mentioning that the existing fractional derivatives lack certain properties that integral derivatives possess. In simpler terms, fractional derivatives do not behave in the same way as regular derivatives when it comes to certain aspects.

Khalil et al. [20] introduced an alternative approach to define fractional derivatives by extending the familiar limit definition, termed as conformable-type fractional derivatives.

Let $u(t)$ be any continuous real-valued function defined on a positive real number. For any fixed ψ, where ψ lies in $(0, 1]$, according to Khalil et al. [20] the function $u(t)$ is called conformable ψ-differentiable at any point p that is ${}_pD_t^{\psi}(u(t))$ if

$$\lim_{\epsilon \to 0} \frac{u(t + \epsilon(t-p)^{1-\psi}) - u(t)}{\epsilon}$$

exists.

2 Methodology

For any nonlinear conformable partial differential equation of the unknown function $u = u(x,t)$, where x and t are independent variables

$$P(u, u_x, u_t^a, u_{xx}) = 0. \tag{1}$$

The main steps of the $(\frac{G'}{G^2})$-expansion method for constructing an exact solution for (1) as discussed in [3] can be given as follows:

Step 1: Convert nonlinear conformable partial differential equation (1) into an ordinary differential equation (ODE) via the traveling wave transformation, i. e.

$$u(x,t) = U(\zeta), \quad \text{where } \zeta = x - \frac{ct^a}{a}, \tag{2}$$

where c is a nonzero constant that will be determined at a later step. Applying transformation (1) and then integrating the resulting equation with respect to ζ as many times as possible, we obtain the following ODE in $U = U(\zeta) : Q(U, U', U'', U''', \ldots) = 0$, where Q is a polynomial function of $U(\zeta)$, and its various integer-order derivatives. The prime notation ($'$) denotes the ordinary derivative with respect to ζ.

Step 2: Suppose that the general solution of the above ODE can be expressed in terms of $(\frac{G'}{G^2})$ as

$$U(\zeta) = a_0 + \sum_{i=1}^{N}\left[a_i\left(\frac{G'}{G^2}\right)^i + b_i\left(\frac{G'}{G^2}\right)^{-i}\right], \tag{3}$$

where $G = G(x)$ satisfies the simple Riccati equation

$$\left(\frac{G'}{G^2}\right)' = \mu + \lambda\left(\frac{G'}{G^2}\right)^2, \tag{4}$$

such that $\mu \neq 1$ and $\lambda \neq 0$ are arbitrary constants. The unknown constants a_N or b_N may be zero, but both of them cannot be zero simultaneously. The coefficients a_0, a_j, b_j ($j = 1, 2, \ldots, N$) are unknown constants to be determined in Step 3. The value of the positive integer N can be computed using the homogeneous balance principle, in other words, by balancing between the highest-order derivatives and the nonlinear terms appearing in equation (1). More precisely, if the degree of $U(\zeta)$ is $\text{Deg}[U(\zeta)] = N$, then the degree of the subsequent terms can be calculated using the following formulas:

$$\text{Deg}\left[\frac{d^q U(\zeta)}{d\zeta^q}\right] = N + q; \quad \text{Deg}\left[U(\zeta)^p \frac{d^q (U(\zeta))^s}{d\zeta^q}\right] = Np + s(N + q).$$

Step 3: Substituting equation (3), along with equation (4), into equation (1), we have a polynomial in $(\frac{G'}{G^2})$. Collecting all coefficients of the same power of $(\frac{G'}{G^2})^i$, where $i =$

$0, \pm 1, \pm 2, \ldots, \pm M$, where M is some positive integer and then setting all of the obtained coefficients to zero, we obtain a system of nonlinear algebraic equations for the unknown constants a_0, a_j, b_j $(j = 1, 2, \ldots, N)$. Assume that the resulting algebraic system can be possibly solved for the unknown constants using symbolic software packages such as Maple and Mathematica.

Step 4: The general solutions of (1) can be separated into the following three cases depending on the values of μ and λ:

If $\mu\lambda > 0$, then (1) has the trigonometric function solution given as

$$\left(\frac{G'}{G^2}\right) = \sqrt{\frac{\mu}{\lambda}}\left(\frac{C\cos\sqrt{\mu\lambda}\zeta + D\sin\sqrt{\mu\lambda}\zeta}{D\cos\sqrt{\mu\lambda}\zeta - C\sin\sqrt{\mu\lambda}\zeta}\right).$$

If $\mu\lambda < 0$, then (1) has the hyperbolic function solution given as

$$\left(\frac{G'}{G^2}\right) = -\frac{\sqrt{|\mu\lambda|}}{\lambda}\left(\frac{C\sinh\sqrt{|\mu\lambda|}\zeta + C\cosh\sqrt{|\mu\lambda|}\zeta + D}{C\sinh\sqrt{|\mu\lambda|}\zeta + C\cosh\sqrt{|\mu\lambda|}\zeta - D)}\right).$$

If $\mu = 0, \lambda \neq 0$ then (1) has the rational function solution given as

$$\left(\frac{G'}{G^2}\right) = -\frac{C}{\lambda(C\zeta + D)'},$$

where C, D are arbitrary nonzero constants.

The exact traveling wave solutions of equation (1) can be obtained by inserting the obtained values of a_0, a_j, b_j $(j = 1, 2, \ldots, N)$.

3 Application of the $(\frac{G'}{G^2})$-expansion method

3.1 Time-fractional modified equal-width equation

In several scientific disciplines, including fluid mechanics, plasma physics, solid-state physics, nonlinear Schrödinger equations, and optical fibers, nonlinear equations are frequently employed to explain complex phenomena [19, 26].

Fluid mechanics uses the equal-width equation as a model to explain how a bore occurs when a large wave of height moves through shallow water and steepens. The bore is undular if the change in the wave's surface elevation is less than 0.28 of the initial depth of the water. The gradual transition between a uniform flow and quiet water is caused by the development of an undular bore from a long wave [11].

A model of equal-width equations is used to handle nonlinear systems, which is the next step in this development's physical explanation. In recent years, there has been a lot of focus on the modified equal-width (MEW) equation resulting from the nonlinear media with a dispersion process [21, 33]

$$u_t^{\psi} - a(u^3)_x - bu_{xxt}^{\psi} = 0, \tag{5}$$

with $0 < \psi \le 1, t \le 0$, and a, b, ψ as nonzero parameters

$$u_t^{a} + a(u^3)_x - bu_{xxt}^{a} = 0. \tag{6}$$

Using the wave transformation in equation (2), equation (6) becomes

$$-cU' + a(U^3)' - bU''' = 0. \tag{7}$$

Integrating equation (7) and ignoring the integrating constant, we find

$$-cU + a(U^3) - bU'' = 0. \tag{8}$$

According to the $(\frac{G'}{G^2})$-expansion method

$$U(\zeta) = a_0 + \sum_{i=1}^{N}\left[a_i\left(\frac{G'}{G^2}\right)^i + b_i\left(\frac{G'}{G^2}\right)^{-i}\right],$$

where $G = G(\zeta)$ satisfies the simple Riccati equation

$$\left(\frac{G'}{G^2}\right)' = \mu + \lambda\left(\frac{G'}{G^2}\right)^2. \tag{9}$$

Then clearly, the maximum degree of $(\frac{G'}{G^2})$ in $\frac{d^m(U)}{d\zeta^m}$ is $N + m$ and the maximum degree of $(\frac{G'}{G^2})$ in U^p is Np. Comparing the highest-degree term and nonlinear term, we find $3n = n + 2$ gives $n = 1$. Therefore, the solution is

$$U(\zeta) = a_0 + a_1\left(\frac{G'}{G^2}\right) + b_1\left(\frac{G'}{G^2}\right)^{-1}. \tag{10}$$

Using equations (9) and (10), we find

$$U''(\zeta) = 2a_1\lambda^2\left(\frac{G'}{G^2}\right)^3 + 2a_1\lambda\mu\left(\frac{G'}{G^2}\right) + 2b_1\mu^2\left(\frac{G'}{G^2}\right)^{-3} + 2b_1\lambda\mu\left(\frac{G'}{G^2}\right)^{-1}. \tag{11}$$

Using equation (10) and equation (11) in equation (8), collecting all terms with the same powers of $(\frac{G'}{G^2})$ and setting each coefficient to zero, we obtain a system of algebraic equations for a_0, a_1, c, λ, and μ as follows:

$$\left(\frac{G'}{G^2}\right)^0 : \quad 6aa_1a_0b_1 + aa_0^3 - a_0c = 0;$$

$$\left(\frac{G'}{G^2}\right)^1 : \quad 3aa_1^2b_1 + 3aa_1a_0^2 - 2a_1b\lambda\mu - a_1c = 0;$$

$$\left(\frac{G'}{G^2}\right)^2 : \quad 3aa_0a_1^2 = 0;$$

$$\left(\frac{G'}{G^2}\right)^3 : \quad aa_1^3 - 2ba_1\lambda^2 = 0;$$

$$\left(\frac{G'}{G^2}\right)^{-1} : \quad 3ab_1^2a_1 + 3ab_1a_0^2 - 2b_1b\lambda\mu - b_1c = 0;$$

$$\left(\frac{G'}{G^2}\right)^{-2} : \quad 3aa_0b_1^2 = 0;$$

$$\left(\frac{G'}{G^2}\right)^{-3} : \quad ab_1^3 - 2bb_1\mu^2 = 0.$$

Solving this system by Maple gives the following solution in Table 1.

Table 1: Different solution sets for the time-fractional modified equal-width equation.

a	b	c	a_1	b_1	λ	μ	a_1
$\frac{2b\mu^2}{b_1^2}$	b	$-2b\lambda\mu$	0	b_1	λ	μ	0
a	b	$-2\sqrt{2}\sqrt{a}\sqrt{b}b_1\lambda$	$-\frac{\sqrt{2}\sqrt{b}\lambda}{\sqrt{a}}$	b_1	λ	$-\frac{\sqrt{a}b_1}{\sqrt{2}\sqrt{b}}$	0
a	b	$-4\sqrt{2}\sqrt{a}\sqrt{b}b_1\lambda$	$\frac{-\sqrt{2}\sqrt{b}\lambda}{\sqrt{a}}$	b_1	λ	$\frac{\sqrt{a}b_1}{\sqrt{2}\sqrt{b}}$	0
a	b	$4\sqrt{2}\sqrt{a}\sqrt{b}b_1\lambda$	$\frac{\sqrt{2}\sqrt{b}\lambda}{\sqrt{a}}$	b_1	λ	$-\frac{\sqrt{a}b_1}{\sqrt{2}\sqrt{b}}$	0
a	b	$2\sqrt{2}\sqrt{a}\sqrt{b}b_1\lambda$	$\frac{\sqrt{2}\sqrt{b}\lambda}{\sqrt{a}}$	b_1	λ	$\frac{\sqrt{a}b_1}{\sqrt{2}\sqrt{b}}$	0

By taking $a = 1$, $a_1 = 1$, $b = 1$, $\lambda = 1$ in the 2nd solution, we have $c = 4$, $\mu = -\frac{1}{2}$, $a_0 = 0$, $C = 1$, $D = 1$, then we have the value of

$$\left(\frac{G'}{G^2}\right) = -\frac{1}{\sqrt{2}}\left(\frac{\sinh(\sqrt{\frac{1}{\sqrt{2}}})(x - \frac{4t^\alpha}{\alpha}) + \cosh(\sqrt{\frac{1}{\sqrt{2}}})(x - \frac{4t^\alpha}{\alpha}) + 1}{\sinh(\sqrt{\frac{1}{\sqrt{2}}})(x - \frac{4t^\alpha}{\alpha}) + \cosh(\sqrt{\frac{1}{\sqrt{2}}})(x - \frac{4t^\alpha}{\alpha}) - 1}\right).$$

Using $U(\zeta) = a_0 + a_1(\frac{G'}{G^2}) + b_1(\frac{G'}{G^2})^{-1}$, we have the following graphs (Figure 1) over different values of α over the range $t \in [0, 4]$ and $x \in [-4, 4]$.

3.2 Time-fractional Fisher equation

The time derivative is eliminated from the Fisher equation using an FDO of order $0 < \alpha \leq 1$ to create the FPDE that numerically describes a broad range of significant physical phenomena. The TFFE is described as

$$u_t^\alpha = u_{xx} + au(1 - u) = 0,$$

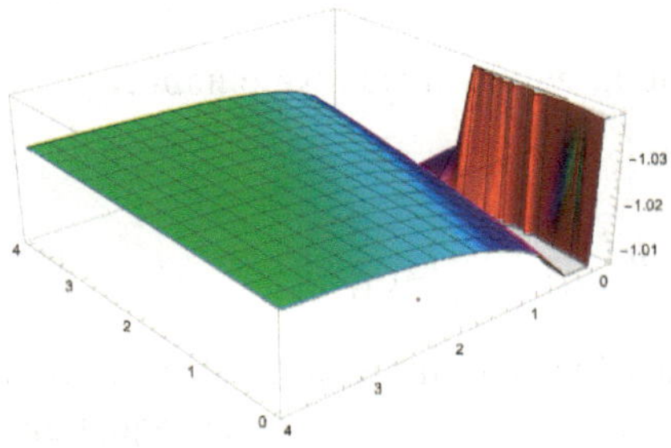

(a) $\alpha = 0.5$

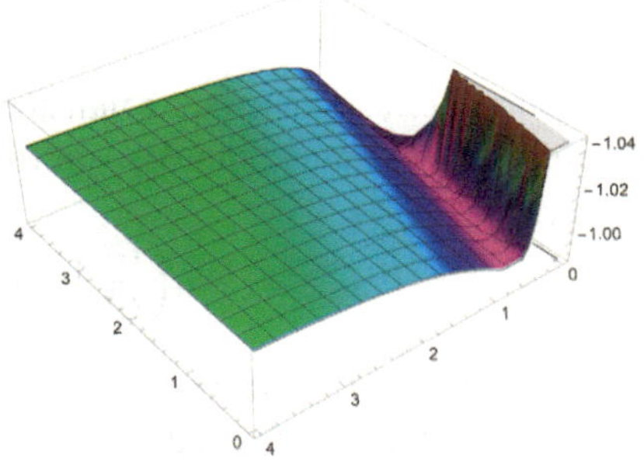

(b) $\alpha = 0.75$

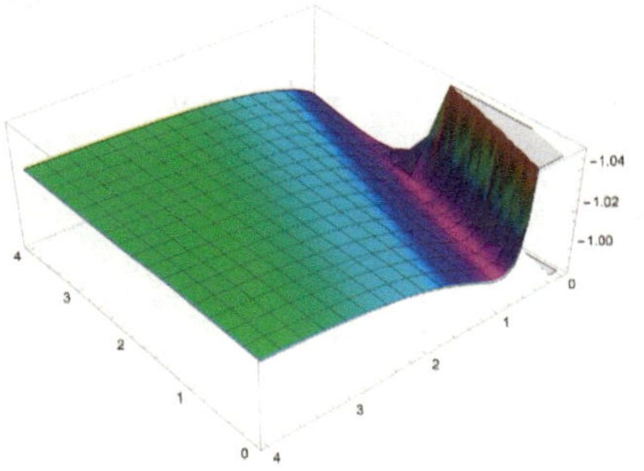

(c) $\alpha = 0.9$

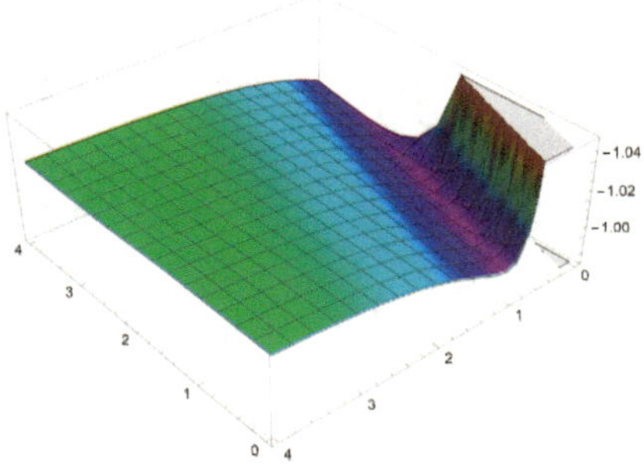

(d) $\alpha = 1$

Figure 1: Graphs showing the changes with different values of α: (a) $\alpha = 0.5$, (b) $\alpha = 0.75$, (c) $\alpha = 0.9$, and (d) $\alpha = 1$.

with $0 < \alpha \leq 1$, $t \geq 0$, and a as nonzero parameters.

TFFE is used in a model equation for the evolution of a neutron population in a nuclear reactor and a prototype model for a spreading flame. The mutant is housed in an indefinitely long habitat. Numerous researchers have examined TFFE theoretically and mathematically; for more information and projections, the reader should see [4, 17]. The world has witnessed an influx of crucial information, such as tsunami wave fields caused by Japan's 2011 earthquakes. These incredible capabilities of ocean waves can be harnessed to provide a variety of energy sources for a variety of uses. We must look at the mathematical underpinnings of such natural problems to lessen the immense force of such massive catastrophes or utilize them as useful energy sources. If we approach these problems differently, we might select the best method for analyzing such possible catastrophes and implement the appropriate safety measures

$$u_t^{\alpha} = u_{xx} + au(1 - u) = 0. \tag{12}$$

Using the wave transformation in equation (2) with equation (12), the following equation is determined:

$$-cU' = U'' + aU(1 - U). \tag{13}$$

Using the homogeneous balancing principle in equation (13), we obtain $N = 2$. Therefore, the assumed solution, for the TFFE is

$$U(\zeta) = a_0 + a_1\left(\frac{G'}{G^2}\right) + a_2\left(\frac{G'}{G^2}\right)^2 + b_1\left(\frac{G'}{G^2}\right)^{-1} + b_2\left(\frac{G'}{G^2}\right)^{-2}. \tag{14}$$

Using equations (4) and (14) in equation (13), collecting all terms with the same powers of $(\frac{G'}{G^2})$ and setting each coefficient to zero, we obtain a system of algebraic equations for $a_0, a_1, a_2, b_1, b_2, c, \lambda$, and μ as follows:

$$\begin{aligned}
\left(\frac{G'}{G^2}\right)^0 &: \quad -2aa_1b_1 - 2aa_2b_2 - aa_0^2 + aa_0 + 2a_2\mu^2 - b_1c\lambda + 2b_2\lambda^2 = 0;\\
\left(\frac{G'}{G^2}\right)^1 &: \quad -2aa_1a_0 + aa_1 - 2aa_2b_1 + 2a_1\lambda\mu + 2a_2c\mu = 0;\\
\left(\frac{G'}{G^2}\right)^2 &: \quad -aa_1^2 - 2aa_2a_0 + aa_2 + a_1c\lambda + 8a_2\lambda\mu = 0;\\
\left(\frac{G'}{G^2}\right)^3 &: \quad -2aa_1a_2 + 2a_1\lambda^2 + 2a_2c\lambda = 0;\\
\left(\frac{G'}{G^2}\right)^4 &: \quad 6a_2\lambda^2 - aa_2^2 = 0;\\
\left(\frac{G'}{G^2}\right)^{-1} &: \quad -2aa_1b_2 - 2aa_0b_1 + ab_1 + 2b_1\lambda\mu - 2b_2c\lambda = 0;\\
\left(\frac{G'}{G^2}\right)^{-2} &: \quad -2aa_0b_2 - ab_1^2 + ab_2 - b_1c\mu + 8b_2\lambda\mu = 0;\\
\left(\frac{G'}{G^2}\right)^{-3} &: \quad -2ab_1b_2 - 2b_2c\mu = 0;\\
\left(\frac{G'}{G^2}\right)^{-4} &: \quad 6b_2\mu^2 - ab_2^2 = 0.
\end{aligned}$$

Solving this system by Maple gives the following solutions in Table 2. In solution 5, if we take $a = 1$, $\mu = 1$ then we have $a_1 = -\frac{1}{16\sqrt{6}}$, $a_2 = \frac{1}{1536}$, $b_1 = -\sqrt{6}$, $b_2 = 6$, $\mu = 1$, $\lambda = \frac{-1}{96}$, $c = \frac{5}{\sqrt{6}}$. Using these values $(\frac{G'}{G^2}) = -\frac{1}{\sqrt{96}}(\frac{\sinh(\sqrt{\frac{1}{\sqrt{96}}})(x-\frac{5}{\sqrt{6}}\frac{t^\alpha}{\alpha})+\cosh(\sqrt{\frac{1}{\sqrt{96}}})(x-\frac{5}{\sqrt{6}}\frac{t^\alpha}{\alpha})+1}{\sinh(\sqrt{\frac{1}{\sqrt{96}}})(x-\frac{5}{\sqrt{6}}\frac{t^\alpha}{\alpha})+\cosh(\sqrt{\frac{1}{\sqrt{96}}})(x-\frac{5}{\sqrt{6}}\frac{t^\alpha}{\alpha})-1})$. Using $U(\zeta) = a_0 + a_1(\frac{G'}{G^2}) + a_2(\frac{G'}{G^2})^2 + b_1(\frac{G'}{G^2})^{-1} + b_2(\frac{G'}{G^2})^{-2}$, then we have the following graphs (Figure 2) over α.

Table 2: Different solution sets for the time-fractional Fisher equation.

a	a_1	a_2	a_0	b_1	b_2	c	λ	μ
a	0	0	$-\frac{1}{2}$	0	$\frac{6\mu^2}{a}$	0	$-\frac{a}{4\mu}$	μ
a	0	$\frac{3a}{128\mu^2}$	$\frac{1}{4}$	0	$\frac{6\mu^2}{a}$	0	$-\frac{a}{16\mu}$	μ
a	0	0	$\frac{1}{4}$	$-\frac{\sqrt{6}\mu}{\sqrt{a}}$	$\frac{6\mu^2}{a}$	$\frac{5\sqrt{a}}{\sqrt{6}}$	$-\frac{a}{24\mu}$	μ
a	0	0	$\frac{1}{4}$	$\frac{\sqrt{6}\mu}{\sqrt{a}}$	$\frac{6\mu^2}{a}$	$-\frac{5\sqrt{a}}{\sqrt{6}}$	$-\frac{a}{24\mu}$	μ
a	$-\frac{\sqrt{a}}{16\sqrt{6}\mu}$	$\frac{a}{1536\mu^2}$	$\frac{3}{8}$	$-\frac{\sqrt{6}\mu}{\sqrt{a}}$	$\frac{6\mu^2}{a}$	$\frac{5\sqrt{a}}{\sqrt{6}}$	$-\frac{a}{96\mu}$	μ
a	$\frac{\sqrt{a}}{16\sqrt{6}\mu}$	$\frac{a}{1536\mu^2}$	$\frac{3}{8}$	$\frac{\sqrt{6}\mu}{\sqrt{a}}$	$\frac{6\mu^2}{a}$	$-\frac{5\sqrt{a}}{\sqrt{6}}$	$-\frac{a}{96\mu}$	μ
a	0	$\frac{3a}{128\mu^2}$	$\frac{3}{4}$	0	$\frac{6\mu^2}{a}$	0	$\frac{a}{16\mu}$	μ
a	0	0	$\frac{3}{2}$	0	$\frac{6\mu^2}{a}$	0	$\frac{a}{4\mu}$	μ
a	0	$\frac{6\lambda^2}{a}$	$-\frac{1}{2}$	0	0	0	λ	$-\frac{a}{4\lambda}$
a	$-\frac{\sqrt{6}\lambda}{\sqrt{a}}$	$\frac{6\lambda^2}{a}$	$\frac{1}{4}$	0	0	$-\frac{5\sqrt{a}}{\sqrt{6}}$	λ	$-\frac{a}{24\lambda}$
a	$\frac{\sqrt{6}\lambda}{\sqrt{a}}$	$\frac{6\lambda^2}{a}$	$\frac{1}{4}$	0	0	$\frac{5\sqrt{a}}{\sqrt{6}}$	λ	$-\frac{a}{24\lambda}$
a	0	$\frac{6\lambda^2}{a}$	$\frac{3}{2}$	0	0	0	λ	$\frac{a}{4\lambda}$

3.3 Time-fractional generalized Burger–Fisher equation

"The time-fractional generalized Burger–Fisher equation (TF-GBFE)" is a convection–diffusion model that has the potential to replicate the most hazardous tsunamis in the coastal zone by simulating underwater landslides [13].

The TF-GBFE with the fractional order α and any real constants a, b, and ψ is defined as follows: $u_t^\alpha - u_{xx} + au^\psi u_x + bu(u^\psi - 1) = 0$, with $0 < \alpha \leq 1$, $t \geq 0$, and a, b, ψ as nonzero parameters. The current study may be expanded to include oceanic modeling, nonlinear mechanical models, and thermal systems

$$u_t^\alpha - u_{xx} + au^\psi u_x + bu(u^\psi - 1) = 0. \tag{15}$$

Using the wave transformation $u(x,t) = u(\zeta)$, $\zeta = x - \frac{ct^\alpha}{\alpha}$. The following equation is determined:

$$-cu' - u'' - au^\psi u' + bu(u^\psi - 1) = 0, \tag{16}$$

where $u' = \frac{du}{d\zeta}$. Let

$$u(\zeta) = v^{\frac{1}{\psi}}(\zeta).$$

Then, equation (16) becomes

$$-\psi vv'' + (\psi - 1)v'^2 + (-c\psi + a\psi v)vv' - b\psi^2 v^2(1 - v) = 0. \tag{17}$$

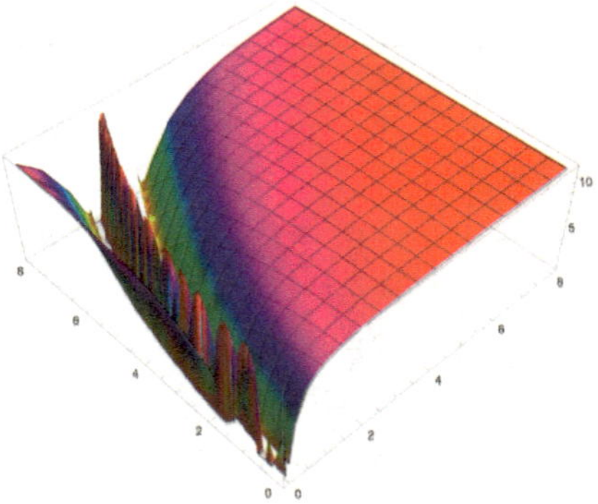

(a) $\alpha = 0.5$

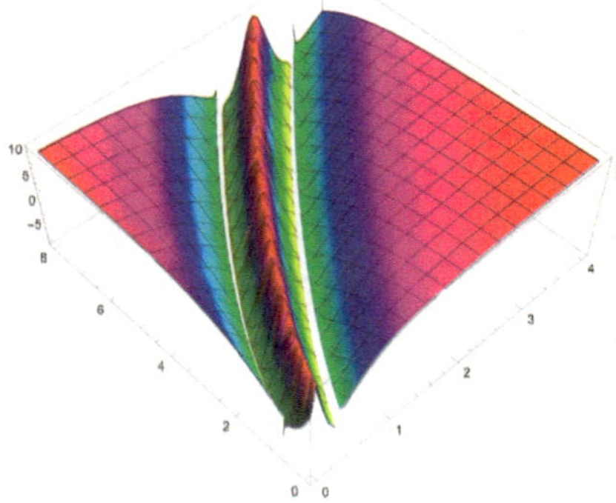

(b) $\alpha = 0.75$

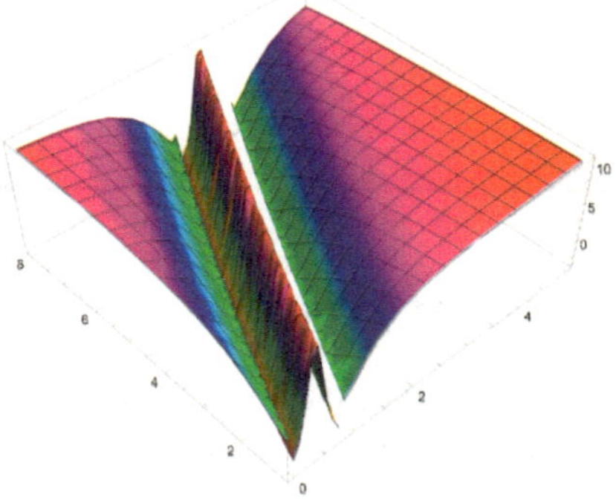

(c) $\alpha = 0.9$

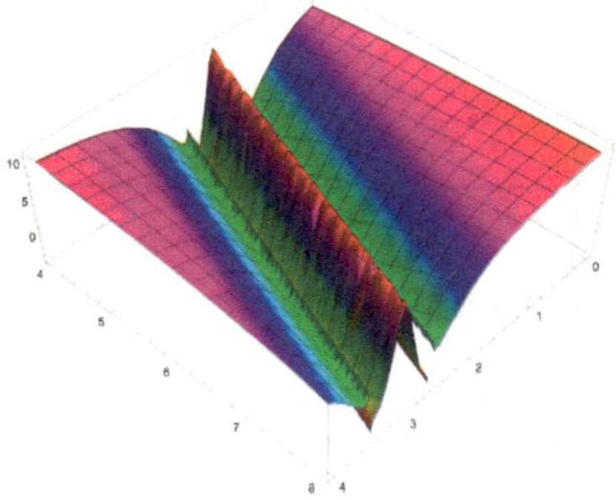

(d) $\alpha = 1$

Figure 2: Graphs showing the changes with different values of α: (a) $\alpha = 0.5$, (b) $\alpha = 0.75$, (c) $\alpha = 0.9$, and (d) $\alpha = 1$.

According to the $(\frac{G'}{G^2})$-expansion method, the solution will be

$$v(\zeta) = a_0 + \sum_{i=1}^{N}\left[a_i\left(\frac{G'}{G^2}\right)^{i} + b_i\left(\frac{G'}{G^2}\right)^{-i}\right],$$

where $G = G(\zeta)$ satisfies the simple Riccati equation

$$\left(\frac{G'}{G^2}\right)' = \mu + \lambda\left(\frac{G'}{G^2}\right)^2. \tag{18}$$

Then clearly, the maximum degree of $(\frac{G'}{G})$ in $\frac{d^m u}{d\zeta^m}$ is $N + m$ and the maximum degree of $(\frac{G'}{G})$ in u^p is Np. Comparing the highest-order derivative and nonlinear term, we find $3n + 1 = 2n + 2$ gives $n = 1$. Therefore, the solution is

$$v(\zeta) = a_0 + a_1\left(\frac{G'}{G^2}\right) + b_1\left(\frac{G'}{G^2}\right)^{-1}. \tag{19}$$

Using equation (19) in equation (17), collecting all terms with the same powers of $(\frac{G'}{G^2})$ and setting each coefficient to zero, we obtain a system of algebraic equations for a_0, a_1, c, λ, and μ as follows:

$$\begin{aligned}
\left(\frac{G'}{G^2}\right)^0 &: \ aa_1^2b_1\psi\mu + aa_1a_0^2\psi\mu - aa_1b_1^2\lambda\psi - aa_0^2b_1\lambda\psi + a_1^2\psi\mu^2 - a_1^2\mu^2 \\
&\quad + 6a_1a_0bb_1\psi^2 - a_1a_0c\psi\mu - 2a_1bb_1\psi^2 - 8a_1b_1\lambda\psi\mu + 4a_1b_1\lambda\mu \\
&\quad + a_0^3b\psi^2 - a_0^2b\psi^2 + a_0b_1c\lambda\psi + b_1^2\lambda^2\psi - b_1^2\lambda^2 = 0; \\
\left(\frac{G'}{G^2}\right)^1 &: \ a_1\psi(-2a_0(\mu(\lambda - aa_1) + b\psi) + a_1(3bb_1\psi - c\mu) + 3a_0^2b\psi) = 0; \\
\left(\frac{G'}{G^2}\right)^2 &: \ a_1(a\psi(a_1^2\mu + a_0^2\lambda) + b_1\lambda(aa_1\psi + \lambda(2 - 4\psi)) + 3a_1a_0b\psi^2 - a_1b\psi^2 \\
&\quad - 2a_1\lambda\mu - a_0c\lambda\psi) = 0; \\
\left(\frac{G'}{G^2}\right)^3 &: \ a_1\psi(2aa_1a_0\lambda + a_1^2b\psi - a_1c\lambda - 2a_0\lambda^2) = 0; \\
\left(\frac{G'}{G^2}\right)^4 &: \ a_1^2\lambda(aa_1\psi - \lambda(\psi + 1)) = 0; \\
\left(\frac{G'}{G^2}\right)^{-1} &: \ b_1\psi(-2aa_0b_1\lambda + b_1(3a_1b\psi + c\lambda) + a_0(3a_0b\psi - 2b\psi - 2\lambda\mu)) = 0; \\
\left(\frac{G'}{G^2}\right)^{-2} &: \ -b_1(a\psi(a_1b_1\mu + a_0^2\mu + b_1^2\lambda) + \mu(4a_1\psi\mu - 2a_1\mu - a_0c\psi + 2b_1\lambda) \\
&\quad + (1 - 3a_0)bb_1\psi^2) = 0; \\
\left(\frac{G'}{G^2}\right)^{-3} &: \ b_1\psi(\mu(-2aa_0b_1 - 2a_0\mu + b_1c) + bb_1^2\psi) = 0; \\
\left(\frac{G'}{G^2}\right)^{-4} &: \ -b_1^2\mu(ab_1\psi + \psi\mu + \mu) = 0.
\end{aligned}$$

Solving this system by Maple gives Table 3. In solution 2, if we take $b = 16$, $a_1 = \frac{1}{4}$, $\lambda = 1$, $\psi = 4$ then we have $b_1 = \frac{1}{4}$, $\mu = -1$, $c = 2$, $a = 5$, $C = 1$, $D = 1$. Using these values, we have $(\frac{G'}{G^2}) = (\frac{\sinh(\sqrt{\frac{1}{\sqrt{96}}})(x - \frac{5}{\sqrt{6}}\frac{t^\alpha}{\alpha}) + \cosh(\sqrt{\frac{1}{\sqrt{96}}})(x - \frac{5}{\sqrt{6}}\frac{t^\alpha}{\alpha}) + 1}{\sinh(\sqrt{\frac{1}{\sqrt{96}}})(x - \frac{5}{\sqrt{6}}\frac{t^\alpha}{\alpha}) + \cosh(\sqrt{\frac{1}{\sqrt{96}}})(x - \frac{5}{\sqrt{6}}\frac{t^\alpha}{\alpha}) - 1})$. Using $U(\zeta) = a_0 + a_1(\frac{G'}{G^2}) + a_2(\frac{G'}{G^2})^2 + b_1(\frac{G'}{G^2})^{-1} + b_2(\frac{G'}{G^2})^{-2}$, then we have the following graphs over α. If we have $c_1 = 2$ and $c_2 = 1$, then we have the following graphs (Figure 3) over different values of α.

Table 3: Different solution sets for the time-fractional generalized Burger–Fisher equation.

a	b	a_1	a_0	b_1	c	λ	μ	ψ
$\frac{\lambda(\psi+1)}{a_1\psi}$	b	a_1	$\frac{1}{2}$	0	$\frac{a_1^2 b\psi^2+\lambda^2}{a_1\lambda\psi}$	λ	$-\frac{\lambda}{4a_1^2}$	ψ
$\frac{\lambda(\psi+1)}{a_1\psi}$	b	a_1	$\frac{1}{2}$	$\frac{1}{16a_1}$	$\frac{a_1^2 b\psi^2+\lambda^2}{a_1\lambda\psi}$	λ	$-\frac{\lambda}{16a_1^2}$	ψ
$\frac{\lambda(\psi+1)}{a_1\psi}$	0	a_1	a_0	0	$\frac{2a_0\lambda}{a_1\psi}$	λ	$-\frac{a_0^2\lambda}{a_1^2}$	ψ
$\frac{2\lambda}{a_1}$	0	a_1	a_0	b_1	$\frac{2a_0\lambda}{a_1}$	λ	$-\frac{b_1\lambda}{a_1}$	1
0	b	a_1	$\frac{1}{2}$	b_1	$\frac{bb_1^2+\mu^2}{b_1\mu}$	$\frac{-\mu}{4b_1^2}$	μ	−1
0	b	a_1	$\frac{1}{2}$	0	$\frac{-a_1^2 b-\lambda^2}{a_1\lambda}$	λ	$\frac{-\lambda}{4a_1^2}$	−1
0	b	a_1	$\frac{1}{2}$	$\frac{1}{16a_1^2}$	$\frac{-a_1^2 b-\lambda^2}{a_1\lambda}$	λ	$\frac{-\lambda}{16a_1^2}$	−1
0	b	a_1	$\frac{1}{2}$	0	$\frac{a_1^2 b+\lambda^2}{a_1\lambda}$	λ	$\frac{-\lambda}{4a_1^2}$	1
0	b	a_1	$\frac{1}{2}$	$\frac{1}{16a_1^2}$	$\frac{a_1^2 b+\lambda^2}{a_1\lambda}$	λ	$\frac{-\lambda}{16a_1^2}$	1

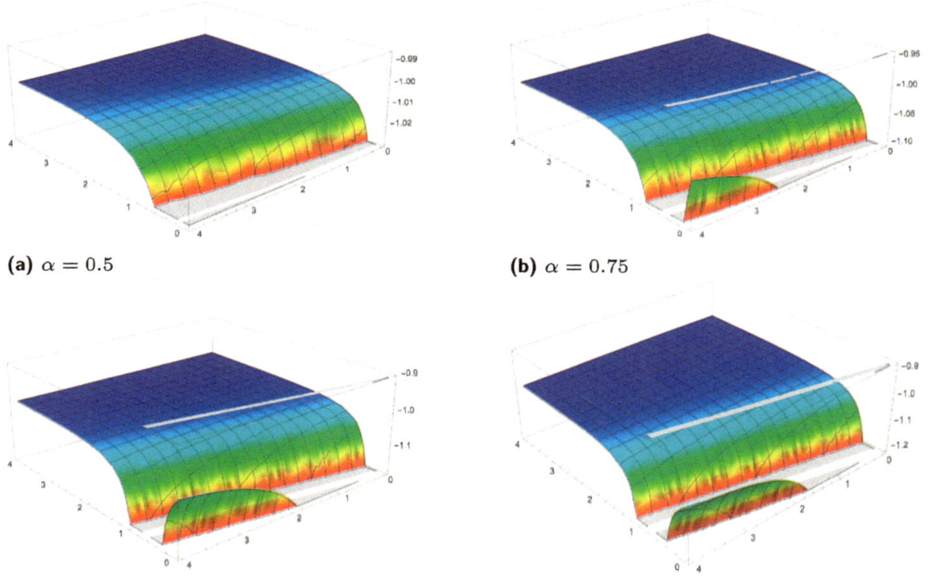

(a) $\alpha = 0.5$ **(b)** $\alpha = 0.75$

(c) $\alpha = 0.9$ **(d)** $\alpha = 1$

Figure 3: Graphs showing the changes with different values of α: (a) $\alpha = 0.5$, (b) $\alpha = 0.75$, (c) $\alpha = 0.9$, and (d) $\alpha = 1$.

4 Conclusion

This study has showcased the robustness of the (G'/G^2)-expansion method in effectively addressing conformable fractional partial differential equations (FPDEs), with a particular focus on the fractional-order modified equal-width (MEW) equation, the time-fractional Fisher equation (TFFE), and the time-fractional generalized Burger–Fisher equation (TFGBFE). Through the derivation of exact solutions for the TFGBFE, the research not only confirms the presence of soliton waves, crucial for understanding nonlinear wave phenomena across scientific and engineering realms, but also unveils additional closed-form solutions for these FPDEs. The examination of graphs illustrating the variation of α underscores the strengthening potential of fractional derivatives as versatile tools in handling these equations in more generalized forms. Furthermore, the proposed solutions for these time-fractional equations offer deeper insights into nonlinear wave dynamics, thereby aiding in the analysis of diverse physical and mathematical models. Leveraging these exact solutions can also enhance the precision of other numerical procedures. In essence, this study paves the way for further exploration and application of the (G'/G^2)-expansion method in tackling a broad spectrum of conformable FPDEs. By doing so, it not only advances our understanding but also facilitates progress in various scientific and engineering disciplines, fostering innovation and knowledge dissemination.

Bibliography

[1] Abdulaziz, O., Hashim, I., & Momani, S. (2008). Solving systems of fractional differential equations by homotopy-perturbation method. *Physics Letters A*, *372*(4), 451–459.

[2] Abualigah, L., & Diabat, A. (2021). Advances in sine cosine algorithm: a comprehensive survey. *Artificial Intelligence Review*, *54*(4), 2567–2608.

[3] Aljahdaly, N. H. (2019). Some applications of the modified (G'/G^2)-expansion method in mathematical physics. *Results in Physics*, *13*, 102272.

[4] Al-Deiakeh, R., Abu Arqub, O., Al-Smadi, M., & Momani, S. (2022). Lie symmetry analysis, explicit solutions, and conservation laws of the time-fractional Fisher equation in two-dimensional space. *Journal of Ocean Engineering and Science*, *7*(4), 345–352.

[5] Aminikhah, H., Refahi Sheikhani, A., & Rezazadeh, H. (2016). Sub-equation method for the fractional regularized long-wave equations with conformable fractional derivatives. *Scientia Iranica*, *23*(3), 1048–1054.

[6] Baleanu, D., Ghafarnehzad, K., Rezapour, S., & Shabibi, M. (2018). On the existence of solutions of a three steps crisis integro-differential equation. *Advances in Difference Equations*, *2018*(1), 1–20.

[7] Baleanu, D., Etemad, S., Pourrazi, S., & Rezapour, S. (2019). On the new fractional hybrid boundary value problems with three-point integral hybrid conditions. *Advances in Difference Equations*, *2019*(1), 1–21.

[8] Baleanu, D., Jajarmi, A., Mohammadi, H., & Rezapour, S. (2020). Analysis of the human liver model with Caputo–Fabrizio fractional derivative. *Chaos, Solitons and Fractals*, *134*, 109705.

[9] Baleanu, D., Etemad, S., & Rezapour, S. (2020). A hybrid Caputo fractional modeling for thermostat with hybrid boundary value conditions. *Boundary Value Problems*, *2020*(1), 1–16.

[10] Bulut, H., Baskonus, H. M., & Belgacem, F. B. M. (2013). The analytical solution of some fractional ordinary differential equations by the Sumudu transform method. *Abstract and Applied Analysis, 2013*.

[11] Danladi, A., Abdullah, A. R., Hussain, E., & Li, Z. (2025). Interactions and soliton dynamics for a (2 + 1)-dimensional nonlinear integrable model arising in shallow water. *Mathematics*, *13*(21), 3474.

[12] Eslami, M., Fathi Vajargah, B., Mirzazadeh, M., & Biswas, A. (2014). Application of first integral method to fractional partial differential equations. *Indian Journal of Physics*, *88*(2), 177–184.

[13] Fakhrusy, Q. Z., Anggraeni, C. P., & Gunawan, P. H. (2019). Simulating water and sediment flow using SWE-convection diffusion model on OpenMP platform. In *Proceedings of the 7th International Conference on Information and Communication Technology (ICoICT)* (pp. 1–6). IEEE.

[14] Gaur, M., Bhushan, B., & Jain, R. (2023). Application of the $(\frac{G'}{G})$-expansion method on conformable time-fractional generalized Burger–Fisher equation. In *Proceedings of the 3rd International Conference on Technological Advancements in Computational Sciences* (pp. 566–571).

[15] Gaur, M. (2023). Symmetry analysis of conformable fractional coupled KdV–Burgers equation. *Palestine Journal of Mathematics*, *12*(1).

[16] Gepreel, K. A. (2011). The homotopy perturbation method applied to the nonlinear fractional Kolmogorov–Petrovskii–Piskunov equations. *Applied Mathematics Letters*, *24*(8), 1428–1434.

[17] Hammond, J., & Bortz, D. (2011). Analytical solutions to Fisher's equation with time-variable coefficients. *Applied Mathematics and Computation*, *218*(6), 2497–2508.

[18] Hamoud, A. A., Ghadle, K., & Atshan, S. (2019). Approximate solutions of fractional integro-differential equations using modified Adomian decomposition method. *Khayyam Journal of Mathematics*, *5*(1), 21–39.

[19] Hussain, E., Abdullah, A. R., Farooq, K., & Younas, U. (2025). Optical multi-peakon dynamics in the fractional cubic–quintic nonlinear pulse propagation model using a novel integral approach. *Fractal and Fractional*, *9*(10), 631.

[20] Khalil, R., Al Horani, M., Yousef, A., & Sababheh, M. (2014). A new definition of fractional derivative. *Journal of Computational and Applied Mathematics*, *264*, 65–70.

[21] Khatun, M. A., Arefin, M. A., Uddin, M. H., Inc, M., & Akbar, M. A. (2022). An analytical approach to the solution of fractional-coupled modified equal width and fractional-coupled Burgers equations. *Journal of Ocean Engineering and Science*.

[22] Kumar, D., Seadawy, A. R., & Joardar, A. K. (2018). Modified Kudryashov method for exact solutions of conformable fractional differential equations. *Chinese Journal of Physics*, *56*(1), 75–85.

[23] Lu, B. (2012). The first integral method for some time fractional differential equations. *Journal of Mathematical Analysis and Applications*, *395*(2), 684–693.

[24] Malik, S., Boulaaras, S. M., Talafha, A. M., & Sağlam, F. M. K. (2025). Soliton dynamics of the fourth-order nonlinear Boussinesq equation arising in shallow-water via advanced analytical techniques. *Scientific Reports*, *15*, 45421.

[25] Miller, K. S., & Ross, B. (1993). *An Introduction to the Fractional Calculus and Fractional Differential Equations*. Wiley.

[26] Murad, M. A. S., Tedjani, A. H., Li, Z., & Hussain, E. (2025). Optical solutions to time-fractional improved (2 + 1)-dimensional nonlinear Schrödinger equation in optical fibers. *Scientific Reports*, *15*(1), 29208.

[27] Pandir, Y., & Duzgun, H. H. (2017). New exact solutions of time fractional Gardner equation using a new version of the F-expansion method. *Communications in Theoretical Physics*, *67*(1), 9.

[28] Prakasha, D., Veeresha, P., & Singh, J. (2019). Fractional approach for water transport in unsaturated porous media with Mittag–Leffler kernel. *Frontiers in Physics*, *7*, 193.

[29] Qazza, A., Burqan, A., Saadeh, R., & Khalil, R. (2022). Applications of double Ara–Sumudu transform in solving fractional partial differential equations. *Symmetry*, *14*(9), 1817.

[30] Rashid, S., Noor, M. A., Nisar, K. S., Baleanu, D., & Rahman, G. (2020). A new dynamic scheme via fractional operators on time scale. *Frontiers in Physics*, *8*, 165.

[31] Sağlam, F. N. K., Movaheedi, Z., Talafha, A. M., Kopçasız, B., & Malik, S. (2025). Different types of soliton solutions for the resonant nonlinear Schrödinger equation with parabolic law nonlinearity via Kumar–Malik approach. *Arab Journal of Basic and Applied Sciences*, *32*(1), 361–373.
[32] Samko, S. G. (1993). *Fractional Integrals and Derivatives: Theory and Applications*. Gordon and Breach.
[33] Seadawy, A. R. (2016). Three-dimensional nonlinear modified Zakharov–Kuznetsov equation of ion-acoustic waves in a magnetized plasma. *Computers & Mathematics with Applications*, *71*(1), 201–212.
[34] Singla, K. (2021). Existence of series solutions for certain nonlinear systems of time fractional partial differential equations. *Journal of Geometry and Physics*, *167*, 104301.
[35] Talafha, A. M., Jhangeer, A., & Kazmi, S. S. (2023). Dynamical analysis of (4 + 1)-dimensional Davey–Stewartson–Kadomtsev–Petviashvili equation using Lie symmetry approach. *Ain Shams Engineering Journal*, *14*(11), 102537.
[36] Wang, G. (2021). Symmetry analysis and analytical solutions of a generalized KdV–Burgers–Kuramoto equation. *Fractals*, *29*(4), 2150101.
[37] Wazwaz, A.-M. (2005). The tanh method: exact solutions of the sine-Gordon and sinh-Gordon equations. *Applied Mathematics and Computation*, *167*(2), 1196–1210.
[38] Wazwaz, A.-M. (2008). The extended tanh method for the Zakharov–Kuznetsov equation and its generalized forms. *Communications in Nonlinear Science and Numerical Simulation*, *13*(6), 1039–1047.

G. Kalaimurugan and R. Thamizhmaran

QSPR modeling with reverse degree topological descriptors for the drugs used in the treatment of Anthrax

Abstract: Anthrax, caused by Bacillus anthracis, continues to pose a serious threat to public health and livestock, highlighting the need for efficient and reliable methods to evaluate therapeutic agents. In this work, a Quantitative Structure–Property Relationship (QSPR) approach based on reverse degree-based topological indices is developed to predict key physicochemical properties of widely used Anthrax drugs, including ciprofloxacin, doxycycline, levofloxacin, and penicillin G. Molecular graphs of these drugs were constructed, and ten reverse topological descriptors—such as reverse Randić, reverse Zagreb, reverse hyper-Zagreb, reverse atom–bond connectivity, reverse forgotten, and reverse redefined Zagreb indices—were analytically computed. Linear regression analysis using SPSS was employed to correlate these descriptors with experimentally reported physicochemical properties, namely boiling point, enthalpy of vaporization, flash point, molar refractivity, polarizability, and molar volume. The results demonstrate strong and statistically significant correlations, where the reverse forgotten index effectively predicts boiling point, flash point, and enthalpy of vaporization (R = 0.885–0.889), the reverse hyper-Zagreb index shows excellent correlation with molar refractivity and polarizability (R = 0.989), and the reverse redefined first Zagreb index exhibits a strong association with molar volume (R = 0.98). The predicted values closely align with experimental data, confirming the robustness of the proposed models. The novelty of this study lies in it being the first systematic identification of descriptor-specific dominance among reverse degree-based indices for Anthrax drugs, demonstrating that different reverse descriptors optimally govern different physicochemical properties. This work establishes a mathematically rigorous and computationally efficient framework that enhances the predictive power of QSPR modeling and supports rational design and optimization of antimicrobial agents.

Keywords: Anthrax drugs, reverse topological descriptors, QSPR, linear regression model

MSC 2020: 05C90, 05C12, 62J05, 62P10, 92E10

G. Kalaimurugan, R. Thamizhmaran, Department of Mathematics, Thiruvalluvar University, Vellore, 632 115, Tamil Nadu, India, e-mail: thamizhmaran2699@gmail.com

https://doi.org/10.1515/9783112234525-004

1 Introduction

Anthrax is an economically significant zoonotic disease that affects both humans and livestock, particularly cattle. The disease is caused by the spore-forming, Gram-positive bacterium Bacillus anthracis. Although Anthrax can infect a wide range of animal species, outbreaks are most commonly associated with herbivorous animals such as cattle, buffalo, sheep, and goats. Globally, Anthrax remains underappreciated in terms of funding allocated for prevention and control, placing both human and animal populations at continued risk. In livestock, the disease is often fatal, and outbreaks can result in substantial economic losses. Moreover, Anthrax has been historically exploited as a biological weapon in certain conflict settings, further emphasizing its public health and security relevance [1].

In livestock, acute or subacute Anthrax infections typically exhibit an incubation period ranging from 36 to 72 h, with only rare instances of prolonged incubation. Human infection most commonly occurs through contact with infected animals or animal-derived products, including meat, hides, and bones. Clinically, human Anthrax presents in multiple forms. Among these, cutaneous and gastrointestinal Anthrax are the most frequently reported, particularly following exposure to contaminated animal products [2, 4]. While gastrointestinal Anthrax is relatively uncommon and associated with the consumption of infected meat, cutaneous Anthrax has historically accounted for more than 95 % of reported human cases.

Vaccination remains the most effective preventive strategy against Anthrax, particularly given the potentially fatal nature of human infection. All currently available Anthrax vaccines contain antigenic components derived from B. anthracis. Several vaccine formulations have been approved for human use across different countries. In animal populations, the implementation of systematic livestock vaccination programs is essential for establishing herd immunity and interrupting disease transmission. Effective control strategies rely on adherence to regular immunization schedules, especially in endemic regions. Continuous field surveillance and laboratory evaluations are necessary to assess vaccine efficacy against circulating regional strains of B. anthracis [3]. According to the World Health Organization, the Sterne vaccine remains the most effective and cost-efficient option for preventing Anthrax in domestic animals, although it requires annual administration. Despite the availability of vaccines, antimicrobial therapy remains a critical component in the management of Anthrax infections. Several antibiotics are widely used for both prophylaxis and treatment, including penicillin G, levofloxacin, doxycycline, and ciprofloxacin, as illustrated in Figure 1. In this context, Quantitative Structure–Property Relationship (QSPR) modeling provides a powerful computational approach for predicting the physicochemical properties of novel Anthrax drug candidates [9]. By employing reverse topological descriptors and linear regression techniques, the trained QSPR model enables the estimation of key molecular properties with reduced experimental cost and time. The reliability of the proposed model is evaluated by visualizing the correlation between predicted and experimentally observed values.

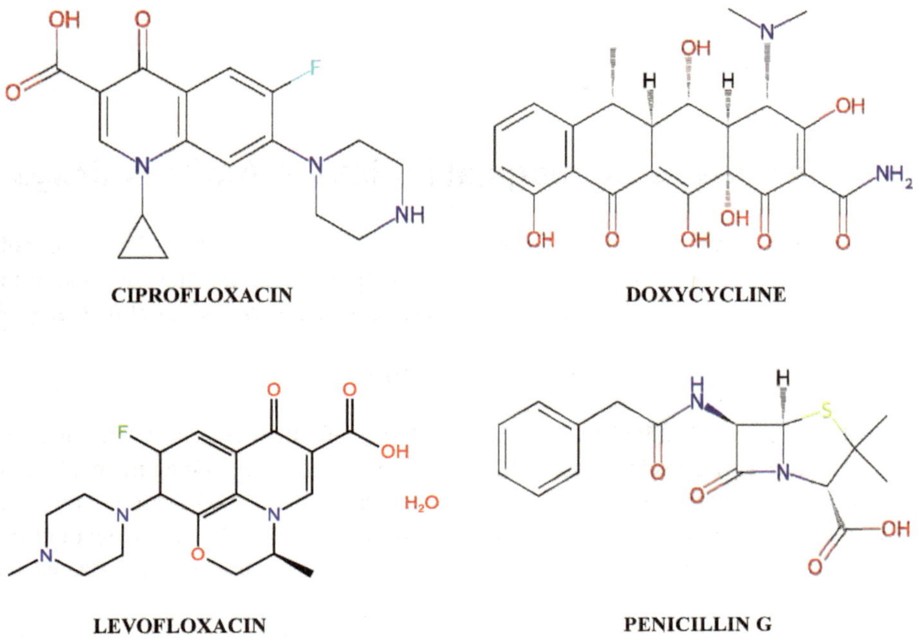

Figure 1: The molecular structure of penicillin G, levofloxacin, doxycycline, and ciprofloxacin.

Furthermore, structural descriptors are utilized to anticipate potential toxicity or adverse effects, thereby supporting the rational design and optimization of safer and more effective Anthrax therapeutics [7, 10].

In QSPR modeling, the two most frequently employed topological indices, in addition to the Wiener index, are the Zagreb and Randić indices. For the molecular graph TUC4[m, n], Jung et al. computed several reverse degree-based descriptors, including the reverse atom–bond connectivity (ABC) index, the reverse geometric–arithmetic (GA) index, the first and second reverse Zagreb indices, and the first and second reverse hyper-Zagreb indices. In the context of antiviral drug design, Wei et al. investigated the Remdesivir compound used in the treatment of Coronavirus disease (COVID-19) and evaluated multiple reverse topological indices, such as the reverse general Randić index, reverse atom–bond connectivity index, reverse geometric–arithmetic index, reverse forgotten index, and reverse Zagreb-type indices [12]. Furthermore, reverse degree-based topological indices for the Remdesivir line graph were reported by Rosary, highlighting their structural significance. The quantitative structure–property relationship (QSPR) of reverse degree-based topological indices was examined by Ravi et al. [5], demonstrating their effectiveness in modeling physicochemical properties. In addition, Haoer and Virk computed reverse Zagreb indices, reverse hyper-Zagreb indices, and their associated polynomials for metal-organic networks. The reverse ABC index and its QSPR applicability were further analyzed by Hashmi et al. for several generations of dendrimers [6], reinforcing the growing importance of reverse degree-based descriptors in predictive chemical modeling.

2 Some reverse topological indices of Anthrax drugs

Let $G = (V, E)$ be a molecular graph, where $|V(G)|$ and $|E(G)|$ denote the order and size of G, respectively. The degree of a vertex $m \in V(G)$ is denoted by $\psi(m)$, while the maximum degree of G is represented by $\triangle(G)$. The reverse degree of a vertex m is defined as

$$R_m = \triangle(G) - \psi(m) + 1.$$

Motivated by the work of Wei et al. [8], Zhao et al. [11] computed several reverse degree-based topological indices for the TM–TCNB molecular structure. Among the evaluated descriptors, the reverse general Randić index with parameter $\alpha = -\frac{1}{2}$ exhibited the best performance. In addition, reverse atom–bond connectivity (ABC), reverse geometric–arithmetic, reverse Zagreb, and reverse hyper-Zagreb indices were also investigated, as summarized in Table 1.

Table 1: Definitions of reverse degree-based topological indices.

Index name	Mathematical expression
Reverse Randic index	$RR_\alpha(G) = \sum_{uv\in E(G)} [R\psi(m) \times R\psi(n)]^\alpha$
Reverse atom–bond connectivity index	$RABC(G) = \sum_{uv\in E(G)} \sqrt{\frac{R\psi(m)+R\psi(n)-2}{R\psi(m)\times R\psi(n)}}$
Reverse geometric arithmetic index	$RGA(G) = \sum_{uv\in E(G)} \frac{2\sqrt{R\psi(m)\times R\psi(n)}}{R\psi(m)+R\psi(n)}$
Reverse first Zagreb index	$RM_1(G) = \sum_{uv\in E(G)} (R\psi(m) + R\psi(n))$
Reverse second Zagreb index	$RM_2(G) = \sum_{uv\in E(G)} (R\psi(m) \times R\psi(n))$
Reverse hyper-Zagreb index	$RHM(G) = \sum_{uv\in E(G)} [R\psi(m) \times R\psi(n)]^2$
Reverse forgotten index	$RF(G) = \sum_{uv\in E(G)} ([R\psi(m)]^2 + [R\psi(n)]^2)$
Reverse redefined first Zagreb index	$RReZG_1(G) = \sum_{uv\in E(G)} \frac{R\psi(m)+R\psi(n)}{R\psi(m)\times R\psi(n)}$
Reverse redefined second Zagreb index	$RReZG_2(G) = \sum_{uv\in E(G)} \frac{R\psi(m)\times R\psi(n)}{R\psi(m)+R\psi(n)}$
Reverse redefined third Zagreb index	$RReZG_3(G) = \sum_{uv\in E(G)} (R\psi(m) + R\psi(n))(R\psi(m) \times R\psi(n))$

Theorem 1. *Let us consider that C is the molecular graph of ciprofloxacin, then*
1. $RR_{-1/2}(C) = 11.1309$
2. $RABC(C) = 18.47$
3. $RGA(C) = 25.3556$

4. $RM_1(C) = 128$
5. $RM_2(C) = 155$
6. $RHM(C) = 644$
7. $RF(C) = 334$
8. $RReZG_1(C) = 23$
9. $RReZG_2(C) = 30.65$
10. $RReZG_3(C) = 800$

Proof. We give the graphical comparison after determining the ciprofloxacin reverse degree-based topological indices. Figure 2 displays the ciprofloxacin graph with vertices and edges. There are 23 vertices and 26 edges in the ciprofloxacin molecular graph, respectively. Based on the reverse degree, the ciprofloxacin edge set is divided into four sets. If $E_{13} = \{mn \in E(C) \mid \psi(m) = 1, \psi(n) = 3\}$, $E_{22} = \{mn \in E(C) \mid \psi(m) = 2, \psi(n) = 2\}$, $E_{23} = \{mn \in E(C) \mid \psi(m) = 2, \psi(n) = 3\}$, $E_{33} = \{mn \in E(C) \mid \psi(m) = 3, \psi(n) = 3\}$.

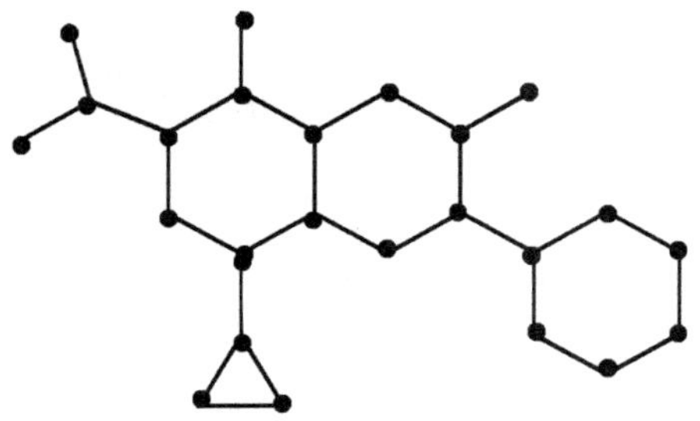

Figure 2: The graphical structure of ciprofloxacin.

We can see that $|E_{13}| = 3$, $|E_{22}| = 5$, $|E_{23}| = 12$, $|E_{33}| = 6$ from the molecular graph of C.

From Table 1, we compute

1. Reverse Randic index of C is

$$RR_\alpha(C) = \sum_{mn \in E(C)} (R\psi(m) \times R\psi(n))^\alpha$$

we take $\alpha = -1/2$,

$$= 3(1 \times 3)^{-1/2} + 5(2 \times 2)^{-1/2} + 12(2 \times 3)^{-1/2} + 6(3 \times 3)^{-1/2}$$

$$= 1.7321 + 2.5 + 4.8989 + 2$$

$$RR_{-1/2}(C) = 11.1309.$$

2. Reverse atom–bond connectivity index of C is

$$\begin{aligned}\mathrm{RABC}(C) &= \sum_{mn\in E(C)} \sqrt{\frac{R\psi(m)+R\psi(n)-2}{R\psi(m)\times R\psi(n)}}\\ &= 3\sqrt{\frac{1+3-2}{3}}+5\sqrt{\frac{2+2-2}{4}}+12\sqrt{\frac{2+3-2}{6}}+6\sqrt{\frac{3+3-2}{9}}\\ \mathrm{RABC}(C) &= 18.47.\end{aligned}$$

3. Reverse geometric arithmetic index of C is

$$\begin{aligned}\mathrm{RGA}(C) &= \sum_{mn\in E(C)} 2\frac{\sqrt{R\psi(m)\times R\psi(n)}}{R\psi(m)+R\psi(n)}\\ &= 6\frac{\sqrt{3}}{2}+10\frac{\sqrt{4}}{4}+24\frac{\sqrt{6}}{5}+12\frac{\sqrt{9}}{6}\\ \mathrm{RGA}(C) &= 128.\end{aligned}$$

4. Reverse first Zagreb index of C is

$$\begin{aligned}\mathrm{RM}_1(C) &= \sum_{mn\in E(C)} (R\psi(m)+R\psi(n))\\ &= 3(1+3)+5(2+2)+12(2+3)+6(3+3)\\ \mathrm{RM}_1(C) &= 128.\end{aligned}$$

5. Reverse second Zagreb index of C is

$$\begin{aligned}\mathrm{RM}_2(C) &= \sum_{mn\in E(C)} (R\psi(m)\times R\psi(n))\\ &= 3(1\times 3)+5(2\times 2)+12(2\times 3)+6(3\times 3)\\ \mathrm{RM}_2(C) &= 155.\end{aligned}$$

6. Reverse hyper-Zagreb index of C is

$$\begin{aligned}\mathrm{RHM}(C) &= \sum_{mn\in E(C)} [R\psi(m)\times R\psi(n)]^2\\ &= 3(1\times 3)^2+5(2\times 2)^2+12(2\times 3)^2+6(3\times 3)^2\\ \mathrm{RHM}(C) &= 644.\end{aligned}$$

7. Reverse Forgotten index of C is

$$\begin{aligned}\mathrm{RF}(C) &= \sum_{mn\in E(C)} [R\psi(m)]^2+[R\psi(n)]^2\\ &= 3(10)+5(8)+12(13)+6(18)\end{aligned}$$

$$\mathrm{RF}(C) = 334.$$

8. Reverse redefined first Zagreb indices of C is

$$\mathrm{RReZG}_1(C) = \sum_{mn\in E(C)} \frac{R\psi(m) + R\psi(n)}{R\psi(m) \times R\psi(n)}$$
$$= 3\frac{4}{3} + 5\frac{4}{4} + 12\frac{5}{6} + 6\frac{6}{9}$$
$$\mathrm{RReZG}_1(C) = 23.$$

9. Reverse redefined second Zagreb indices of C is

$$\mathrm{RReZG}_2(C) = \sum_{mn\in E(C)} \frac{R\psi(m) \times R\psi(n)}{R\psi(m) + R\psi(n)}$$
$$= 3\frac{3}{4} + 5\frac{4}{4} + 12\frac{6}{5} + 6\frac{9}{6}$$
$$\mathrm{RReZG}_2(C) = 30.65.$$

10. Reverse redefined third Zagreb indices of C is

$$\mathrm{RReZG}_3(C) = \sum_{mn\in E(C)} [R\psi(m) \times R\psi(n)][R\psi(m) + R\psi(n)]$$
$$= 3(4)(3) + 5(4)(4) + 12(5)(6) + 6(6)(9)$$
$$\mathrm{RReZG}_3(C) = 800.$$

□

Theorem 2. *Let us consider that D is the molecular graph of doxycycline, then*

1. $\mathrm{RR}_{-1/2}(D) = 15.6042$
2. $\mathrm{RABC}(D) = 37.7058$
3. $\mathrm{RGA}(D) = 34.8141$
4. $\mathrm{RM}_1(D) = 200$
5. $\mathrm{RM}_2(D) = 264$
6. $\mathrm{RHM}(D) = 1136$
7. $\mathrm{RF}(D) = 608$
8. $\mathrm{RReZG}_1(D) = 34$
9. $\mathrm{RReZG}_2(D) = 45.55$
10. $\mathrm{RReZG}_3(D) = 1594$

Proof. We give the graphical comparison after determining the doxycycline reverse degree-based topological indices. Figure 3 displays the doxycycline graph with vertices and edges. There are 34 vertices and 37 edges in the doxycycline molecular graph, respectively. Based on the reverse degree, the doxycycline edge set is divided into seven sets. If $E_{13} = \{mn \in E(D) \mid \psi(m) = 1, \psi(n) = 3\}$, $E_{14} = \{mn \in E(D) \mid \psi(m) = 1, \psi(n) = 4\}$,

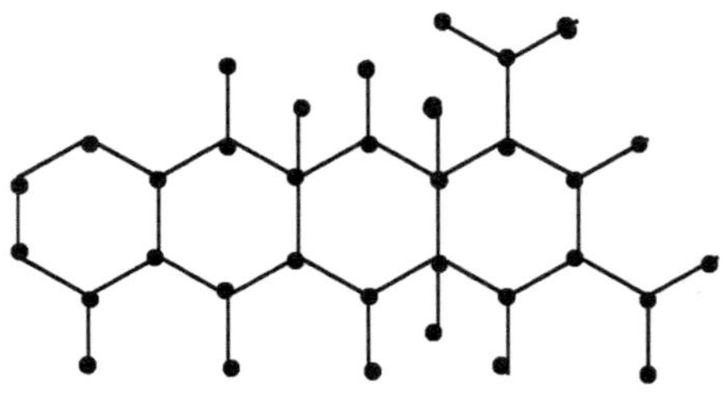

Figure 3: The graphical structure of doxycycline.

$E_{22} = \{mn \in E(D) \mid \psi(m) = 2, \psi(n) = 2\}$, $E_{23} = \{mn \in E(D) \mid \psi(m) = 2, \psi(n) = 3\}$, $E_{33} = \{mn \in E(D) \mid \psi(m) = 3, \psi(n) = 3\}$, $E_{34} = \{mn \in E(D) \mid \psi(m) = 3, \psi(n) = 4\}$, $E_{44} = \{mn \in E(D) \mid \psi(m) = 4, \psi(n) = 4\}$.

We can see that $|E_{13}| = 11$, $|E_{14}| = 3$, $|E_{22}| = 2$, $|E_{23}| = 2$, $|E_{33}| = 11$, $|E_{34}| = 7$, $|E_{44}| = 1$ from the molecular graph of D.

From Table 1, we compute

1. Reverse Randic index of D is

$$\mathrm{RR}_\alpha(D) = \sum_{mn\in E(D)} (R\psi(m) \times R\psi(n))^\alpha$$

we take $\alpha = -1/2$,

$$\begin{aligned} &= 11(1\times 3)^{-1/2} + 3(1\times 4)^{-1/2} + 2(2\times 2)^{-1/2} + 2(2\times 3)^{-1/2} \\ &\quad + 11(3\times 3)^{-1/2} + 7(3\times 4)^{-1/2} + 1(4\times 4)^{-1/2} \\ &= 6.35 + 1.5 + 1 + 0.8165 + 3.6667 + 2.02 + 0.25 \end{aligned}$$

$$\mathrm{RR}_{-1/2}(D) = 15.6042.$$

2. Reverse atom–bond connectivity index of D is

$$\mathrm{RABC}(D) = \sum_{mn\in E(D)} \sqrt{\frac{R\psi(m) + R\psi(n) - 2}{R\psi(m) \times R\psi(n)}}$$

$$= 11\sqrt{\frac{2}{4}} + 3\sqrt{\frac{3}{4}} + 2\sqrt{\frac{1}{2}} + 2\sqrt{\frac{1}{2}} + 11\sqrt{\frac{4}{9}} + 7\sqrt{\frac{5}{12}} + 1\sqrt{\frac{6}{16}}$$

$$\mathrm{RABC}(D) = 37.7058.$$

3. Reverse geometric arithmetic index of D is

$$\mathrm{RGA}(D) = \sum_{mn\in E(D)} 2\frac{\sqrt{R\psi(m) \times R\psi(n)}}{R\psi(m) + R\psi(n)}$$

$$= 22\frac{\sqrt{3}}{4} + 6\frac{\sqrt{4}}{5} + 4\frac{\sqrt{4}}{4} + 4\frac{\sqrt{6}}{5} + 22\frac{\sqrt{9}}{6} + 14\frac{\sqrt{12}}{7} + 2\frac{\sqrt{16}}{8}$$

$$\mathrm{RGA}(D) = 34.8141.$$

4. Reverse first Zagreb index of D is

$$\mathrm{RM}_1(D) = \sum_{mn\in E(D)} (R\psi(m) + R\psi(n))$$
$$= 11(1+3) + 3(1+4) + 2(2+2) + 2(2+3) + 11(3+3) + 7(3+4) + 1(4+4)$$
$$\mathrm{RM}_1(D) = 200.$$

5. Reverse second Zagreb index of D is

$$\mathrm{RM}_2(D) = \sum_{mn\in E(D)} (R\psi(m) \times R\psi(n))$$
$$= 11(1\times 3) + 3(1\times 4) + 2(2\times 2) + 2(2\times 3) + 11(3\times 3) + 17(3\times 4) + 1(4\times 4)$$
$$\mathrm{RM}_2(D) = 155.$$

6. Reverse hyper-Zagreb index of D is

$$\mathrm{RHM}(D) = \sum_{mn\in E(D)} [R\psi(m) \times R\psi(n)]^2$$
$$= 11(1\times 3)^2 + 3(1\times 4)^2 + 2(2\times 2)^2 + 2(2\times 3)^2 + 11(3\times 3)^{2^2} + 17(3\times 4)^2 + 1(4\times 4)^2$$
$$\mathrm{RHM}(D) = 1136.$$

7. Reverse forgotten index of D is

$$\mathrm{RF}(D) = \sum_{mn\in E(D)} [R\psi(m)]^2 + [R\psi(n)]^2$$
$$= 11(10) + 3(17) + 2(8) + 2(13) + 11(18) + 7(25) + 1(32)$$
$$\mathrm{RF}(D) = 608.$$

8. Reverse redefined first Zagreb indices of D is

$$\mathrm{RReZG}_1(D) = \sum_{mn\in E(D)} \frac{R\psi(m) + R\psi(n)}{R\psi(m) \times R\psi(n)}$$
$$= 11\frac{4}{3} + 3\frac{5}{4} + 2\frac{4}{4} + 2\frac{5}{6} + 11\frac{6}{9} + 7\frac{7}{12} + 1\frac{8}{16}$$
$$\mathrm{RReZG}_1(D) = 34.$$

9. Reverse redefined second Zagreb indices of D is

$$\mathrm{RReZG}_2(D) = \sum_{mn\in E(D)} \frac{R\psi(m) \times R\psi(n)}{R\psi(m) + R\psi(n)}$$

$$= 11\frac{3}{4} + 3\frac{4}{5} + 2\frac{4}{4} + 2\frac{6}{5} + 11\frac{9}{6} + 7\frac{12}{7} + 1\frac{16}{8}$$

$$\mathrm{RReZG}_2(D) = 45.55.$$

10. Reverse redefined third Zagreb indices of D is

$$\mathrm{RReZG}_3(D) = \sum_{mn\in E(D)} [R\psi(m) \times R\psi(n)][R\psi(m) + R\psi(n)]$$

$$= 11(3)(4) + 3(4)(5) + 2(4)(4) + 2(6)(5) + 11(6)(9) + 7(12)(7) + 1(16)(8)$$

$$\mathrm{RReZG}_3(D) = 1594.$$ □

Theorem 3. *Let us consider that L is the molecular graph of levofloxacin, then*
1. $\mathrm{RR}_{-1/2}(L) = 12.0461$
2. $\mathrm{RABC}(L) = 20.0913$
3. $\mathrm{RGA}(L) = 26.9941$
4. $\mathrm{RM}_1(L) = 140$
5. $\mathrm{RM}_2(L) = 171$
6. $\mathrm{RHM}(L) = 718$
7. $\mathrm{RF}(L) = 376$
8. $\mathrm{RReZG}_1(L) = 25.3333$
9. $\mathrm{RReZG}_2(L) = 73.5$
10. $\mathrm{RReZG}_3(L) = 906$

Proof. We give the graphical comparison after determining the levofloxacin reverse degree-based topological indices. Figure 4 displays the levofloxacin graph with vertices and edges. There are 26 vertices and 28 edges in the levofloxacin molecular graph, respectively. Based on the reverse degree, the levofloxacin edge set is divided into four sets. If $E_{13} = \{mn \in E(C) \mid \psi(m) = 1, \psi(n) = 3\}$, $E_{22} = \{mn \in E(C) \mid \psi(m) = 2, \psi(n) = 2\}$, $E_{23} = \{mn \in E(C) \mid \psi(m) = 2, \psi(n) = 3\}$, $E_{33} = \{mn \in E(C) \mid \psi(m) = 3, \psi(n) = 3\}$.

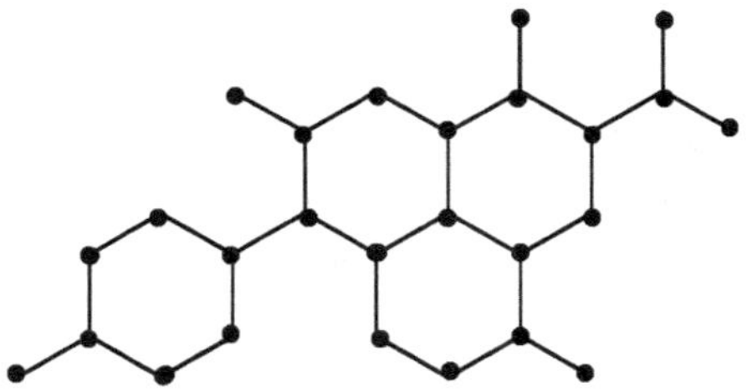

Figure 4: The graphical structure of levofloxacin.

We can see that $|E_{13}| = 6$, $|E_{22}| = 3$, $|E_{23}| = 10$, $|E_{33}| = 9$ from the molecular graph of L.

From Table 1, we compute

1. Reverse Randic index of L is

$$\mathrm{RR}_{\alpha}(L) = \sum_{mn \in E(L)} (R\psi(m) \times R\psi(n))^{\alpha}$$

we take $\alpha = -1/2$,

$$= 6(1 \times 3)^{-1/2} + 3(2 \times 2)^{-1/2} + 10(2 \times 3)^{-1/2} + 9(3 \times 3)^{-1/2}$$

$$= 3.4641 + 1.5 + 4.082 + 3$$

$$\mathrm{RR}_{-1/2}(L) = 12.0461.$$

2. Reverse atom–bond connectivity index of L is

$$\mathrm{RABC}(L) = \sum_{mn \in E(L)} \sqrt{\frac{R\psi(m) + R\psi(n) - 2}{R\psi(m) \times R\psi(n)}}$$

$$= 6\sqrt{\frac{1+3-2}{3}} + 3\sqrt{\frac{2+2-2}{4}} + 10\sqrt{\frac{2+3-2}{6}} + 9\sqrt{\frac{3+3-2}{9}}$$

$$\mathrm{RABC}(L) = 20.0913.$$

3. Reverse geometric arithmetic index of L is

$$\mathrm{RGA}(L) = \sum_{mn \in E(L)} 2\frac{\sqrt{R\psi(m) \times R\psi(n)}}{R\psi(m) + R\psi(n)}$$

$$= 12\frac{\sqrt{3}}{4} + 6\frac{\sqrt{4}}{4} + 20\frac{\sqrt{6}}{5} + 18\frac{\sqrt{9}}{6}$$

$$\mathrm{RGA}(L) = 26.9941.$$

4. Reverse first Zagreb index of L is

$$\mathrm{RM}_1(L) = \sum_{mn \in E(L)} (R\psi(m) + R\psi(n))$$

$$= 6(1+3) + 3(2+2) + 10(2+3) + 9(3+3)$$

$$\mathrm{RM}_1(L) = 140.$$

5. Reverse second Zagreb index of L is

$$\mathrm{RM}_2(L) = \sum_{mn \in E(L)} (R\psi(m) \times R\psi(n))$$

$$= 6(1 \times 3) + 3(2 \times 2) + 10(2 \times 3) + 9(3 \times 3)$$

$$\mathrm{RM}_2(L) = 171.$$

6. Reverse hyper-Zagreb index of L is

$$\begin{aligned}\mathrm{RHM}(L) &= \sum_{mn\in E(L)} [R\psi(m)\times R\psi(n)]^2\\ &= 6(1\times 3)^2 + 3(2\times 2)^2 + 10(2\times 3)^2 + 9(3\times 3)^2\\ \mathrm{RHM}(L) &= 718.\end{aligned}$$

7. Reverse forgotten index of L is

$$\begin{aligned}\mathrm{RF}(L) &= \sum_{mn\in E(L)} [R\psi(m)]^2 + [R\psi(n)]^2\\ &= 6(10) + 3(8) + 10(13) + 9(18)\\ \mathrm{RF}(L) &= 376.\end{aligned}$$

8. Reverse redefined first Zagreb indices of L is

$$\begin{aligned}\mathrm{RReZG}_1(L) &= \sum_{mn\in E(L)} \frac{R\psi(m)+R\psi(n)}{R\psi(m)\times R\psi(n)}\\ &= 6\frac{4}{3} + 3\frac{4}{4} + 10\frac{5}{6} + 9\frac{6}{9}\\ \mathrm{RReZG}_1(L) &= 25.3333.\end{aligned}$$

9. Reverse redefined second Zagreb indices of L is

$$\begin{aligned}\mathrm{RReZG}_2(L) &= \sum_{mn\in E(L)} \frac{R\psi(m)\times R\psi(n)}{R\psi(m)+R\psi(n)}\\ &= 6\frac{3}{4} + 3\frac{4}{4} + 10\frac{6}{5} + 9\frac{9}{6}\\ \mathrm{RReZG}_2(L) &= 73.5.\end{aligned}$$

10. Reverse redefined third Zagreb indices of L is

$$\begin{aligned}\mathrm{RReZG}_3(L) &= \sum_{mn\in E(L)} [R\psi(m)\times R\psi(n)][R\psi(m)+R\psi(n)]\\ &= 6(4)(3) + 3(4)(4) + 10(5)(6) + 9(6)(9)\\ \mathrm{RReZG}_3(L) &= 906.\end{aligned}$$

□

Theorem 4. *Let us consider that P is the molecular graph of penicillin G, then*

1. $\mathrm{RR}_{-1/2}(P) = 10.8772$
2. $\mathrm{RABC}(P) = 18.307$
3. $\mathrm{RGA}(P) = 23.6079$
4. $\mathrm{RM}_1(P) = 127$
5. $\mathrm{RM}_2(P) = 152$

6. $\mathrm{RHM}(P) = 667$
7. $\mathrm{RF}(P) = 363$
8. $\mathrm{RReZG}_1(P) = 23.4167$
9. $\mathrm{RReZG}_2(P) = 28.6952$
10. $\mathrm{RReZG}_3(P) = 832$

Proof. We give the graphical comparison after determining the penicillin G reverse degree-based topological indices. Figure 5 displays the penicillin G graph with vertices and edges. There are 24 vertices and 25 edges in the penicillin G molecular graph, respectively. Based on the reverse degree, the penicillin G edge set is divided into seven sets. If $E_{13} = \{mn \in E(P) \mid \psi(m) = 1, \psi(n) = 3\}$, $E_{14} = \{mn \in E(P) \mid \psi(m) = 1, \psi(n) = 4\}$, $E_{22} = \{mn \in E(P) \mid \psi(m) = 2, \psi(n) = 2\}$, $E_{23} = \{mn \in E(P) \mid \psi(m) = 2, \psi(n) = 3\}$, $E_{24} = \{mn \in E(P) \mid \psi(m) = 2, \psi(n) = 4\}$, $E_{33} = \{mn \in E(P) \mid \psi(m) = 3, \psi(n) = 3\}$, $E_{34} = \{mn \in E(P) \mid \psi(m) = 3, \psi(n) = 4\}$.

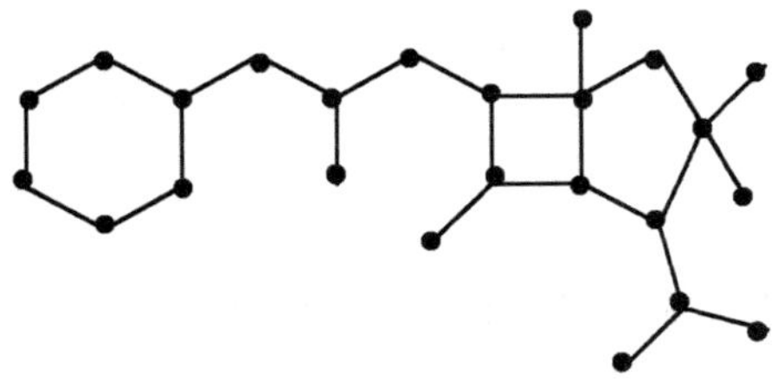

PENICILLIN G

Figure 5: The graphical structure of penicillin G.

We can see that $|E_{13}| = 4$, $|E_{14}| = 3$, $|E_{22}| = 4$, $|E_{23}| = 6$, $|E_{24}| = 2$, $|E_{33}| = 4$, $|E_{34}| = 2$ from the molecular graph of P.

From Table 1, we compute

1. Reverse Randic index of P is

$$\mathrm{RR}_\alpha(P) = \sum_{mn\in E(P)} (R\psi(m) \times R\psi(n))^\alpha$$

we take $\alpha = -1/2$,

$$= 4(1\times 3)^{-1/2} + 3(1\times 4)^{-1/2} + 4(2\times 2)^{-1/2} + 6(2\times 3)^{-1/2} + 2(2\times 4)^{-1/2} + 4(3\times 3)^{-1/2} + 2(3\times 4)^{-1/2}$$

$$= 2.31 + 1.5 + 2 + 2.4495 + 0.7071 + 1.3333 + 0.5773$$

$$\mathrm{RR}_{-1/2}(P) = 10.8772.$$

2. Reverse atom–bond connectivity index of P is

$$\mathrm{RABC}(P) = \sum_{mn\in E(P)} \sqrt{\frac{R\psi(m) + R\psi(n) - 2}{R\psi(m) \times R\psi(n)}}$$
$$= 4\sqrt{\frac{2}{3}} + 3\sqrt{\frac{3}{4}} + 4\sqrt{\frac{2}{4}} + 6\sqrt{\frac{3}{6}} + 2\sqrt{\frac{4}{8}} + 4\sqrt{\frac{4}{9}} + 2\sqrt{\frac{5}{12}}$$
$$\mathrm{RABC}(P) = 18.307.$$

3. Reverse geometric arithmetic index of P is

$$\mathrm{RGA}(P) = \sum_{mn\in E(P)} 2\frac{\sqrt{R\psi(m) \times R\psi(n)}}{R\psi(m) + R\psi(n)}$$
$$= 8\frac{\sqrt{3}}{4} + 6\frac{\sqrt{4}}{5} + 8\frac{\sqrt{4}}{4} + 12\frac{\sqrt{6}}{5} + 4\frac{\sqrt{8}}{6} + 8\frac{\sqrt{9}}{6} + 4\frac{\sqrt{12}}{7}$$
$$\mathrm{RGA}(P) = 23.6079.$$

4. Reverse first Zagreb index of P is

$$\mathrm{RM}_1(P) = \sum_{mn\in E(P)} (R\psi(m) + R\psi(n))$$
$$= 4(1+3) + 3(1+4) + 4(2+2) + 6(2+3) + 2(2+4) + 4(3+3) + 2(3+4)$$
$$\mathrm{RM}_1(P) = 127.$$

5. Reverse second Zagreb index of P is

$$\mathrm{RM}_2(P) = \sum_{mn\in E(P)} (R\psi(m) \times R\psi(n))$$
$$= 4(1\times 3) + 3(1\times 4) + 4(2\times 2) + 6(2\times 3) + 2(2\times 4) + 4(3\times 3) + 2(3\times 4)$$
$$\mathrm{RM}_2(P) = 152.$$

6. Reverse hyper-Zagreb index of P is

$$\mathrm{RHM}(P) = \sum_{mn\in E(P)} [R\psi(m) \times R\psi(n)]^2$$
$$= 4(1\times 3)^2 + 3(1\times 4)^2 + 4(2\times 2)^2 + 6(2\times 3)^2 + 2(2\times 4)^2 + 4(3\times 3)^2 + 2(3\times 4)^2$$
$$\mathrm{RHM}(P) = 667.$$

7. Reverse forgotten index of P is

$$\mathrm{RF}(P) = \sum_{mn\in E(P)} [R\psi(m)]^2 + [R\psi(n)]^2$$
$$= 4(10) + 3(17) + 4(8) + 6(13) + 2(20) + 4(18) + 2(25)$$

$\mathrm{RF}(P) = 363.$

8. Reverse redefined first Zagreb indices of P is

$$\mathrm{RReZG}_1(P) = \sum_{mn \in E(P)} \frac{R\psi(m) + R\psi(n)}{R\psi(m) \times R\psi(n)}$$
$$= 4\frac{4}{3} + 3\frac{5}{4} + 4\frac{4}{4} + 6\frac{5}{6} + 2\frac{6}{8} + 4\frac{6}{9} + 2\frac{7}{12}$$
$$\mathrm{RReZG}_1(P) = 23.4167.$$

9. Reverse redefined second Zagreb indices of P is

$$\mathrm{RReZG}_2(P) = \sum_{mn \in E(P)} \frac{R\psi(m) \times R\psi(n)}{R\psi(m) + R\psi(n)}$$
$$= 4\frac{3}{4} + 3\frac{4}{5} + 4\frac{4}{4} + 6\frac{6}{5} + 2\frac{8}{6} + 4\frac{9}{6} + 2\frac{12}{7}$$
$$\mathrm{RReZG}_2(P) = 28.6952.$$

10. Reverse redefined third Zagreb indices of P is

$$\mathrm{RReZG}_3(P) = \sum_{mn \in E(P)} [R\psi(m) \times R\psi(n)][R\psi(m) + R\psi(n)]$$
$$= 4(3)(4) + 3(4)(5) + 4(4)(4) + 6(6)(5) + 2(6)(8) + 4(9)(6) + 2(12)(7)$$
$$\mathrm{RReZG}_3(P) = 1594.$$ □

3 Results and discussion

By establishing quantitative structure–property relationships (QSPR) between selected reverse degree-based topological indices and key physicochemical properties, we evaluate the predictive effectiveness of these descriptors for potential Anthrax drug candidates. The experimental physicochemical data for the selected drugs were obtained from the chemspider website and the physicochemical properties considered in this study comprise boiling point (BP), enthalpy of vaporization (E), flash point (FP), molar refractivity (MR), polarizability (P), and molar volume (MV).

In this study, ten reverse degree-based topological indices are employed. These include the reverse first and second Zagreb indices, the reverse Randić index, the reverse hyper-Zagreb index, the reverse atom–bond connectivity (ABC) index, the reverse forgotten index, the reverse geometric–arithmetic index, the reverse redefined first and second Zagreb indices, and the reverse third Zagreb index. All selected indices considered in this analysis are summarized in Table 3.

Table 2: Physicochemical properties of selected drugs.

Drug	BP	E	FP	MR	P	MV
Ciprofloxacin	581.8	91.5	305.6	83.3	33.0	226.8
Doxycycline	762.6	116.5	415.0	109.0	43.2	271.1
Levofloxacin	571.5	90.1	299.4	91.1	36.1	244.0
Penicillin G	663.3	102.5	355.0	86.3	34.2	235.2

Table 3: Computed topological index values for selected drugs.

Drug	RR	RABC	RGA	RM_1	RM_2	RHM	RF	$RReZG_1$	$RReZG_2$	$RReZG_3$
Ciprofloxacin	11.1309	18.4700	25.3556	128	155	644	334	23.0000	30.6500	800
Doxycycline	15.6042	37.7058	34.8141	200	264	1136	608	34.0000	45.5500	1594
Levofloxacin	12.0461	20.0913	26.9941	140	171	718	376	25.3333	73.5000	906
Penicillin G	10.8772	18.3070	23.6079	127	152	667	363	23.4167	28.6952	832

The data presented in Tables 2 and 3, together with statistical analysis performed using SPSS software, were used to construct Table 4. Table 4 reports the correlation coefficients between the physicochemical properties of the selected Anthrax drugs and the corresponding reverse degree-based topological indices. These relationships are further illustrated through graphical representations shown in Figures 6, 7, and 8. In the QSPR analysis, each reverse topological index is statistically evaluated using the following parameters: sample size (N), regression slope (b), correlation coefficient (R), coefficient of determination (R^2), root mean square error (RMSE), and the associated p-value. The p-value is used to assess the statistical significance of each predictor, where a higher p-value indicates that variations in the independent variable are not significantly related to changes in the dependent variable.

Table 4: Correlation values of different indices with physicochemical properties.

Indices	BP	E	FP	MR	P	MV
$RR_{-1/2}$	0.783	0.790	0.783	0.987	0.987	0.969
RABC	0.857	0.863	0.857	0.978	0.978	0.953
RGA	0.731	0.739	0.731	0.968	0.968	0.944
RM_1	0.502	0.510	0.502	0.901	0.901	0.917
RM_2	0.824	0.830	0.824	0.986	0.986	0.965
RHM	0.858	0.864	0.858	**0.989**	**0.989**	0.970
RF	**0.885**	**0.889**	**0.885**	0.987	0.987	0.971
$RreZG_1$	0.827	0.834	0.827	0.995	0.955	**0.980**
$RreZG_2$	−0.258	−0.252	−0.258	0.283	0.283	0.340
$RreZG_3$	0.862	0.868	0.862	0.986	0.986	0.966

Each regression coefficient is tested against the null hypothesis. It is observed that, for all examined properties, the correlation coefficient satisfies $R > 0.6$ and the corresponding p-value is less than 0.05, indicating strong correlations and statistical significance. However, in such cases, the predictive capability of the model may be limited due to potential overfitting. To further assess the contribution of each predictor, an additional test is performed by comparing the estimated coefficients with a model containing zero predictor variables, as reported in Table 5.

Table 5: Statistical parameters of regression analysis for physicochemical properties.

Phys. prop.	N	A	B	R	R^2	F	Standard error of estimate	p	Indicator
BP	4	384.0542	0.6205	0.885	0.783	0.5746	72.102	0.013	SIGNIFICANT
E	4	63.9697	0.0860	0.889	0.792	0.0030	70.649	0.001	SIGNIFICANT
FP	4	186.0300	0.3753	0.885	0.782	0.1920	72.170	0.015	SIGNIFICANT
MR	4	53.6020	0.0490	0.989	0.977	0.00041	42.453	0.0004	SIGNIFICANT
P	4	21.2240	0.0190	0.989	0.977	0.00002	42.584	0.0003	SIGNIFICANT
MV	4	147.4390	3.6628	0.980	0.961	0.0570	1.238	0.0000001	SIGNIFICANT

Models of regression for the reverse-based topological index from Table 4:
- Boiling Point (BP) = 384.0542 + 0.6205[RF(G)]
- Enthalpy (E) = 63.9697 + 0.086[RF(G)]
- Flash Point (FP) = 186.03 + 0.3753[RF(G)]
- Molar Refraction (MR) = 53.602 + 0.049[RHM(G)]
- Polarizability (P) = 21.224 + 0.019[RHM(G)]
- Molar Volume (MV) = 147.439 + 3.6628[$\mathrm{RReZG_1}(G)$]

A linear regression model is employed in the present QSPR investigation. Seven physicochemical properties and four selected reverse degree-based topological indices are used to evaluate the performance of the proposed models, as illustrated in Figures 6, 7, and 8.

1. Figure 6 demonstrates a strong correlation between the reverse forgotten index RF(G) and the boiling point (BP) and flash point (FP), with a correlation coefficient of $R = 0.885$, indicating a superior relationship compared to other physicochemical variables.
2. Figure 7 illustrates an excellent correlation between the reverse hyper-Zagreb index RHM(G) and molar refractivity (MR) and polarizability (P), achieving a correlation coefficient of $R = 0.989$, which is the highest among the examined properties.
3. Figure 8 reveals a strong correlation between the reverse forgotten index RF(G) and enthalpy of vaporization (E), with $R = 0.889$. In addition, the reverse redefined first Zagreb index $\mathrm{RreZG_1}(G)$ exhibits an excellent correlation with molar volume (MV), yielding a correlation coefficient of $R = 0.98$.

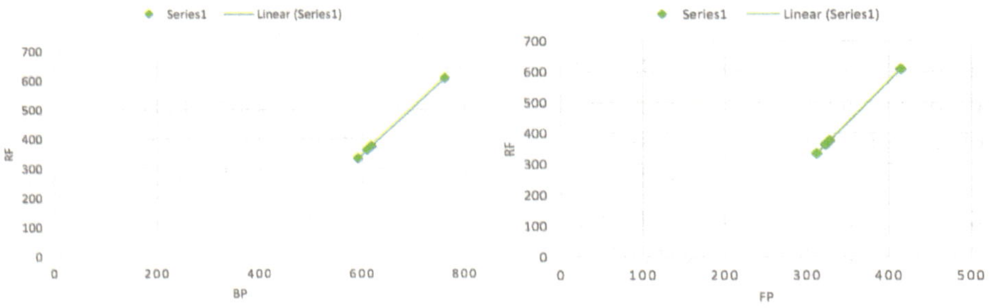

Figure 6: Reverse forgotten index (RF) versus boiling point (BP) and flash point (FP).

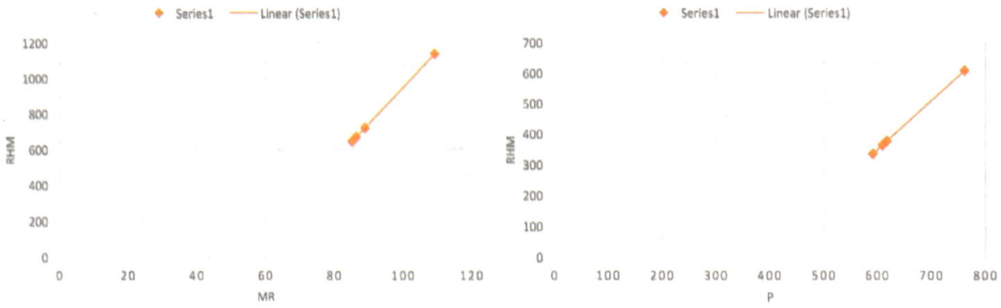

Figure 7: Reverse hyper-Zagreb index (RHM) versus molar refractivity (MR) and polarizability (P).

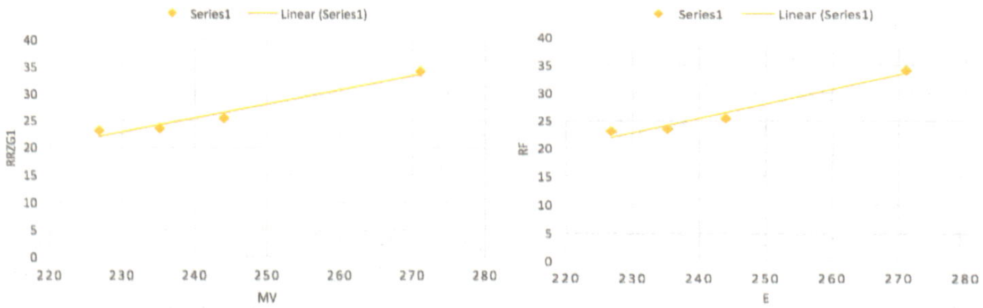

Figure 8: Reverse forgotten index (RF) versus enthalpy of vaporization (E), and reverse redefined third Zagreb indices ($RReZG_1$) versus molar volume (MV).

Table 6 illustrates the influence of molecular structure on the pharmacokinetic behavior and therapeutic efficacy of Anthrax drugs by comparing their physicochemical properties with the predicted values obtained from reverse topological indices. This comparison enables an assessment of the predictive capability of reverse topological descriptors for key drug-related characteristics that govern bioavailability and efficacy, including solubility, lipophilicity, molecular weight, and polar surface area.

Table 6: Discrepancy between the drug's actual and revised values as determined by QSPR analysis using topological indices.

Parameter		Drug			
		Ciprofloxacin	Doxycycline	Levofloxacin	Penicillin G
Boiling point	Actual values	581.8	762.6	571.5	663.3
	Revised	RF(G)	RF(G)	RF(G)	RF(G)
	Values by indices	591.3	761.3	617.4	609.3
Enthalpy	Actual values	91.5	116.5	90.1	102.5
	Revised	RF(G)	RF(G)	RF(G)	RF(G)
	Values by indices	92.6	116.5	96.3	95.2
Flash point	Actual values	305.6	415.0	299.4	355.0
	Revised	RF(G)	RF(G)	RF(G)	RF(G)
	Values by indices	311.4	414.2	327.1	322.3
Molar refractivity	Actual values	83.3	109.0	91.1	86.3
	Revised	RHM(G)	RHM(G)	RHM(G)	RHM(G)
	Values by indices	85.5	109.2	88.9	86.3
Polarizability	Actual values	33.0	43.2	36.1	34.2
	Revised	RHM(G)	RHM(G)	RHM(G)	RHM(G)
	Values by indices	35.5	42.8	34.8	33.8
Molar volume	Actual values	226.8	271.1	244.0	235.2
	Revised	$RReZG_1(G)$	$RReZG_1(G)$	$RReZG_1(G)$	$RReZG_1(G)$
	Values by indices	231.6	271.9	240.2	233.1

4 Conclusion

This study establishes an effective QSPR framework based on reverse degree-based topological indices for predicting the physicochemical properties of Anthrax therapeutic drugs. The strong and statistically significant correlations obtained between selected reverse descriptors and key properties confirm that these indices successfully encode molecular structural information relevant to drug behavior. In particular, the reverse forgotten, reverse hyper-Zagreb, and reverse redefined first Zagreb indices emerged as reliable predictors for thermodynamic and pharmacokinetic parameters, with predicted values closely matching experimental data. These results demonstrate the robustness, simplicity, and computational efficiency of the proposed approach. Future work may extend this framework to larger classes of antimicrobial agents, incorporate nonlinear and machine-learning-based models to enhance predictive accuracy, and combine reverse degree-based indices with other molecular descriptors to develop more generalized and scalable QSPR models for drug discovery and optimization.

Bibliography

[1] Ogunleye, Seto C., Olorunshola, Mercy M., Fasina, Kolapo A., Aborode, Abdullahi T., Akinsulie, Olalekan C., Amoo, Abimbola, Olatoye, Boluwatife J., Bakare, Akeem, Lawal, Mariam A., Adekanye, Oluwabori, & Chinyere, Ezemba C. (18 January 2024). Anthrax outbreak: exploring its biological agents and public health implications. https://doi.org/10.3389/fitd.2023.1297896.

[2] Subedi, Deepak, Pantha, Saurav, Jyoti, Sumit, Gautam, Bickal, Kaphle, Krishna, Yadav, Rakesh Kumar, Ghimire, Shristi, & Dhakal, Santosh (2024). Anthrax in humans, animals, and the environment and the one health strategies for anthrax control. https://doi.org/10.3390/pathogens13090773.

[3] Georgopoulos, Apostolos P., & James, Lisa M. (2024). Anthrax vaccines in the 21st century. https://doi.org/10.3390/vaccines12020159.

[4] Chanda, Mohammed Mudassar, Campbell, Lindsay, Walke, Henry, Salzer, Johanna S., Hemadri, Divakar, Patil, Sharanagouda S., Purse, Bethan V., & Shivachandra, Sathish Bhadravati (2024). A thirty-year time series analyses identifies coherence between oscillations in Anthrax outbreaks and El Niño in Karnataka, India. https://doi.org/10.1038/s41598-024-67736-w.

[5] Ali, Parvez, Kirmani, Syed Ajaz K., & Kumar, Alok (2022). Computation of reverse degrees of some antiviral drugs targeting COVID-19 with QSPR analysis. https://doi.org/10.33263/BRIAC131.066.

[6] Arockiaraj, Micheal, Greeni, A. Berin, & Kalaam, AR. Abul (2023). Linear versus cubic regression models for analyzing generalized reverse degree based topological indices of certain latest corona treatment drug molecules. https://doi.org/10.1002/qua.27136.

[7] Singh, Omendra, Garg, Pravin, Kansal, Neha, & Sharma, Gaurav (2023). On analysis of temperature based topological indices of anti-cancer drugs. *European Chemical Bulletin*, *12*(10), 9522–9541.

[8] Poojary, Prasanna, Shenoy, Gautham B, Swamy, Narahari Narasimha, Ananthapadmanabha, Raghavendra, Sooryanarayana, Badekara, & Poojary, Nandini (2022). Reverse topological indices of some molecules in drugs used in the treatment of H1N1. https://doi.org/10.33263/BRIAC131.071.

[9] Roy, Kunal, Kar, Supratik, Das, Rudra Narayan. A primer on QSAR/QSPR modeling fundamental concepts. https://doi.org/10.1007/978-3-319-17281-1.

[10] Ahmed, Wakeel, Ali, Kashif, Zaman, Shahid, & Raza, Asma. Molecular insights into anti-Alzheimer's drugs through predictive modeling using linear regression and QSPR analysis. *Modern Physics Letters B*. https://doi.org/10.1142/S0217984924502609.

[11] Ibrahim, Hassan, Shewa, Gladys Amos, Aderibigbe, Babatunde Adewale. On the QSPr models of some anti-malaria drugs using some reversed degree-based topological indices. https://doi.org/10.62050/ljsir2024.v2n2.296.

[12] Havare, Özge Çolakoglu (2021). Quantitative structure analysis of some molecules in drugs used in the treatment of COVID-19 with topological indices. https://doi.org/10.1080/10406638.2021.1934045.

V. G. Deepa, S. Aparna Lakshmanan, and V. N. Sreeja

Understanding social networks: an overview of essential parameters for analysis

Abstract: Two or more people interacting with one another constitute a social network. Relationships, sharing of information, or a common interest could all influence how social entities engage with one another. Social networks all share the same fundamental characteristics, regardless of whether they are virtual or actual. Consequently, the modeling of interacting networks will almost definitely involve social network analysis. A graph model of the social system is used to analyze the social network structure that emerges from these interactions. Mathematical representations of social networks have been made utilizing relations, matrices, and graphs. In this chapter, we provide a summary of several graph theoretic parameters that are frequently used to analyze social networks.

Keywords: Social network, graph theoretic parameters, centrality, reciprocity, sharing network

MSC 2020: 05C82, 05C90, 91D30, 68R10

1 Introduction

Currently, the term "Social Network" refers to the social media platforms, for example, LinkedIn, Facebook, and so on; nonetheless, the term also refers to the minute changes among social groups for common interests. Researchers are increasingly interested in learning about the massive information networks in social media, and more work has been done around this since the twentieth century. Although social networks may be used to grasp a specific area in general, most of the focus in this area has not been accomplished. To understand the behavior of a society deep down, we actually consider thinking of it as social groups, and the review is led in these groups. The formation of group events cut off from a community is dependent on a variety of relationships. They can be an association between at least two persons or groups, animosity, or an ideology shared by at least two people. Social network research provides insights into why entities seen in reality accept explicit structures, why some connections are used more frequently than others and some are dropped entirely, and how relations can be influenced through the flow of information and

V. G. Deepa, V. N. Sreeja, Department of Mathematics, Sree Krishna College, Guruvayur, Kerala-680541, India, e-mail: skcdeepa@gmail.com
S. Aparna Lakshmanan, Department of Mathematics, Cochin University of Science and Technology, Cochin, Kerala-682022, India

https://doi.org/10.1515/9783112234525-005

knowledge. The heterogeneity of theories that have been developed complement classical social network analysis. The one who is interested in social network may read [1, 3, 4, 8, 10].

1.1 Objective

We are aiming to provide a first-hand look at some of the graph theoretic parameters that help to understand and analyze a social network extensively in this work. A substantial effort has been made to demonstrate this theory through the academic sharing of aspirants in a classroom that includes teenagers.

2 Social network

Social networks are communities that allow people to communicate with one another. Numerous social networks revolve around us: interpersonal relationships such as friendship, mentorship, etc. Social networks provide a variety of opportunities to reconnect with relatives and friends, meet new people, and develop new relationships. The proper network creation is the most crucial stage of social network analysis. However, it is about the nature of the relationship and which interactions need more consideration. Also, in Social Network Analysis, what matters is not what the network is in the Social Network or how it is formed, what is needed to be analyzed is their inherent features and it is about comprehending a condition that has arisen [7].

The structural qualities of a network are measured and analyzed using social network analysis [9]. It enables the evaluation of interactions and relationships between groups, organizations, and other entities that are linked. Complete network analysis and ego network analysis are the two basic categories used to categorize social networks. Ego network analysis is the process of determining the relationship between entities. The analysis is performed on a specific sample of persons drawn from the entire society. To evaluate the correlation, this selection is done at random. This ego network analysis takes into account the neighborhood of a vertex, diversity, and so on. The analysis that is adopted in all networks is complete network analysis. It examines the relationship between the actors drawn from a large population. It provides the foundation for subgroup analysis, centrality measure, and equivalence analysis. The ego network collects relationship of an actor in the network, whereas the complete network collects relationships among the actors in the network. The nuances of this approach may be lost when considering a large social network. As a result, the scope of comprehending network features may be less significant than the quality of information. Social networks are examined at the scale appropriate to the researcher's theoretical objective. Networks can be classified into three types based on their level: micro, meso, and macro. The micro level denotes that the network begins with a single person or a small group of people in a specific social setting. Leading levels are dyadic (connection between two entities), triadic (relationship between an entity and a dyad), and so on. The population size at the meso level is between the micro and macro levels. The treatment of interpersonal interactions in a large population is done at the macro level.

2.1 Formation of a social network

The formation of a community into something like a social network by taking into account the specific relationship is the most significant step in social network study. A network is an idealistic depiction of a collection of objects that are linked together via relationships. The interconnected entities are referred to as vertices (actors, nodes, and so on), and the linkages that interconnect the vertices are referred to as edges. Edges can be either directed or undirected. An ordered pair $G = (V, E)$ represents a graph with V as the set of nodes and E as the set of edges [5]. A weighted graph is a special type of graph by assigning a weight for each edge.

Through this study, the authors collected the data of a classroom consisting of 34 adolescence by taking each individual as vertices by naming $v_1, v_2, \ldots, v_{34}$ and each vertex v_i is connected to the vertex v_j by edge e_k if v_i is sharing his academic details to $v_j, i \neq j$. We give the weight of an edge as

$$w(e_k) = \begin{cases} 1 & \text{if } v_i \text{ shared to } v_j \text{ occasionally} \\ 2 & \text{if } v_i \text{ shared to } v_j \text{ frequently.} \end{cases}$$

Obviously, if v_i has not shared his academic details to v_j, there does not exist a link from v_i to v_j. Also, we avoid the situation when an individual shares information to himself. To construct the directed weighted graph for the above network, the authors used the methodology as in [2]. The resultant directed weighted graph of the 34 individuals in the classroom network is given in Figure 1.

2.2 Analysis

The core part of SNA is its 'analysis' part. After forming a social network, we must use analysis tools to determine 'why' this is happening in networks. For the purpose of studying social networks and drawing conclusions from them, analysis tools can be quite helpful. Analysis tools can assist in visualizing the social network's structure, including the connections between members, the frequency of interactions, and the sorts of content shared. This data can be used to locate important communities and influencers inside the network. It may detect trends and patterns in user behavior by examining the content shared on social networks, such as the most well-liked subjects or the hours of the

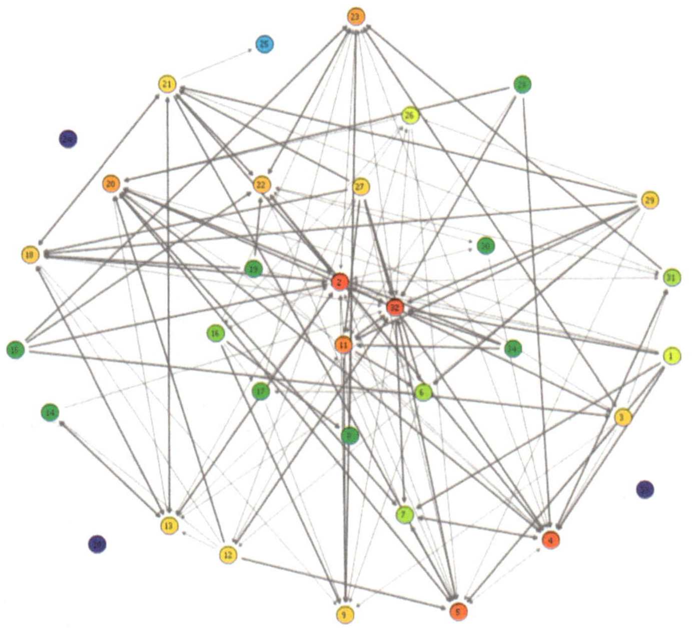

Figure 1: Figure showing the academic sharing of teenagers in a classroom network.

day when users are most active. This data can help direct content strategy and marketing initiatives. Analysis tools can be used to track mentions, sentiment, and the tone of social media conversations about a brand. This data can be used to spot possible problems and react immediately to client comments. It can provide comprehensive numbers on social media effectiveness, including as engagement rates, reach, and conversions. This data can be used to evaluate the success of social media initiatives and improve future tactics. Analysis tools are crucial for comprehending social network dynamics and for selecting the right social media strategy and engagement tactics.

In this section, we will provide a survey of several parameters that are used to analyze a social network.

2.2.1 Matrix analysis

Matrix analysis of a social network involves the use of mathematical matrices to represent the relationships between nodes in a network. The adjacency matrix, a square matrix that depicts the relationships between network nodes, is the most typical type of matrix used in network research. A social network's patterns and structures, such as groups of strongly connected nodes or core nodes with a high degree of connectivity to other nodes, can be found using matrix analysis. Matrix analysis is an effective method for comprehending the dynamics and structure of social networks and can offer insightful information for a variety of applications, including marketing, community-building, and public health.

2.2.1.1 Adjacency matrix

An effective tool for examining social networks and comprehending their dynamics and relationships is the adjacency matrix. As it offers a means of representing the connections between network nodes, the adjacency matrix is a crucial tool in the analysis of social networks.

The matrix $A = [a_{(i,j)}]$, where $a_{(i,j)}$ indicates the weight of the edge from the vertex v_i to vertex v_j and it is equal to zero if there is no connection. The adjacency matrix of our study is given in the Appendix.

2.2.1.2 Cocitation matrix

The product of transposing the adjacency matrices A and A is a crucial tool in social network analysis, giving essential insights into the dynamics and structure of social networks as well as the connections between their nodes.

The cocitation matrix $C = [c_{(i,j)}]$, which is obtained as the product of the transpose of the adjacency matrix A and A. Its diagonal element $c_{(i,i)}$ the number of incoming edges of the vertex v_i and nondiagonal elements $c_{(i,j)}, i \neq j$ indicates the number of vertices with outgoing edges to both i and j. The cocitation matrix of Figure 1 is given in the Appendix.

2.2.1.3 Degree matrix

The degree of each node in the network is represented by a diagonal matrix called a degree matrix. The number of edges that link to a node, or alternatively, the number of other nodes that the node is connected to, determines that node's degree in a network. In social network analysis, it is frequently combined with the adjacency matrix to calculate different network metrics.

The diagonal Matrix $D = [d_{(i,j)}]$ is called a degree matrix if

$$d_{(i,j)} = \begin{cases} \text{degree of } v_i & \text{if } i = j \\ 0 & \text{if } i \neq j. \end{cases}$$

The degree matrix of Figure 1 is given in the Appendix.

2.2.1.4 Laplacian matrix

The Laplacian matrix, also known as the graph Laplacian, is a matrix derived from the adjacency matrix of a graph or network. It offers crucial details regarding the network's connection and organizational structure.

The Laplacian matrix $L = [l_{(i,j)}]$ is obtained by subtracting A (which is explained in Section 2.2.1.1) from D (which is explained in Section 2.2.1.3), that is, $L = D - A$. The Laplacian matrix of Figure 1 is given in the Appendix.

2.2.2 Cohesion measures

Insights into the strength and intensity of relationships inside the network are revealed by cohesion measures, which are crucial in the research of social networks. These metrics aid in determining the degree of cohesiveness or companionship between individuals or groups inside the network. Cohesion metrics offer important insights into the dynamics and structure of social networks, assisting in the comprehension of relationship patterns, the prediction of behavior, and the direction of actions for community- and network-building.

2.2.2.1 Reciprocity

Reciprocity R in a directed network is a quantitative measurement that vertices are linked to one another. Arc and dyad reciprocities are the two primary categories [6]. The fraction of reciprocal ties divided by the total number of ties in the network is known as arc reciprocity. The ratio of actor pairs that have reciprocated to all actor pairs with any relationship is known as the dyad reciprocity. If $R = 1$, the adjacency matrix A will be symmetric. In an undirected graph, all edges are assumed to be reciprocated and $R = 1$.

In our study, we found that the arc reciprocity is 0.396018 and the dyad reciprocity is 0.209677, which means 39.60 % of all actual ties in the network are reciprocated and 20.97 % of all pairs of actors with any relationships have a reciprocating connection.

2.2.2.2 Distance

The length of the shortest distance in between two network nodes is known as the geodesic distance. For example, in Figure 1, the geodesic distance between v_1 to v_2 is 2 (they are connected) and the geodesic distance between v_1 to v_{24} is ∞ (they are not connected).

2.2.2.3 Average distance

This is the average length of geodesics between all pairs of connected vertices. For an undirected or directed network that is strongly connected, the Average Graph Distance is calculated as follows:

$$d^{-} = \frac{\sum_{i \neq j}\{d(v_i, v_j), \forall v_i, v_j \in V}{n(n-1)}.$$

In a connected network, the lowest upper bound of the average distance is $n + \frac{1}{3}$. The total number of existing links between connected nodes is used as denominator in the case of disconnected networks. A natural measure of compactness of a graph is the average distance. In our example, the average distance is 2.3404.

2.2.2.4 Eccentricity

The eccentricity of the vertex v_i measures how far v_i is from other vertices. For a connected network, eccentricity is the largest geodesic distance among a vertex and any other vertex in the network

$$E_{v_i} = \max\{d(v_i, v_j), \forall v_j \in V\}.$$

In a disconnected graph, the eccentricity of every vertex is considered as ∞. The eccentricity of each vertex in Figure 1 is ∞, since it is disconnected.

2.2.2.5 Diameter

Diameter is the greatest eccentricity among the vertices in a connected network, or the largest distance between any two linked vertices

$$d = \max\{d(v_i, v_j), \forall v_i, v_j \in V\}.$$

In Figure 1, it is 5.

2.2.2.6 Clustering coefficient

The clustering coefficient shows the probability of a node to be cluster. It is frequently used in social network analysis to characterize a network's transitivity. How close a node and its neighbors are to forming a clique can be measured by looking at the Clustering Coefficient of that node. The Global Clustering Coefficient (GC) and Local Clustering Coefficient are the two main types of clustering coefficient (LC). GC is based on triplets of actors and serves as an indicator of overall network clustering. A triplet is made up of three interconnected nodes. LC is used to determine network transitivity and to show how embedded a single node is.

In our study, maximum LC = 1 (for node 3) and minimum LC = 0 (for node 1).

The Global Clustering Coefficient GC = 0.281443. Generally, GC varies between 0 (if there is no clique) and 1 (if every node and its neighborhood are complete cliques).

2.2.2.7 Centrality

The centrality of a node in a network indicates how 'prominent' or influential it is. Centrality measures the attempt to assess the importance of each individual within the social network. In order to do so, these metrics usually quantify the relationships between nodes along with their geodesic distances from one another. The following are the most commonly used measures of centrality in network analysis:

2.2.2.7.1 Degree centrality

A node's degree centrality (DC) is the most basic centrality and counts the connections a node has with other nodes in the network. This index is frequently used in social network theory to quantify actor activity. It can be estimated in both types of networks, but undirected ones are normally more practical. The degree centrality index of each vertex in an undirected graph is the number of edges connecting to, but in a directed network it is the total number of outgoing edges coming from. It is the sum of the weights of all outgoing edges associated with in weighted graphs. The following formula is used to calculate the degree centrality of a network with n vertices:

$$\mathrm{DC} = \frac{\sum_{i=1}^{n}[\mathrm{DC}(v^*) - \mathrm{DC}(v_i)]}{\sum_{j=1}^{m}[\mathrm{DC}(w^*) - \mathrm{DC}(w_j)]},$$

where v^* is the vertex having the highest DC in G and w^* is the vertex highest degree centrality in the connected subgraph $G^* = (W, E)$ of G with m vertices. In Figure 1, the maximum degree centrality is 13 (for node 2).

2.2.2.7.2 Closeness centrality

Closeness centrality (CC) is a measure of how close a vertex is to the rest of the network's vertices. Nodes with a high closeness centrality are the ones that can quickly reach a large number of other nodes. A node is considered to be more central if it can readily communicate with more other nodes. The reciprocal of the sum of geodesic distances from that node to all other nodes is the vertex's closeness centrality

$$\mathrm{CC}(v_i) = \frac{(n-1)}{\sum_{v_j \in V} d(v_i, v_j)}.$$

In a strongly connected directed graph, CC can be calculated. If the network has isolation vertices, they are discarded by default. The normal CC is undefined if the graph is weakly linked. The Influence Range Closeness Centrality index could be applied in that situation. In Figure 1, the highest value of CC is 0.02439 (node 2).

2.2.2.7.3 Betweenness centrality

Betweenness centrality (BC) is a vertex measure. A vertex that is included in many geodesic paths has higher betweenness than those do not have this property. It is extremely sensitive to network dynamics, i. e., it changes significantly when a vertex or an edge is added or removed. This index may be determined in both undirected graphs and directed graphs, however, it is often most useful in undirected graphs. A vertex v_i's

Betweenness Centrality is the sum of $\eta_{(s,v_i,t)}$ for all $s, t \in V$, where $\eta_{(s,v_i,t)}$ is the ratio of all shortest paths between s and t that passes through the vertex v_i.

2.2.2.7.4 Stress centrality

The Stress Centrality (SC) of a vertex v_i in a network is the total number of geodesics connecting all other vertices that flow through v_i. The Stress Centrality is very high when one vertex intersects all geodesics connecting the remaining $(n - 1)$ nodes, that is, for a star-like network structure. The SC index indicates how frequently a vertex lies on the geodesics that link other vertices.

2.2.2.7.5 Eccentricity centrality

The eccentricity centrality (EC) of a vertex v_i in a network is the inverse of the maximum geodesic distance $\{d(v_i, v_j), \forall v_j \in V\}$. The eccentricity centrality score indicates how near each node is to one another and hence to the network's center. High EC nodes are located near other nodes in the graph and are therefore more likely to be in the middle of the network. Low EC nodes are farther from one another in the graph. This index is also called Harary graph centrality. In our study, EC is zero.

2.2.2.7.6 Power centrality

This is a generalized degree centrality metric, which adds the degree (with weight 1), the size of the 2nd-order neighborhood (with weight 2), and the size of the kth-order neighborhood for each vertex v_i (with weight k). Thus, the most significant other nodes for vertex v_i are its immediate neighbors, following by nodes in its second-order neighborhood, third-order neighborhood, and so on. The total nodes in the very same components minus one is used to normalize the sum computed for each node. This index may be calculated for both undirected graphs as well as directed graphs, however, it is often most useful in undirected graphs.

2.2.2.7.7 Eigenvector centrality

Eigenvector centrality (EC) of a node v_i is defined as the ith element of the adjacency matrix's main eigenvector. The eigenvector with highest positive eigenvalue is the leading eigenvector. Eigenvector centrality extends degree centrality by assigning a value to each actor that is proportional to the values of its neighbors. It is also called the Gould index.

2.2.2.7.8 In degree centrality

In degree centrality (IDC) counts the number of incoming arcs at each vertex v_i. In directed graphs, for each node, this index is useful as a measure of reputation. As a result, it is known as degree prestige. It is worth noting that the IDC index is the same as degree centrality in undirected graphs. Actors having a higher IDC are seen to be more prominent than others since they earn more nominations or picks. The higher the index value, the more renowned the actor. From Figure 1, the maximum IDC is 17 (for node 32).

3 Conclusion

In the current period, social network analysis has evolved as a critical approach. People have utilized the concept of a social network to explore interpersonal relationships between individuals in a social systems. Social network analysis has progressed from a suggestive analogy to an analytic technique to a paradigm, complete with its own theoretical declarations and procedures. Only when the parameters of Social Network Analysis are exact does it have a scientific basis. Despite the fact that there are several parameters for evaluating a network, the use of precise parameters may vary depending on the network's characteristic. Through this chapter, we tried to take a survey of some parameters that is essential for the understanding of a network. Also, we tried to understand these parameters in the analysis of a network, which is formed in a classroom of teenagers by considering their academic sharing. The tools, which are listed here, can be used to automate the calculation of matrix and cohesion measures and such methodologies are more accessible to researchers in various disciplines.

Appendix

2.2.1.1 Adjacency matrix

Actor/Actor	1	2	3	4	5	6	7	8	9	10	11	12	13	14	15	16	17	18	19	20	21	22	23	24	25	26	27	28	29	30	31	32	33	34
1	0	0	0	2	2	0	2	0	0	0	0	0	0	0	0	0	0	0	0	1	0	0	0	0	0	0	0	0	0	0	0	2	0	0
2	1	0	2	1	1	0	0	0	0	0	0	0	2	0	0	1	0	2	0	0	2	2	1	0	0	0	0	0	0	1	1	2	0	0
3	0	2	0	1	1	0	0	0	1	0	0	0	0	0	0	0	0	0	0	0	0	1	2	0	0	0	0	0	0	0	1	1	0	0
4	0	1	0	0	1	0	2	0	0	0	0	0	0	0	0	0	0	0	0	0	0	2	1	0	0	0	0	0	0	0	2	0	0	0
5	0	1	0	2	0	0	2	0	0	0	0	0	0	0	0	0	0	0	0	2	0	0	0	0	0	0	0	0	0	0	0	2	0	0
6	0	1	0	0	0	0	0	0	1	0	0	0	0	0	0	0	1	0	0	0	0	0	0	0	0	0	0	0	0	0	0	0	0	0
7	0	0	0	2	2	0	0	0	0	0	0	0	0	0	0	0	0	0	0	2	0	1	0	0	0	0	0	0	0	0	0	0	0	0
8	0	1	0	0	0	0	0	0	0	0	0	0	0	0	0	0	0	0	0	0	0	0	1	0	0	0	0	0	0	0	0	0	0	0
9	0	1	0	0	0	0	0	0	0	0	0	0	0	0	0	0	0	0	0	0	0	0	0	0	0	0	0	0	0	0	0	1	0	0
10	0	0	0	0	0	0	0	0	0	0	0	0	0	0	0	0	0	0	0	0	0	0	0	0	0	0	0	0	0	0	0	0	0	0
11	0	0	0	1	1	2	0	0	2	0	0	1	0	0	0	0	0	0	0	1	0	0	0	0	0	1	0	0	0	0	0	2	0	0
12	0	0	0	0	2	0	0	0	0	0	2	0	1	1	0	0	0	1	0	2	0	0	0	0	0	0	0	0	0	0	0	2	0	0
13	0	2	0	0	0	0	0	0	0	0	0	0	0	2	0	0	0	2	0	1	2	0	0	0	0	0	1	0	0	0	0	1	0	0
14	0	0	0	0	0	0	0	0	0	0	0	0	2	0	0	0	0	0	0	0	0	0	0	0	0	0	0	0	0	0	0	1	0	0
15	0	2	2	0	0	0	0	0	0	0	0	0	0	0	0	0	0	0	0	0	0	2	2	0	0	0	0	0	0	0	0	0	0	0
16	0	1	0	0	0	0	0	2	2	0	0	0	0	0	0	0	1	0	0	0	0	0	1	0	0	0	0	0	0	0	0	0	0	0
17	0	1	0	0	0	0	0	0	1	0	0	0	0	0	0	0	0	0	0	0	0	0	1	0	0	0	0	0	0	0	0	0	0	0
18	0	2	0	0	0	0	0	0	1	0	0	0	2	0	0	0	0	0	0	0	2	0	0	0	0	0	0	0	0	0	0	1	0	0
19	1	0	0	1	0	0	0	0	0	0	0	0	0	0	0	0	0	2	0	0	0	2	0	0	0	0	0	0	0	0	0	0	0	0
20	0	0	0	2	2	0	2	0	0	0	0	0	0	0	0	0	0	0	0	0	0	0	0	0	0	0	0	0	0	0	0	2	0	0
21	0	2	0	0	2	0	0	0	0	0	0	0	2	0	0	0	0	2	0	0	0	2	0	0	1	0	0	0	0	0	1	0	0	0
22	0	0	0	2	0	0	0	0	0	0	0	0	0	0	0	0	0	0	0	0	0	0	2	0	0	0	0	0	0	1	1	0	0	0
23	0	1	2	1	0	0	0	0	2	0	0	0	0	0	0	0	0	0	0	0	0	1	0	0	0	0	0	0	0	0	0	1	0	0
24	0	0	0	0	0	0	0	0	0	0	0	0	0	0	0	0	0	0	0	0	0	0	0	0	0	0	0	0	0	0	0	0	0	0
25	0	0	0	0	0	0	0	0	0	0	0	0	0	0	0	0	0	0	0	0	0	0	0	0	0	0	0	0	0	0	0	0	0	0
26	0	0	0	0	1	0	0	0	0	0	1	0	0	0	0	1	0	0	0	1	0	0	0	0	0	0	0	0	0	0	0	1	0	0
27	0	0	0	0	1	2	0	0	0	0	2	1	1	0	0	0	0	2	0	0	2	0	0	0	0	1	0	0	0	0	0	2	0	0
28	0	0	0	2	0	0	0	0	0	0	1	0	0	0	0	0	0	0	0	2	0	0	0	0	0	0	0	0	0	0	0	2	0	0
29	0	0	0	0	1	2	0	0	0	0	2	1	1	0	0	0	0	2	0	0	2	0	0	0	0	1	0	0	0	0	0	2	0	0
30	0	2	0	1	0	0	0	0	0	0	0	0	0	0	0	0	0	0	0	0	0	2	0	0	0	0	0	0	0	0	0	0	0	0
31	0	2	0	2	0	0	0	0	0	0	0	0	0	0	0	0	0	0	0	0	0	1	2	0	0	0	0	0	0	0	0	0	0	0
32	0	1	0	2	2	0	2	0	0	0	0	0	0	0	0	0	0	0	0	2	0	0	1	0	0	0	0	0	0	0	0	0	0	0
33	0	0	0	0	0	0	0	0	0	0	0	0	0	0	0	0	0	0	0	0	0	0	0	0	0	0	0	0	0	0	0	0	0	0
34	0	0	0	0	0	0	0	0	0	0	2	0	0	0	0	0	0	0	0	2	0	0	0	0	0	0	0	0	0	0	0	2	0	0

2.2.1.2 Cocitation matrix

Actor/Actor	1	2	3	4	5	6	7	8	9	10	11	12	13	14	15	16	17	18	19	20	21	22	23	24	25	26	27	28	29	30	31	32	33	34
1	2	0	2	2	1	0	0	0	0	0	0	0	2	0	0	1	0	4	0	0	2	4	1	0	0	0	0	0	0	1	1	2	0	0
2	0	37	6	13	9	0	6	2	10	0	0	0	8	4	0	0	2	8	0	6	8	19	17	0	2	0	2	0	0	0	6	10	0	0
3	2	6	12	4	2	0	0	0	4	0	0	0	4	0	0	2	0	4	0	0	4	10	6	0	0	0	0	0	0	2	2	6	0	0
4	2	13	4	38	19	2	16	0	5	0	2	1	2	0	0	1	0	4	0	19	2	12	13	0	0	1	0	0	0	3	4	22	0	0
5	1	9	2	19	31	6	14	0	3	0	9	3	10	2	0	2	0	12	0	16	6	11	6	0	2	3	0	0	0	1	6	22	0	0
6	0	0	0	2	6	12	0	0	4	0	8	6	4	0	0	0	0	8	0	2	8	0	0	0	0	6	0	0	0	0	0	12	0	0
7	0	6	0	16	14	0	20	0	0	0	0	0	0	0	0	0	0	0	0	10	0	4	4	0	0	0	0	0	0	0	4	12	0	0
8	0	2	0	0	0	0	0	4	4	0	0	0	0	0	0	0	2	0	0	0	0	0	2	0	0	0	0	0	0	0	0	0	0	0
9	0	10	4	5	3	4	0	4	16	0	0	2	2	0	0	0	3	0	0	2	2	3	5	0	0	2	0	0	0	0	1	8	0	0
10	0	0	0	0	0	0	0	0	0	0	0	0	0	0	0	0	0	0	0	0	0	0	0	0	0	0	0	0	0	0	0	0	0	0
11	0	0	0	2	9	8	0	0	0	0	18	4	6	2	0	1	0	10	0	11	8	0	0	0	0	4	0	0	0	0	0	19	0	0
12	0	0	0	1	3	6	0	0	2	0	4	3	2	0	0	0	0	4	0	1	4	0	0	0	0	3	0	0	0	0	0	6	0	0
13	2	8	4	2	10	4	0	0	2	0	6	2	19	1	0	2	0	13	0	2	12	8	2	0	2	2	0	0	0	2	4	14	0	0
14	0	4	0	0	2	0	0	0	0	0	2	0	1	5	0	0	0	5	0	4	4	0	0	0	0	0	2	0	0	0	0	4	0	0
15	0	0	0	0	0	0	0	0	0	0	0	0	0	0	0	0	0	0	0	0	0	0	0	0	0	0	0	0	0	0	0	0	0	0
16	1	0	2	1	2	0	0	0	0	0	1	0	2	0	0	2	0	2	0	1	2	2	1	0	0	0	0	0	0	1	1	3	0	0
17	0	2	0	0	0	0	0	2	3	0	0	0	0	0	0	0	2	0	0	0	0	0	1	0	0	0	0	0	0	0	0	0	0	0
18	4	8	4	4	12	8	0	0	0	0	10	4	13	5	0	2	0	25	0	4	16	12	2	0	2	4	2	0	0	2	4	16	0	0
19	0	0	0	0	0	0	0	0	0	0	0	0	0	0	0	0	0	0	0	0	0	0	0	0	0	0	0	0	0	0	0	0	0	0
20	0	6	0	19	16	2	10	0	2	0	11	1	2	4	0	1	0	4	0	28	2	2	2	0	0	1	1	0	0	0	0	22	0	0
21	2	8	4	2	6	8	0	0	2	0	8	4	12	4	0	2	0	16	0	2	20	4	2	0	0	4	2	0	0	2	2	16	0	0
22	4	19	10	12	11	0	4	0	3	0	0	0	8	0	0	2	0	12	0	2	4	28	12	0	2	0	0	0	0	2	9	6	0	0
23	1	17	6	13	6	0	4	2	5	0	0	0	2	0	0	1	1	2	0	2	2	12	22	0	0	0	0	0	0	3	7	4	0	0
24	0	0	0	0	0	0	0	0	0	0	0	0	0	0	0	0	0	0	0	0	0	0	0	0	0	0	0	0	0	0	0	0	0	0
25	0	2	0	0	2	0	0	0	0	0	0	0	2	0	0	0	0	2	0	0	0	2	0	0	1	0	0	0	0	0	1	0	0	0
26	0	0	0	1	3	6	0	0	2	0	4	3	2	0	0	0	0	4	0	1	4	0	0	0	0	3	0	0	0	0	0	6	0	0
27	0	2	0	0	0	0	0	0	0	0	0	0	0	2	0	0	0	2	0	1	2	0	0	0	0	0	1	0	0	0	0	1	0	0
28	0	0	0	0	0	0	0	0	0	0	0	0	0	0	0	0	0	0	0	0	0	0	0	0	0	0	0	0	0	0	0	0	0	0
29	0	0	0	0	0	0	0	0	0	0	0	0	0	0	0	0	0	0	0	0	0	0	0	0	0	0	0	0	0	0	0	0	0	0
30	1	0	2	3	1	0	0	0	0	0	0	0	2	0	0	1	0	2	0	0	2	2	3	0	0	0	0	0	0	2	2	2	0	0
31	1	6	2	4	6	0	4	0	1	0	0	0	4	0	0	1	0	4	0	0	2	9	7	0	1	0	0	0	0	2	8	3	0	0
32	2	10	6	22	22	12	12	0	8	0	19	6	14	4	0	3	0	16	0	22	16	6	4	0	0	6	1	0	0	2	3	47	0	0
33	0	0	0	0	0	0	0	0	0	0	0	0	0	0	0	0	0	0	0	0	0	0	0	0	0	0	0	0	0	0	0	0	0	0
34	0	0	0	0	0	0	0	0	0	0	0	0	0	0	0	0	0	0	0	0	0	0	0	0	0	0	0	0	0	0	0	0	0	0

2.2.1.3 Degree matrix

Actor/Actor	1	2	3	4	5	6	7	8	9	10	11	12	13	14	15	16	17	18	19	20	21	22	23	24	25	26	27	28	29	30	31	32	33	34
1	9	0	0	0	0	0	0	0	0	0	0	0	0	0	0	0	0	0	0	0	0	0	0	0	0	0	0	0	0	0	0	0	0	0
2	0	19	0	0	0	0	0	0	0	0	0	0	0	0	0	0	0	0	0	0	0	0	0	0	0	0	0	0	0	0	0	0	0	0
3	0	0	10	0	0	0	0	0	0	0	0	0	0	0	0	0	0	0	0	0	0	0	0	0	0	0	0	0	0	0	0	0	0	0
4	0	0	0	9	0	0	0	0	0	0	0	0	0	0	0	0	0	0	0	0	0	0	0	0	0	0	0	0	0	0	0	0	0	0
5	0	0	0	0	9	0	0	0	0	0	0	0	0	0	0	0	0	0	0	0	0	0	0	0	0	0	0	0	0	0	0	0	0	0
6	0	0	0	0	0	3	0	0	0	0	0	0	0	0	0	0	0	0	0	0	0	0	0	0	0	0	0	0	0	0	0	0	0	0
7	0	0	0	0	0	0	7	0	0	0	0	0	0	0	0	0	0	0	0	0	0	0	0	0	0	0	0	0	0	0	0	0	0	0
8	0	0	0	0	0	0	0	2	0	0	0	0	0	0	0	0	0	0	0	0	0	0	0	0	0	0	0	0	0	0	0	0	0	0
9	0	0	0	0	0	0	0	0	2	0	0	0	0	0	0	0	0	0	0	0	0	0	0	0	0	0	0	0	0	0	0	0	0	0
10	0	0	0	0	0	0	0	0	0	0	0	0	0	0	0	0	0	0	0	0	0	0	0	0	0	0	0	0	0	0	0	0	0	0
11	0	0	0	0	0	0	0	0	0	0	11	0	0	0	0	0	0	0	0	0	0	0	0	0	0	0	0	0	0	0	0	0	0	0
12	0	0	0	0	0	0	0	0	0	0	0	11	0	0	0	0	0	0	0	0	0	0	0	0	0	0	0	0	0	0	0	0	0	0
13	0	0	0	0	0	0	0	0	0	0	0	0	11	0	0	0	0	0	0	0	0	0	0	0	0	0	0	0	0	0	0	0	0	0
14	0	0	0	0	0	0	0	0	0	0	0	0	0	3	0	0	0	0	0	0	0	0	0	0	0	0	0	0	0	0	0	0	0	0
15	0	0	0	0	0	0	0	0	0	0	0	0	0	0	6	0	0	0	0	0	0	0	0	0	0	0	0	0	0	0	0	0	0	0
16	0	0	0	0	0	0	0	0	0	0	0	0	0	0	0	7	0	0	0	0	0	0	0	0	0	0	0	0	0	0	0	0	0	0
17	0	0	0	0	0	0	0	0	0	0	0	0	0	0	0	0	3	0	0	0	0	0	0	0	0	0	0	0	0	0	0	0	0	0
18	0	0	0	0	0	0	0	0	0	0	0	0	0	0	0	0	0	8	0	0	0	0	0	0	0	0	0	0	0	0	0	0	0	0
19	0	0	0	0	0	0	0	0	0	0	0	0	0	0	0	0	0	0	6	0	0	0	0	0	0	0	0	0	0	0	0	0	0	0
20	0	0	0	0	0	0	0	0	0	0	0	0	0	0	0	0	0	0	0	8	0	0	0	0	0	0	0	0	0	0	0	0	0	0
21	0	0	0	0	0	0	0	0	0	0	0	0	0	0	0	0	0	0	0	0	12	0	0	0	0	0	0	0	0	0	0	0	0	0
22	0	0	0	0	0	0	0	0	0	0	0	0	0	0	0	0	0	0	0	0	0	6	0	0	0	0	0	0	0	0	0	0	0	0
23	0	0	0	0	0	0	0	0	0	0	0	0	0	0	0	0	0	0	0	0	0	0	6	0	0	0	0	0	0	0	0	0	0	0
24	0	0	0	0	0	0	0	0	0	0	0	0	0	0	0	0	0	0	0	0	0	0	0	0	0	0	0	0	0	0	0	0	0	0
25	0	0	0	0	0	0	0	0	0	0	0	0	0	0	0	0	0	0	0	0	0	0	0	0	0	0	0	0	0	0	0	0	0	0
26	0	0	0	0	0	0	0	0	0	0	0	0	0	0	0	0	0	0	0	0	0	0	0	0	0	5	0	0	0	0	0	0	0	0
27	0	0	0	0	0	0	0	0	0	0	0	0	0	0	0	0	0	0	0	0	0	0	0	0	0	0	14	0	0	0	0	0	0	0
28	0	0	0	0	0	0	0	0	0	0	0	0	0	0	0	0	0	0	0	0	0	0	0	0	0	0	0	7	0	0	0	0	0	0
29	0	0	0	0	0	0	0	0	0	0	0	0	0	0	0	0	0	0	0	0	0	0	0	0	0	0	0	0	14	0	0	0	0	0
30	0	0	0	0	0	0	0	0	0	0	0	0	0	0	0	0	0	0	0	0	0	0	0	0	0	0	0	0	0	5	0	0	0	0
31	0	0	0	0	0	0	0	0	0	0	0	0	0	0	0	0	0	0	0	0	0	0	0	0	0	0	0	0	0	0	7	0	0	0
32	0	0	0	0	0	0	0	0	0	0	0	0	0	0	0	0	0	0	0	0	0	0	0	0	0	0	0	0	0	0	0	10	0	0
33	0	0	0	0	0	0	0	0	0	0	0	0	0	0	0	0	0	0	0	0	0	0	0	0	0	0	0	0	0	0	0	0	0	0
34	0	0	0	0	0	0	0	0	0	0	0	0	0	0	0	0	0	0	0	0	0	0	0	0	0	0	0	0	0	0	0	0	0	6

2.2.1.4 Laplacian matrix

Actor/Actor	1	2	3	4	5	6	7	8	9	10	11	12	13	14	15	16	17
1	9	0	0	-2	-2	0	-2	0	0	0	0	0	0	0	0	0	0
2	-1	19	-2	-1	-1	0	0	0	0	0	0	0	-2	0	0	-1	0
3	0	-2	10	-1	-1	0	0	0	-1	0	0	0	0	0	0	0	0
4	0	-1	0	9	-1	0	-2	0	0	0	0	0	0	0	0	0	0
5	0	-1	0	-2	9	0	-2	0	0	0	0	0	0	0	0	0	0
6	0	-1	0	0	0	3	0	0	-1	0	0	0	0	0	0	0	-1
7	0	0	0	-2	-2	0	7	0	0	0	0	0	0	0	0	0	0
8	0	-1	0	0	0	0	0	2	0	0	0	0	0	0	0	0	0
9	0	-1	0	0	0	0	0	0	2	0	0	0	0	0	0	0	0
10	0	0	0	0	0	0	0	0	0	0	0	0	0	0	0	0	0
11	0	0	0	-1	-1	-2	0	0	-2	0	11	-1	0	0	0	0	0
12	0	0	0	0	-2	0	0	0	0	0	-2	11	-1	-1	0	0	0
13	0	-2	0	0	0	0	0	0	0	0	0	0	11	-2	0	0	0
14	0	0	0	0	0	0	0	0	0	0	0	0	-2	3	0	0	0
15	0	-2	-2	0	0	0	0	0	0	0	0	0	0	0	8	0	0
16	0	-1	0	0	0	0	0	-2	-2	0	0	0	0	0	0	7	-1
17	0	-1	0	0	0	0	0	0	-1	0	0	0	0	0	0	0	3
18	0	-2	0	0	0	0	0	0	-1	0	0	0	-2	0	0	0	0
19	-1	0	0	-1	0	0	0	0	0	0	0	0	0	0	0	0	0
20	0	0	0	-2	-2	0	-2	0	0	0	0	0	0	0	0	0	0
21	0	-2	0	0	-2	0	0	0	0	0	0	0	-2	0	0	0	0
22	0	0	0	-2	0	0	0	0	0	0	0	0	0	0	0	0	0
23	0	-1	-2	-1	0	0	0	0	-2	0	0	0	0	0	0	0	0
24	0	0	0	0	0	0	0	0	0	0	0	0	0	0	0	0	0
25	0	0	0	0	0	0	0	0	0	0	0	0	0	0	0	0	0
26	0	0	0	0	-1	0	0	0	0	0	-1	0	0	0	0	-1	0
27	0	0	0	0	-1	-2	0	0	0	0	-2	-1	-1	0	0	0	0
28	0	0	0	-2	0	0	0	0	0	0	-1	0	0	0	0	0	0
29	0	0	0	0	-1	-2	0	0	0	0	-2	-1	-1	0	0	0	0
30	0	-2	0	-1	0	0	0	0	0	0	0	0	0	0	0	0	0
31	0	-2	0	-2	0	0	0	0	0	0	0	0	0	0	0	0	0
32	0	-1	0	-2	-2	0	-2	0	0	0	0	0	0	0	0	0	0
33	0	0	0	0	0	0	0	0	0	0	0	0	0	0	0	0	0
34	0	0	0	0	0	0	0	0	0	0	-2	0	0	0	0	0	0

Actor/Actor	18	19	20	21	22	23	24	25	26	27	28	29	30	31	32	33	34
1	0	0	-1	0	0	0	0	0	0	0	0	0	0	0	-2	0	0
2	-2	0	0	-2	-2	-1	0	0	0	0	0	0	-1	-1	-2	0	0
3	0	0	0	0	-1	-2	0	0	0	0	0	0	0	-1	-1	0	0
4	0	0	0	0	-2	-1	0	0	0	0	0	0	0	-2	0	0	0
5	0	0	-2	0	0	0	0	0	0	0	0	0	0	0	-2	0	0
6	0	0	0	0	0	0	0	0	0	0	0	0	0	0	0	0	0
7	0	0	-2	0	-1	0	0	0	0	0	0	0	0	0	0	0	0
8	0	0	0	0	0	-1	0	0	0	0	0	0	0	0	0	0	0
9	0	0	0	0	0	0	0	0	0	0	0	0	0	0	-1	0	0
10	0	0	0	0	0	0	0	0	0	0	0	0	0	0	0	0	0
11	0	0	-1	0	0	0	0	0	-1	0	0	0	0	0	-2	0	0
12	-1	0	-2	0	0	0	0	0	0	0	0	0	0	0	-2	0	0
13	-2	0	-1	-2	0	0	0	0	0	-1	0	0	0	0	-1	0	0
14	0	0	0	0	0	0	0	0	0	0	0	0	0	0	-1	0	0
15	0	0	0	0	-2	-2	0	0	0	0	0	0	0	0	0	0	0
16	0	0	0	0	0	-1	0	0	0	0	0	0	0	0	0	0	0
17	0	0	0	0	0	-1	0	0	0	0	0	0	0	0	0	0	0
18	8	0	0	-2	0	0	0	0	0	0	0	0	0	0	-1	0	0
19	-2	6	0	0	-2	0	0	0	0	0	0	0	0	0	0	0	0
20	0	0	8	0	0	0	0	0	0	0	0	0	0	0	-2	0	0
21	-2	0	0	12	-2	0	0	0	-1	0	0	0	0	-1	0	0	0
22	0	0	0	0	6	-2	0	0	0	0	0	0	-1	-1	0	0	0
23	0	0	0	0	-1	6	0	0	0	0	0	0	0	0	-1	0	0
24	0	0	0	0	0	0	0	0	0	0	0	0	0	0	0	0	0
25	0	0	0	0	0	0	0	0	0	0	0	0	0	0	0	0	0
26	0	0	-1	0	0	0	0	0	5	0	0	0	0	0	-1	0	0
27	-2	0	0	-2	0	0	0	0	-1	14	0	0	0	0	-2	0	0
28	0	0	-2	0	0	0	0	0	0	0	7	0	0	0	-2	0	0
29	-2	0	0	-2	0	0	0	0	-1	0	0	14	0	0	-2	0	0
30	0	0	0	0	-2	0	0	0	0	0	0	0	5	0	0	0	0
31	0	0	0	0	-1	-2	0	0	0	0	0	0	0	7	0	0	0
32	0	0	-2	0	0	-1	0	0	0	0	0	0	0	0	10	0	0
33	0	0	0	0	0	0	0	0	0	0	0	0	0	0	0	0	0
34	0	0	-2	0	0	0	0	0	0	0	0	0	0	0	-2	0	6

Bibliography

[1] Carrington, P. J., & Scott, J. (2011). *The SAGE Handbook of Social Network Analysis*. SAGE Publications.

[2] Deepa, V. G., Aparna Lakshmanan, S., & Sreeja, V. N. (2019). Centrality and reciprocity in directed social networks: a case study. *Malaya Journal of Matematik*, *S*(1), 479–484.

[3] Watts, D. J., & Strogatz, S. H. (1998). Collective dynamics of small-world networks. *Nature*, *393*, 440–442. https://doi.org/10.1038/30918.

[4] Harary, F., & Norman, R. Z. (1953). *Graph Theory as a Mathematical Model in Social Science*. Institute for Social Research.

[5] Golumbic, M. C. (1980). *Algorithmic Graph Theory and Perfect Graphs*. Academic Press.

[6] Newman, M. (2018). *Networks* (2nd ed.). Oxford University Press.

[7] Durland, M. M., & Fredericks, K. A. (2005). An introduction to social network analysis. *New Directions for Evaluation*, *2005*(107), 5–13. https://doi.org/10.1002/ev.157.

[8] Wasserman, S., & Faust, K. (1994). *Social Network Analysis: Methods and Applications*. Cambridge University Press.

[9] Wellman, B. (1988). Structural analysis: from method and metaphor to theory and substance. In B. Wellman & S. D. Berkowitz (Eds.), *Social Structures: A Network Approach* (pp. 19–61). Cambridge University Press.

[10] White, H. C. (1963). Uses of mathematics in sociology. In J. C. Charlesworth (Ed.), *Mathematics and the Social Sciences*. American Academy of Political and Social Science.

Pragya and Sukhpreet Kaur Sidhu

Fuzzy Big-M approach for solving fuzzy linear programming using a horizontal membership function

Abstract: In real-life optimization problems, due to uncertainty and hesitancy, it might be challenging for the decider to identify the precise values of the parameters. To deal with this situation, fuzzy set theory came into existence. There exist many methods to solve fuzzy linear programming. Where an identity matrix is not present, then the fuzzy Big-M method is used. Here, we propose an algorithm for the fuzzy Big-M method using horizontal membership functions. The approach transforms fuzzy linear programming problems (FLPP) into crisp models by converting triangular and trapezoidal fuzzy numbers into horizontal representations. By applying a ranking function, the fuzzy parameters are defuzzified, enabling the use of the Big-M method to find crisp optimal solutions. These are then substituted back to determine the final fuzzy optimal solution. Numerical illustrations demonstrate that this horizontal membership approach provides a computationally efficient and robust framework for decision-making under uncertainty.

Keywords: Fuzzy linear programming, horizontal membership function, fuzzy optimal solution, horizontal fuzzy number, fuzzy Big-M approach

MSC 2020: 90C70, 03E72, 90C05, 90C90

1 Introduction

The linear programming problem (LPP) is a mathematical technique to optimize a linear objective function subjected to linear constraints. The constraints may be equalities or inequalities. A solution that satisfies all the linear constraints and nonnegative restrictions of problem is called a feasible solution (FS) and a FS that gives the best possible outcomes of objective functions within the constraint problem is called an optimal feasible solution.

In 1939, Leonid Kantorovich of Russia developed the LPP and many researchers utilized it into other fields. In 1947, George B. Dantzing developed the simplex method of solving LPP.

Pragya, Sukhpreet Kaur Sidhu, Department of Mathematics, Akal University, Talwandi Sabo, India, e-mail: sukhpreetkaursran@gmail.com

https://doi.org/10.1515/9783112234525-006

The most general form of LPP is

$$\begin{aligned} \max(\min)\ z &= c^T x \\ \text{s.t.} \\ Ax &\leq, =, \geq b, \\ x &\geq 0, \end{aligned} \tag{1}$$

where $c \in R^n, A \in R^{m\times n}, x \in R^n, b \in R^m$

In linear programming (LP), uncertainty is not allowed, i. e., in linear programming, all the values of parameters and variables are certainly known. However, in reality, the values of parameters and variables are not certain, then the LP may not provide the relevant solutions. To address this, Zadeh [10] introduced the concept of fuzzy set theory in 1965, which allows for the representation of uncertainty through fuzzy numbers. The concept of fuzzy linear programming problem (FLPP) was introduced by H. J. Zimmerman during 1974–1976. The fuzzy linear programming problem (FLPP) is a problem in which parameters (coefficients) or variables are represented by fuzzy numbers.

The fully fuzzy linear programming problem (FFLPP) [2, 3] is a problem in which all the variables and parameters are expressed as fuzzy numbers.

The general form of FFLPP is

$$\begin{aligned} \max(\min)\ \tilde{z} &\approx \tilde{c}^T \tilde{x} \\ \text{s.t.} \\ \tilde{A}\tilde{x} &\preceq, \approx, \succeq \tilde{b}, \\ \tilde{x} &\succeq \tilde{0}, \end{aligned} \tag{2}$$

where $\tilde{b} \in F(R^m), \tilde{A} \in F(R^{m\times n}), \tilde{c} \in F(R^n), \tilde{x} \in F(R^n)$.

The concept of a horizontal membership function was introduced by Piegat and Landowski [8] in 2015. They define a fuzzy set not in the form of a vertical membership function of type $\mu = f(x)$ but in the horizontal type $x = f(\mu)$. The idea of a relative distance measure variable (RDM), $\beta_x \in [0,1]$ was proposed by Piegat and Landowski [7, 9]. This variable enables us to obtain any value lying between the left border $x_L(\mu)$ and the right border $x_R(\mu)$. On the left border β_x is zero and on the right border β_x is one.

Nasseri et al. [6] proposed a new method for solving FLPP, by use of a linear ranking function. Hatami and Kazemipoor [4] proposed a fuzzy Big-M method for solving FLPP in which all the elements of the coefficient matrix of the constraints are represented by real numbers and the rest of the parameters are represented by symmetric trapezoidal fuzzy numbers.

Piegat and Landowski [8] proposed a horizontal membership function that defined a fuzzy set not in the form of vertical membership functions but in horizontal form. It can be constructed with the use of relative distance measure variables and examples of its application were given.

Landowski [5] proposed a method for solving a real fuzzy system with the use of horizontal fuzzy numbers and the properties of basic algebraic operations on these numbers.

Edalatpanah [1] proposed a new approach to solve FLPP using the theory of horizontal membership functions and a multidimensional solution based on the primal simplex method. After reviewing the literature, it is found that no work has been done on the Big-M method for solving FLPP using a horizontal membership function. In this work, a new algorithm is proposed to solve the fuzzy Big-M approach using a horizontal membership function.

The main motivation of this research is to improve fuzzy math problem solving using ≥ or = rules. Typically, these problems are difficult to solve because there is no clear starting point for the solution to begin. The fuzzy Big-M method is an effective tool for dealing with these complex rules. Additionally, we use the Horizontal Membership Function (HMF) for greater accuracy. When fuzzy ideas are converted into fixed numbers, the majority of traditional methods lose some important information. Using HMF, we can preserve the full shape and details of triangular and trapezoidal fuzzy numbers. This increases the reliability of our final results and makes them easier for computers to calculate.

This chapter is constituted as follows: In Section 2, some related definitions and arithmetic operations of horizontal fuzzy numbers are presented. In Section 3, a fuzzy Big-M approach is presented for solving FLPP. In Section 4, a new algorithm is proposed to apply the fuzzy Big-M approach to solve FLPP using HMF. In Section 5, numerical examples are given to illustrate the new algorithm. Finally, conclusions are provided in Section 6.

2 Preliminaries

In this section, some basic definitions and arithmetic operations of horizontal fuzzy numbers are presented [1]:

Definition 1. Let $Y = [Y^l, Y^u]$. Then, each $y \in Y$ using RDM variable $\beta_y \in [0,1]$ can be described $y = Y^l + \beta_y(Y^u - Y^l)$ and Y can be defined as $X = \{y \mid y = Y^l + \beta_y(Y^u - Y^l)\}$.

Definition 2. The horizontal fuzzy number V^r for an arbitrary fuzzy number $v = (\underline{v}(r), \overline{v}(r))$ is defined in terms of a set of numbers v^r as $V^r = \{v^r \in R \mid v^r = \underline{v}(r) + \beta_v(\underline{v}(r) - \overline{v}(r)), r, \beta_v \in [0,1]\}$. The horizontal fuzzy number V^r can be expressed as $v(r, \beta_v)$.

Remark 1. The horizontal fuzzy number V^r can be expressed as $v(r, \beta_v) = \underline{v}(r) + \beta_v(\underline{v}(r) - \overline{v}(r)), r, \beta_v \in [0,1]$.

Remark 2. The horizontal fuzzy number of the triangular fuzzy number $\tilde{A} = (a, b, c)$ is as follows:

$$A = [a + (b - a)r] + [(c - a)(1 - r)]\beta_{\tilde{A}}. \tag{3}$$

Remark 3. The horizontal fuzzy number of the trapezoidal fuzzy number $\tilde{B} = (a, b, c, d)$ is as follows:

$$B = [a + (b - a)r] + [(d - a) - r(d - a + b - c)]\beta_{\tilde{B}}. \tag{4}$$

Definition 3. Let $\tilde{A}$, $\tilde{B}$ be two fuzzy numbers whose horizontal fuzzy numbers are $A(r, \beta_A)$ and $B(r, \beta_B)$, respectively. Then, the four basic arithmetic operations $o \in \{+, -, ., \div\}$ on $\tilde{A}$ and $\tilde{B}$ are as $\tilde{A}\ o\ \tilde{B} = A(r, \beta_A)\ o\ B(r, \beta_B)$.

Remark 4. Based on Definition 3, the following relations hold:
a) $\tilde{A} - \tilde{A} = 0$;
b) $\tilde{A} - \tilde{B} = -(\tilde{B} - \tilde{A})$;
c) if $\tilde{A} \neq 0$, then $\tilde{A} \div \tilde{A} = 1$;
d) $(\tilde{A} + \tilde{B}).\tilde{C} = \tilde{A}\tilde{C} + \tilde{B}\tilde{C}$.

Definition 4. A function $R : F(R) \to R$, that associates each fuzzy number to the real line, is called a ranking function. For any two fuzzy numbers $\tilde{C}$ and $\tilde{D}$, we have
(i) $\tilde{C} \leq \tilde{D}$ if and only if $R(\tilde{C}) \leq R(\tilde{D})$;
(ii) $\tilde{C} \geq \tilde{D}$ if and only if $R(\tilde{C}) \geq R(\tilde{D})$.

Definition 5. Let $\tilde{B}$ be any fuzzy number such that its horizontal fuzzy number (HFN) is $B(r, \beta B)$, then the ranking function $\tilde{A}$ can be defined as $R(\tilde{A}) = \int_0^1 \int_0^1 B(r, \beta_B)$.

3 Fuzzy Big-M approach for solving FLPPs

The fuzzy Big-M approach is a mathematical way to solve FLPP with inequality constraints. To convert all the inequality constraints into equality constraints, we use fuzzy slack and fuzzy surplus variables. If the identity submatrix are not present in matrix A, then fuzzy artificial variables ($\tilde{a}$) are added. After adding the artificial variables ($\tilde{a}$), the constraints of the considered system are reformulated as $\tilde{A}\tilde{x} + \tilde{a} = \tilde{b}$, where $\tilde{x} \geq \tilde{0}$ (indicating nonnegativity of fuzzy decision variables) and $\tilde{a} \geq \tilde{0}$ (indicating nonnegativity of the fuzzy artificial variables).

Furthermore, a large penalty value $\tilde{M}$ is assigned to each artificial variable in the objective function. After assigning the large penalty and adding a fuzzy artificial variable the problem (2) is converted into standard FLPP

$$\begin{aligned} \max\,(\tilde{z}) &\approx \tilde{c}^T\tilde{x} - \tilde{M}\tilde{a} \\ &\text{s.t.} \\ \tilde{A}\tilde{x} + \tilde{a} &\approx \tilde{b}, \\ \tilde{x},\ \tilde{a} &\succeq \tilde{0}, \end{aligned} \tag{5}$$

where $\tilde{M} = (M, M)$.

4 The proposed algorithm

In this section, an algorithm is proposed to solve FLPP using a horizontal membership function.

Consider the fuzzy parameter linear programming problem as

$$\begin{aligned} \max(\min)\ \tilde{z} &\approx \tilde{c}^T x \\ &\text{s.t.} \\ \tilde{A}x &\preceq, \approx, \succeq \tilde{b}, \\ x &\geq 0, \end{aligned} \tag{6}$$

where $\tilde{b} \in F(R^m), \tilde{A} \in F(R^{m\times n}), \tilde{c} \in F(R^n), x \in R^n$.

Step 1. *Horizontal representation of fuzzy parameters*

Convert all fuzzy parameters into their respective horizontal fuzzy numbers. For any fuzzy number $\tilde{b}$, the horizontal fuzzy number $b(r, \beta_{\tilde{b}})$ is obtained.

Step 2. *Apply the ranking function*

Apply the ranking function $R : F(R) \to R$ to the horizontal fuzzy numbers obtained in Step 1. This process maps the fuzzy horizontal expressions into crisp real numbers.

Step 3. *Formulation of the crisp LLP*

Using Step 1 and Step 2, obtain the following crisp LPP from FLPP (6):

$$\begin{aligned} \max(\min)\ \tilde{z} &= R(\tilde{c})x \\ &\text{s.t.} \\ R(\tilde{A})x &\leq, =, \geq R(\tilde{b}) \\ x &\geq 0. \end{aligned} \tag{7}$$

Step 4. *Computation of optimal crisp solution*

By applying the Big-M approach to problem (7), obtain the crisp feasible and crisp optimal solutions.

Step 5. *Evaluation of fuzzy optimal objective value*

Put the value of the crisp feasible solution in FLPP and obtain the fuzzy optimal solution $\tilde{z} = (a, b, c)$.

5 Numerical illustrations

In this section, some examples are provided to illustrate the fuzzy Big-M approach using a horizontal membership function:

Example 5.1. Consider the following FLPP:

$$\begin{aligned}\text{Maximize}(\tilde{z}) &\approx (2,6,7)x_1 + (5,7,9)x_2\\ &\text{s.t.}\\ (3,7,8)x_1 + (3,4,7)x_2 &\succeq (43,45,48)\\ (4,6,8)x_1 + (5,6,9)x_2 &\preceq (50,53,56)\\ x_1, x_2 &\geq 0.\end{aligned} \tag{8}$$

Step 1. *Horizontal representation of fuzzy parameters*

$$\begin{aligned}
&\tilde{A} = (2,6,7), \quad A(r,\beta_{\tilde{A}}) = (2+4r) + \beta_{\tilde{A}}(5-5r);\\
&\tilde{B} = (5,7,9), \quad B(r,\beta_{\tilde{B}}) = (5+2r) + \beta_{\tilde{B}}(4-4r);\\
&\tilde{C} = (3,7,8), \quad C(r,\beta_{\tilde{C}}) = (3+4r) + \beta_{\tilde{C}}(5-5r);\\
&\tilde{D} = (3,4,7), \quad D(r,\beta_{\tilde{D}}) = (3+r) + \beta_{\tilde{D}}(4-4r);\\
&\tilde{E} = (4,6,8), \quad E(r,\beta_{\tilde{E}}) = (4+2r) + \beta_{\tilde{E}}(4-4r);\\
&\tilde{F} = (5,6,9), \quad F(r,\beta_{\tilde{F}}) = (5+r) + \beta_{\tilde{F}}(4-4r);\\
&\tilde{G} = (43,45,48), \quad G(r,\beta_{\tilde{G}}) = (43+2r) + \beta_{\tilde{G}}(4-4r);\\
&\tilde{H} = (50,53,56), \quad H(r,\beta_{\tilde{H}}) = (50+3r) + \beta_{\tilde{H}}(6-6r).
\end{aligned}$$

Step 2. *Apply the ranking function*

$$\begin{aligned}
&R(\tilde{A}) = 5.25, \quad R(\tilde{B}) = 7, \quad R(\tilde{C}) = 6.25;\\
&R(\tilde{D}) = 4.5, \quad R(\tilde{E}) = 6, \quad R(\tilde{F}) = 6.5;\\
&R(\tilde{G}) = 43, \quad R(\tilde{H}) = 53.
\end{aligned}$$

Step 3. *Formulation of the crisp LLP*
Using Step 1 and Step 2, we obtain the crisp LPP from fuzzy LPP(8)

$$\begin{aligned}\text{Maximize}(z) &= 5.25x_1 + 7x_2\\ 6.25x_1 + 4.5x_2 &\geq 43\\ 6x_1 + 6.5x_2 &\leq 53\\ x_1, x_2 &\geq 0.\end{aligned} \tag{9}$$

Step 4. *Computation of optimal crisp solution*

By applying the Big-M method and using software TORA on problem (9), we obtain the crisp feasible and crisp optimal solutions $x_1 = 3.01$, $x_2 = 5.38$, $z = 53.43$.

Step 5. *Evaluation of fuzzy optimal objective value*

Putting the values of x_1 and x_2 in problem (8), we obtain a fuzzy optimal solution $\tilde{z} = (32.92, 55.72, 69.49)$.

Example 5.2. Consider the following FLPP:

$$\begin{aligned} &\text{Minimize}(\tilde{z}) \approx (2,4,5,7)x_1 + (3,6,8,9)x_2 \\ &\text{s.t.} \\ &(1,5,7,9)x_1 + (5,7,10,12)x_2 \succeq (45,47,49,53) \\ &(4,6,9,11)x_1 + (3,5,7,8)x_2 \succeq (46,48,51,55) \\ &x_1, x_2 \geq 0. \end{aligned} \tag{10}$$

Step 1. *Horizontal representation of fuzzy parameters*

$$\begin{aligned} &\tilde{A} = (2,4,5,7), \quad A(r,\beta_{\tilde{A}}) = (2+2r) + \beta_{\tilde{A}}(5-4r); \\ &\tilde{B} = (3,6,8,9), \quad B(r,\beta_{\tilde{B}}) = (3+3r) + \beta_{\tilde{B}}(6-4r); \\ &\tilde{C} = (1,5,7,9), \quad C(r,\beta_{\tilde{C}}) = (1+4r) + \beta_{\tilde{C}}(8-6r); \\ &\tilde{D} = (5,7,10,12), \quad D(r,\beta_{\tilde{D}}) = (5+2r) + \beta_{\tilde{D}}(7-4r); \\ &\tilde{E} = (4,6,9,11), \quad E(r,\beta_{\tilde{E}}) = (4+2r) + \beta_{\tilde{E}}(7-4r); \\ &\tilde{F} = (3,5,7,8), \quad F(r,\beta_{\tilde{F}}) = (3+2r) + \beta_{\tilde{F}}(5-3r); \\ &\tilde{G} = (45,47,49,53), \quad G(r,\beta_{\tilde{G}}) = (45+2r) + \beta_{\tilde{G}}(8-6r); \\ &\tilde{H} = (46,48,51,55), \quad H(r,\beta_{\tilde{H}}) = (46+2r) + \beta_{\tilde{H}}(9-6r). \end{aligned}$$

Step 2. *Apply the ranking function*

$$\begin{aligned} &R(\tilde{A}) = 4.5, \quad R(\tilde{B}) = 6.5, \quad R(\tilde{C}) = 5.5; \\ &R(\tilde{D}) = 8.5, \quad R(\tilde{E}) = 7.5, \quad R(\tilde{F}) = 5.75; \\ &R(\tilde{G}) = 48.5, \quad R(\tilde{H}) = 50. \end{aligned}$$

Step 3. *Formulation of the crisp LLP*

Using Step 1 and Step 2, we obtain the crisp LPP from fuzzy LPP(10)

$$\begin{aligned} &\text{Minimize}(z) = 4.5x_1 + 6.5x_2 \\ &5.5x_1 + 8.5x_2 \geq 48.5 \\ &7.5x_1 + 5.75x_2 \geq 50 \\ &x_1, x_2 \geq 0. \end{aligned} \tag{11}$$

Step 4. *Computation of optimal crisp solution*

By applying the Big-M approach on problem (11), we obtain the crisp feasible and crisp optimal solutions $x_1 = 4.55$, $x_2 = 2.76$, $z = 38.43$.

Step 5. *Evaluation of fuzzy optimal objective value*

Putting the values of x_1 and x_2 in problem (10), we obtain a fuzzy optimal solution $\tilde{z} = (17.38, 34.76, 44.83, 56.69)$.

5.1 Comparison between "fuzzy Big-M approach" and "fuzzy Big-M approach with HMF"

The suggested Big-M method utilizes Horizontal Membership Functions (HMF), which provides numerous significant advantages over traditional fuzzy approaches. While the traditional approach concentrates only on the height of fuzzy numbers, which often loses essential data, the HMF-based approach considers the complete geometric body of the fuzzy number, ensuring that no information is lost. Traditional approaches can also become complex and complicated during calculations, especially when multiplying fuzzy numbers, but the HMF methodology reduces fuzzy parameters to simple math that acts like regular numbers. Furthermore, this new technique is more stable, eliminating common math paradoxes and errors that might occur with standard fuzzy logic. Finally, although traditional methods provide only approximate results, the incorporation of HMF into the Big-M process produces a highly precise result. Finally, although traditional methods produce only approximate results, including HMF into the Big-M process produces a highly exact and reliable ideal solution.

6 Conclusions

In this work, a fuzzy Big-M approach is proposed to solve FLPP in which decision variables are represented by real numbers and all the parameters are expressed as triangular and trapezoidal fuzzy numbers. The method utilizes horizontal membership functions and ranking functions to transform the FLPP into a classical LPP. This methodological shift enables the application of well-established linear programming algorithms to find optimal solutions in scenarios where traditional methods might struggle due to the presence of fuzziness or ambiguity in the data.

This study simplifies fuzzy optimization by applying the Horizontal Membership Function (HMF) to convert fuzzy numbers into a flat representation. This speeds up and simplifies the calculation of the Big-M approach for both people and machines. The approach is extremely adaptable, as it employs a single simple formula to handle many uncertainty shapes, such as triangular and trapezoidal fuzzy numbers, without giving up critical data.

The proposed HMF-based Big-M technique is both stable and accurate, but it has significant limitations. First, when fuzzy constraints increase, repeated ranking calculations can become lengthy. Second, this strategy is currently limited to linear problems and would require modifications for nonlinear models. Finally, the final solution is dependent on the precise ranking function $R(\tilde{V})$, as different functions may provide somewhat different crisp outcomes. We intend to apply our approach to nonlinear and multiobjective situations in the future.

Bibliography

[1] Edalatpanah, S. A. (2023). Multidimensional solution of fuzzy linear programming. *PeerJ Computer Science*. https://doi.org/10.7717/peerj-cs.1646.

[2] Ezzati, R., Khorram, E., & Enayati, R. (2015). A new algorithm to solve fully fuzzy linear programming problems using the MOLP problem. *Applied Mathematical Modelling*, *39*(12), 3183–3193.

[3] Ghanbari, R., Ghorbani-Moghadam, K., Mahdavi-Amiri, N., & De Baets, B. (2020). Fuzzy linear programming problems: models and solutions. *Soft Computing*, *24*(13), 10043–10073.

[4] Hatami, A., & Kazemipoor, H. (2013). Fuzzy big-M method for solving fuzzy linear programs with trapezoidal fuzzy numbers. *International Journal of Research in Industrial Engineering*, *2*(3), 1–9.

[5] Landowski, M. (2019). Method with horizontal fuzzy numbers for solving fuzzy linear systems. *Soft Computing*, *23*(12), 3921–3933.

[6] Nasseri, S. H., Ardil, E., Yazdani, A., & Zaefarian, R. (2005). Simplex method for solving linear programming problems with fuzzy numbers. *World Academy of Science, Engineering and Technology*, *10*, 284–288.

[7] Piegat, A., & Landowski, M. (2012). Is the conventional interval arithmetic correct? *Journal of Theoretical and Applied Computer Science*, *6*(2), 27–44.

[8] Piegat, A., & Landowski, M. (2015). Horizontal membership function and examples of its applications. *International Journal of Fuzzy Systems*, *17*(1), 22–30.

[9] Tomaszewska, K. (2014). The application of horizontal membership functions to fuzzy arithmetic operations. *Journal of Theoretical and Applied Computer Science*, *8*(2), 3–10.

[10] Zadeh, L. A. (1965). Fuzzy sets. *Information and Control*, *8*(3), 338–353.

Anjaneyulu Naik Kalavath, Satyanarayana Bavanari, and
Bhuvaneswari Dhanala

Neutrosophic falling shadows applied to commutative ideals in BCK-algebras

Abstract: The idea of a commutative falling neutrosophic fuzzy ideal of a BCK-algebra is introduced and its associated properties are systematically studied. Also, we verify that every commutative neutrosophic fuzzy ideal is a commutative falling neutrosophic fuzzy ideal, and provide examples to show that a commutative falling neutrosophic fuzzy ideal is not a commutative neutrosophic fuzzy ideal. We explore the link between a falling neutrosophic fuzzy ideal and a commutative falling neutrosophic fuzzy ideal, and a condition for a falling neutrosophic fuzzy ideal to be a commutative falling neutrosophic fuzzy ideal is provided. Introducing a commutative falling neutrosophic fuzzy ideal extends fuzzy logic applications to BCK-algebras, bridging gaps in existing literature. Links between falling neutrosophic fuzzy ideals and a commutative falling neutrosophic fuzzy ideal are examined, with conditions for a falling neutrosophic fuzzy ideal to be commutative.

Keywords: Commutative (falling) neutrosophic fuzzy ideal, (falling) neutrosophic fuzzy ideal, BCK-algebras

MSC 2020: 03G25, 06F35, 08A72, 11Y50

1 Introduction

Research on combining probability and fuzzy set theory to model uncertainty has led to significant advancements. Goodman [5] demonstrated the equivalence between fuzzy sets and classes of random sets. Building on this, Wang and Sanchez [20] introduced the theory of falling shadows, bridging probability concepts with fuzzy set membership functions. This theory's mathematical structure was formalized in [21]. Tan et al. [18, 19] developed a theoretical framework for fuzzy inference relations and operations based on falling shadows. Meanwhile, Meng [13] explored commutative ideals in BCK-algebras, and Jun and Roh [7] discussed their fuzzification. Jun et al. [9] generalized FCIs

Acknowledgement: K. Anjaneyulu Naik is thankful for the support from the National Fellowship and Scholarship for Higher Education of ST Students (NFST), Ministry of Tribal Affairs, Govt. of India (Agreement No. 11019/07/2018-Sch, Award No. 202223-NFST-AND-00689).

Anjaneyulu Naik Kalavath, Satyanarayana Bavanari, Bhuvaneswari Dhanala, Department of Mathematics, University College of Sciences, Acharya Nagarjuna University, Nagarjuna Nagar 522 510, Guntur, Andhra Pradesh, India, e-mail: nelu.fuzzy@gmail.com

https://doi.org/10.1515/9783112234525-007

to intuitionistic fuzzy versions. Later, Jun and Kang [10] developed the notion of fFCIs in BCK-As, examining key characteristics. In 2023, Brzooei et al. [3, 4] introduced an interesting study of falling shadows applied to commutative ideals by using intuitionistic sets within the BCK-algebras and also their theory and applications in Hoops.

BCK-algebras have found diverse applications in various fields, showcasing their versatility. They play a role in Logic and Foundations by modeling logical systems and algebraic logic, providing a framework for understanding logical structures. In Fuzzy Logic, BCK-algebras facilitate fuzzy set theory and reasoning, handling uncertain information. Cryptography takes advantage of these algebras for secure communication and encryption. In Computer Science, they contribute to formal languages and automata theory, while in Artificial Intelligence, they support knowledge representation and logical inference. BCK-algebras also have potential uses in Quantum Computing, particularly in quantum logic and information processing. Their broad applicability highlights their significance in modern mathematics and technology.

Neutrosophic sets, a generalization of fuzzy sets, offer a more flexible approach to modeling indeterminacy. Although commutative ideals in BCK-algebras have been studied, applying neutrosophic falling shadows to these structures remains unexplored. This study introduces commutative falling neutrosophic fuzzy ideals (CfNFI) in BCK-algebras, bridging gaps in existing literature and extending fuzzy logic applications. By establishing relationships and properties, we pave the way for novel approaches to uncertainty handling in algebraic structures.

Recently, Anjaneyulu Naik and Satyanarayana et al. [2, 15–17] introduced novel studies within the BCK and hyper-BCK-algebras. Jun et al. [12] introduced intuitionistic random sets and falling intuitionistic shadows. They applied these concepts to define fISAs and ideals in BCI/BCK-As, examining properties and relationships. This work also characterized fIIs.

This study presents the CfNFI of BCK-A, and a study of the associated properties. It is proven that every CNFI is a CfNFI, and we provide an example to show that a CfNFI is not a CNFI. We provide relations between an fNFI and a CfNFI. We consider a condition for an fNFI to be a CfNFI.

In this chapter, we used the following abbreviations:

- BCK-A: BCK-algebra
- CfNFI: commutative falling neutrosophic fuzzy ideal
- CNFI: commutative neutrosophic fuzzy ideal
- fNFI: falling neutrosophic fuzzy ideal
- CI: commutative ideal
- IFS: intuitionistic fuzzy set
- IFSA: intuitionistic fuzzy subalgebra
- IFI: intuitionistic fuzzy ideal
- CIFI: commutative intuitionistic fuzzy ideal
- IRS: intuitionistic random set
- IfSh: intuitionistic falling shadow

- fISA: falling intuitionistic subalgebra
- fII: falling intuitionistic ideal
- NFRS: neutrosophic fuzzy random set
- NFfSh: neutrosophic fuzzy falling shadow
- C-BCK-A: commutative BCK-A

2 Preliminaries

This section outlines key concepts used in this chapter. A set $\mathfrak{A}$, together with a binary operation $\between$ and a distinguished element 0, forms a BCI-A, if

(I) $((f \between g) \between (f \between h)) \between (h \between g) = 0$;

(II) $(f \between (f \between g)) \between g = 0$;

(III) $f \between f = 0$;

(IV) If $f \between g = 0$, and $g \between f = 0 \Rightarrow f = g$ for all $f, g, h \in \mathfrak{A}$. A BCI-A $\mathfrak{A}$; specifically, we have the identity:

(V) $0 \between f = 0$ for all $f \in \mathfrak{A}$ is called a "BCK-algebra".

In any BCI/BCK-A $\mathfrak{A}$, we have

(P_1) $f \between 0 = f$;

(P_2) $f \leq g \Rightarrow f \between h \leq g \between h, h \between g \leq h \between f$;

(P_3) $(f \between g) \between h = (f \between h) \between g$;

(P_4) $(f \between h) \between (g \between h) \leq f \between g, \forall f, g, h \in \mathfrak{A}$, where $f \leq g \Leftrightarrow f \between g = 0$.

Let $\mathfrak{S}$ be a $\neq \phi$ subset of a BCI/BCK-A $\mathfrak{A}$. Then, "$\mathfrak{S}$" is said to be a "subalgebra" of $\mathfrak{A}$, if $f \between g \in \mathfrak{S}$ for all $f, g \in \mathfrak{S}$.

A subset $\mathfrak{I}$ of a BCI/BCK-A $\mathfrak{A}$ is said to be an "ideal" of $\mathfrak{A}$ if it fulfils: (I_1) $0 \in \mathfrak{I}$, (I_2) $g \in \mathfrak{I}$, $f \between g \in \mathfrak{I} \Rightarrow f \in \mathfrak{I}$ for all $f, g \in \mathfrak{A}$.

Let $\mathfrak{I}$ be a subset of BCK-A $\mathfrak{A}$. Then, $\mathfrak{I}$ is said to be a "commutative ideal (CI)" of $\mathfrak{A}$, if (I_1) is true and (I_3) $h \in \mathfrak{I}$, $(f \between g) \between h \in \mathfrak{I} \Rightarrow f \between (g \between (g \between f)) \in \mathfrak{I}$ for all $f, g, h \in \mathfrak{A}$.

Importantly, while all "commutative ideals (CIs)" are ideals, the converse does not hold: ideals are not necessarily CIs [14].

For those seeking further insight into BCI/BCK-A, we recommend consulting [6, 14].

Let $\mathfrak{A}$ be a $(\neq \phi)$. An "intuitionistic fuzzy set (IFS)" in $\mathfrak{A}$ ([1]) is $\mathfrak{M} := \{(f; \mathfrak{M}_\xi(f), \mathfrak{M}_\varpi(f)) \mid f \in \mathfrak{A}, 0 \leq \mathfrak{M}_\xi(f) + \mathfrak{M}_\varpi(f) \leq 1\}$, where $\mathfrak{M}_\xi : \mathfrak{A} \to [0, 1]$ is a Degree of membership and $\mathfrak{M}_\varpi : \mathfrak{A} \to [0, 1]$ is a Degree of nonmembership. To simplify notation, we define $\mathfrak{M} = (\mathfrak{M}_\xi, \mathfrak{M}_\varpi)$ to represent the "intuitionistic fuzzy set (IFS)" $\mathfrak{M} := \{(f; \mathfrak{M}_\xi(f), \mathfrak{M}_\varpi(f)) \mid f \in \mathfrak{A}, 0 \leq \mathfrak{M}_\xi(f) + \mathfrak{M}_\varpi(f) \leq 1\}$.

Define an IFS $\mathfrak{M} = (\mathfrak{M}_\xi, \mathfrak{M}_\varpi)$ in $\mathfrak{A}$, and $\xi, \varpi \in [0,1]$. The following sets are hereby introduced: $\mathcal{U}_\epsilon(\mathfrak{M}, \xi) := \{f \in \mathfrak{A} \mid \mathfrak{M}_\xi(f) \geq \xi\}$, $\mathcal{L}_\epsilon(\mathfrak{M}, \varpi) := \{f \in \mathfrak{A} \mid \mathfrak{M}_\varpi(f) \leq \varpi\}$. We say $\mathcal{U}_\epsilon(\mathfrak{M}, \xi)$, and $\mathcal{L}_\epsilon(\mathfrak{M}, \varpi)$ are "intuitionistic fuzzy ϵ-subsets."

An IFS $\mathfrak{M} = (\mathfrak{M}_\xi, \mathfrak{M}_\varpi)$ in a BCI/BCK-A $\mathfrak{A}$ is said to be an IFSA of $\mathfrak{A}$ [8] if it fulfils:
(IFSA_1) $\mathfrak{M}_\xi(f \between g) \geq \min\{\mathfrak{M}_\xi(f), \mathfrak{M}_\xi(g)\}$;
(IFSA_2) $\mathfrak{M}_\varpi(f \between g) \leq \max\{\mathfrak{M}_\varpi(f), \mathfrak{M}_\varpi(g)\}$, for all $f, g \in \mathfrak{A}$.

An IFS $\mathfrak{M} = (\mathfrak{M}_\xi, \mathfrak{M}_\varpi)$ in a BCI/BCK-A $\mathfrak{A}$ is said to be an IFI of $\mathfrak{A}$ [8] if it fulfils:
(IFI_1) $\mathfrak{M}_\xi(0) \geq \mathfrak{M}_\xi(f)$, $\mathfrak{M}_\varpi(0) \leq \mathfrak{M}_\varpi(f)$;
(IFI_2) $\mathfrak{M}_\xi(f) \geq \min\{\mathfrak{M}_\xi(f \between g), \mathfrak{M}_\xi(g)\}$;
(IFI_3) $\mathfrak{M}_\varpi(f) \leq \max\{\mathfrak{M}_\varpi(f \between g), \mathfrak{M}_\varpi(g)\}$, for all $f, g \in \mathfrak{A}$.

An IFS $\mathfrak{M} = (\mathfrak{M}_\xi, \mathfrak{M}_\varpi)$ in a BCK-A $\mathfrak{A}$ is said to be a "commutative intuitionistic fuzzy ideal (CIFI)" of $\mathfrak{A}$ [9] if it satisfies (IFI_1), and
(CIFI_1) $\mathfrak{M}_\xi(f \between (g \between (g \between f))) \geq \min\{\mathfrak{M}_\xi((f \between g) \between h), \mathfrak{M}_\xi(h)\}$;
(CIFI_2) $\mathfrak{M}_\varpi(f \between (g \between (g \between f))) \leq \max\{\mathfrak{M}_\varpi((f \between g) \between h), \mathfrak{M}_\varpi(h)\}$, for all $f, g, h \in \mathfrak{A}$.

Consider $\mathfrak{A}$ is a BCI/BCK-A. For every $f \in \mathfrak{A}$, and $\mathfrak{B} \in 2^{\mathfrak{A}}$, let $\overline{f} := \{\mathfrak{C} \in 2^{\mathfrak{A}} \mid f \in \mathfrak{C}\}$, and $\overline{\mathfrak{B}} := \{\overline{f} \mid f \in \mathfrak{B}\}$.

Assume the ordered pair $(2^{\mathfrak{A}}, \mathcal{P})$. If $\mathcal{P}$ is a σ-field in $2^{\mathfrak{A}}$, and $\overline{\mathfrak{A}} \subseteq \mathcal{P}$, We refer to $(2^{\mathfrak{A}}, \mathcal{P})$ as a "hypermeasurable structure" on $\mathfrak{A}$.

Define an "intuitionistic random set (IRS)" on $\mathfrak{A}$ as a pair $\mathfrak{D} := (\mathfrak{D}_\xi, \mathfrak{D}_\varpi)$, where $\mathfrak{D}_\xi$, and $\mathfrak{D}_\varpi$ are $\mathfrak{E} - \mathcal{P}$ measurable functions from ψ to $2^{\mathfrak{A}}$, within the probability-space $(\psi, \mathfrak{E}, \mathfrak{Q})$ and "hypermeasurable structure" $(2^{\mathfrak{A}}, \mathcal{P})$.

That is,
(PS_1) $\mathfrak{D}_\xi^{(-1)}(\mathfrak{C}) = \{\mathcal{R}_\xi \in \psi \mid \mathfrak{D}_\xi(\mathcal{R}_\xi) \in \mathfrak{C}\} \in \mathfrak{E}$,
(PS_2) $\mathfrak{D}_\varpi^{(-1)}(\mathfrak{C}) = \{\mathcal{R}_\varpi \in \psi \mid \mathfrak{D}_\varpi(\mathcal{R}_\varpi) \in \mathfrak{C}\} \in \mathfrak{E}$, for all $\mathfrak{C} \in \mathcal{P}$.

Given an "intuitionistic random set (IRS)" $\mathfrak{D} := (\mathfrak{D}_\xi, \mathfrak{D}_\varpi)$ on $\mathfrak{A}$, consider the functions: $\tilde{\mathcal{G}}_\xi : \mathfrak{A} \to [0,1], f_\xi \mapsto \mathfrak{Q}(\mathcal{R}_\xi \mid f_\xi \in \mathfrak{D}_\xi(\mathcal{R}_\xi))$, $\tilde{\mathcal{G}}_\varpi : \mathfrak{A} \to [0,1], f_\varpi \mapsto \mathfrak{Q}(\mathcal{R}_\varpi \mid f_\varpi \in \mathfrak{D}_\varpi(\mathcal{R}_\varpi))$. Then, $\tilde{\mathcal{G}} := (\tilde{\mathcal{G}}_\xi, \tilde{\mathcal{G}}_\varpi)$ is an IFS on $\mathfrak{A}$, and we call it the "intuitionistic falling shadow (IfSh)" of the IRS $\mathfrak{D} := (\mathfrak{D}_\xi, \mathfrak{D}_\varpi)$, and $\mathfrak{D} := (\mathfrak{D}_\xi, \mathfrak{D}_\varpi)$ is said to be an intuitionistic cloud of $\tilde{\mathcal{G}} := (\tilde{\mathcal{G}}_\xi, \tilde{\mathcal{G}}_\varpi)$.

Consider, for instance, a probability-space $(\psi, \mathfrak{E}, \mathfrak{Q}) = ([0,1], \mathfrak{E}, n)$ with a Borel field $\mathfrak{E}$ on $[0,1]$ and the usual Lebesgue measure n. Let $\tilde{\mathcal{G}} := (\tilde{\mathcal{G}}_\xi, \tilde{\mathcal{G}}_\varpi)$ be an "intuitionistic fuzzy set (IFS)" in $\mathfrak{A}$. Then, a couple $\mathfrak{D} := (\mathfrak{D}_\xi, \mathfrak{D}_\varpi)$ in which $\mathfrak{D}_\xi : [0,1] \to 2^{\mathfrak{A}}, \xi \mapsto \mathcal{U}_\epsilon(\tilde{\mathcal{G}}; \xi)$, $\mathfrak{D}_\varpi : [0,1] \to 2^{\mathfrak{A}}, \varpi \mapsto \mathcal{L}_\epsilon(\tilde{\mathcal{G}}; \varpi)$ is an IRS and $\mathfrak{D} := (\mathfrak{D}_\xi, \mathfrak{D}_\varpi)$ is an intuitionistic cloud of $\tilde{\mathcal{G}} := (\tilde{\mathcal{G}}_\xi, \tilde{\mathcal{G}}_\varpi)$. We call $\mathfrak{D} := (\mathfrak{D}_\xi, \mathfrak{D}_\varpi)$ previously introduced as the intuitionistic cut cloud of $\tilde{\mathcal{G}} := (\tilde{\mathcal{G}}_\xi, \tilde{\mathcal{G}}_\varpi)$.

For a probability-space $(\psi, \mathfrak{E}, \mathfrak{Q})$, let $\mathfrak{O} := (\mathfrak{O}_\xi, \mathfrak{O}_\varpi)$ be an IRS on a BCI/BCK-A $\mathfrak{A}$. Then, the IfSh $\tilde{\tilde{\mathfrak{F}}} := (\tilde{\tilde{\mathfrak{F}}}_\xi, \tilde{\tilde{\mathfrak{F}}}_\varpi)$ of $\mathfrak{O} := (\mathfrak{O}_\xi, \mathfrak{O}_\varpi)$ is called a fISA (resp., fII) of $\mathfrak{A}$ if $\mathfrak{O}_\xi(\mathcal{R}_\xi)$, and $\mathfrak{O}_\varpi(\mathcal{R}_\varpi)$ are subalgebras (resp., ideals) of $\mathfrak{A}$ for all $\mathcal{R}_\xi, \mathcal{R}_\varpi \in \psi$ (see [11, 12]).

3 Main results

In this section, we establish and examine the concept of CfNFI in BCK-A, along with its properties. Additionally, illustrative examples demonstrate the connection between fNFI and CfNFI. Throughout, $\mathfrak{A}$ is a BCK-algebra unless specified.

Definition 1. Let $(\psi, \mathfrak{E}, \mathfrak{Q})$ be a probability-space and a NFRS $\mathfrak{O} := (\mathfrak{O}_\xi, \mathfrak{O}_\zeta, \mathfrak{O}_\varpi)$ on $\mathfrak{A}$, the NFfSh $\tilde{\mathcal{G}} := (\tilde{\mathcal{G}}_\xi, \tilde{\mathcal{G}}_\zeta, \tilde{\mathcal{G}}_\varpi)$ of $\mathfrak{O} := (\mathfrak{O}_\xi, \mathfrak{O}_\zeta, \mathfrak{O}_\varpi)$ is said to be a CfNFI of $\mathfrak{A}$ if $\mathfrak{O}_\xi(\mathcal{R}_\xi)$, $\mathfrak{O}_\zeta(\mathcal{R}_\zeta)$, and $\mathfrak{O}_\varpi(\mathcal{R}_\varpi)$ are "commutative ideals (CIs)" of $\mathfrak{A}$ for all $\mathcal{R}_\xi, \mathcal{R}_\zeta, \mathcal{R}_\varpi \in \psi$.

Example 1. Define the set $\mathfrak{A} = \{0, \mathfrak{k}_1, \mathfrak{k}_2, \mathfrak{k}_3, \mathfrak{k}_4\}$ and equip it with $\between$ a binary operation, whose operation table is shown in Table 1.

Table 1: BCK-algebra.

$\between$	0	$\mathfrak{k}_1$	$\mathfrak{k}_2$	$\mathfrak{k}_3$	$\mathfrak{k}_4$
0	0	0	0	0	0
$\mathfrak{k}_1$	$\mathfrak{k}_1$	0	0	$\mathfrak{k}_1$	$\mathfrak{k}_1$
$\mathfrak{k}_2$	$\mathfrak{k}_2$	$\mathfrak{k}_1$	0	$\mathfrak{k}_2$	$\mathfrak{k}_2$
$\mathfrak{k}_3$	$\mathfrak{k}_3$	$\mathfrak{k}_3$	$\mathfrak{k}_3$	0	$\mathfrak{k}_3$
$\mathfrak{k}_4$	$\mathfrak{k}_4$	$\mathfrak{k}_4$	$\mathfrak{k}_4$	$\mathfrak{k}_4$	0

Then, $(\mathfrak{A}; \between, 0)$ is a BCK-A [14]. Consider $(\psi, \mathfrak{E}, \mathfrak{Q}) = ([0,1], \mathfrak{E}, n)$, and let $\mathfrak{O} := (\mathfrak{O}_\xi, \mathfrak{O}_\zeta, \mathfrak{O}_\varpi)$ be a NFRS on $\mathfrak{A}$ that is given as follows:

$$\mathfrak{O}_\xi : [0,1] \to \mathcal{W}(\mathfrak{A}), \quad f \mapsto \begin{cases} \{0, \mathfrak{k}_3\} & \text{if } f \in [0, 0.35), \\ \{0, \mathfrak{k}_4\} & \text{if } f \in [0.35, 0.65), \\ \{0, \mathfrak{k}_1, \mathfrak{k}_2\} & \text{if } f \in [0.65, 0.95), \\ \{0, \mathfrak{k}_3, \mathfrak{k}_4\} & \text{if } f \in [0.95, 1), \end{cases}$$

$$\mathfrak{O}_\zeta : [0,1] \to \mathcal{W}(\mathfrak{A}), \quad f \mapsto \begin{cases} \{0, \mathfrak{k}_3\} & \text{if } f \in [0, 0.15), \\ \{0, \mathfrak{k}_4\} & \text{if } f \in [0.15, 0.45), \\ \{0, \mathfrak{k}_1, \mathfrak{k}_2\} & \text{if } f \in [0.45, 0.75), \\ \{0, \mathfrak{k}_3, \mathfrak{k}_4\} & \text{if } f \in [0.75, 1), \end{cases} \quad \text{and}$$

$$\mathfrak{D}_{\varpi} : [0,1] \to \mathcal{W}(\mathfrak{A}), \quad f \mapsto \begin{cases} \{0\} & \text{if } f \in (0.9, 1], \\ \{0, \mathfrak{k}_3\} & \text{if } f \in (0.8, 0.9], \\ \{0, \mathfrak{k}_4\} & \text{if } f \in (0.6, 0.8], \\ \{0, \mathfrak{k}_1, \mathfrak{k}_2, \mathfrak{k}_3\} & \text{if } f \in (0.4, 0.6], \\ \mathfrak{A} & \text{if } f \in [0, 0.4]. \end{cases}$$

Then, $\mathfrak{D}_{\xi}(\mathfrak{s})$, $\mathfrak{D}_{\zeta}(\mathfrak{s})$, and $\mathfrak{D}_{\varpi}(\mathfrak{s})$ are "commutative ideals (CIs)" of $\mathfrak{A}$ for all $\mathfrak{s} \in [0,1]$. Hence, the NFfSh $\tilde{\mathcal{G}} := (\tilde{\mathcal{G}}_{\xi}, \tilde{\mathcal{G}}_{\zeta}, \tilde{\mathcal{G}}_{\varpi})$ of $\mathfrak{D} := (\mathfrak{D}_{\xi}, \mathfrak{D}_{\zeta}, \mathfrak{D}_{\varpi})$ is a CfNFI of $\mathfrak{A}$, and it is given as follows:

$$\tilde{\mathcal{G}}_{\xi}(f) = \begin{cases} 1 & \text{if } f = 0, \\ 0.4 & \text{if } f = \{\mathfrak{k}_1, \mathfrak{k}_2\}, \\ 0.5 & \text{if } f = \mathfrak{k}_3, \\ 0.55 & \text{if } f = \mathfrak{k}_4, \end{cases} \qquad \tilde{\mathcal{G}}_{\zeta}(f) = \begin{cases} 1 & \text{if } f = 0, \\ 0.2 & \text{if } f = \{\mathfrak{k}_1, \mathfrak{k}_2\}, \\ 0.3 & \text{if } f = \mathfrak{k}_3, \\ 0.35 & \text{if } f = \mathfrak{k}_4 \end{cases}$$

and

$$\tilde{\mathcal{G}}_{\varpi}(f) = \begin{cases} 0 & \text{if } f = 0, \\ 0.6 & \text{if } f = \{\mathfrak{k}_1, \mathfrak{k}_2, \mathfrak{k}_4\}, \\ 0.4 & \text{if } f = \mathfrak{k}_3. \end{cases}$$

For a probability-space $(\psi, \mathfrak{E}, \mathfrak{Q})$, let $\tilde{\mathcal{G}} := (\tilde{\mathcal{G}}_{\xi}, \tilde{\mathcal{G}}_{\zeta}, \tilde{\mathcal{G}}_{\varpi})$ be an NFfSh of an NFRS $\mathfrak{D} := (\mathfrak{D}_{\xi}, \mathfrak{D}_{\zeta}, \mathfrak{D}_{\varpi})$. For $f \in \mathfrak{A}$, let $\psi(f; \mathfrak{D}_{\xi}) := \{\mathcal{R}_{\xi} \in \psi \mid f \in \mathfrak{D}_{\xi}(\mathcal{R}_{\xi})\}$, $\psi(f; \mathfrak{D}_{\zeta}) := \{\mathcal{R}_{\zeta} \in \psi \mid f \in \mathfrak{D}_{\zeta}(\mathcal{R}_{\zeta})\}$, and $\psi(f; \mathfrak{D}_{\varpi}) := \{\mathcal{R}_{\varpi} \in \psi \mid f \in \mathfrak{D}_{\varpi}(\mathcal{R}_{\varpi})\}$. Then, $\psi(f; \mathfrak{D}_{\xi})$, $\psi(f; \mathfrak{D}_{\zeta})$, $\psi(f; \mathfrak{D}_{\varpi}) \in \mathfrak{E}$ (see [12]).

Proposition 1. *Let $\tilde{\mathcal{G}} := (\tilde{\mathcal{G}}_{\xi}, \tilde{\mathcal{G}}_{\zeta}, \tilde{\mathcal{G}}_{\varpi})$ be an NFfSh of the NFRS $\mathfrak{D} := (\mathfrak{D}_{\xi}, \mathfrak{D}_{\zeta}, \mathfrak{D}_{\varpi})$ on $\mathfrak{A}$. If $\tilde{\mathcal{G}} := (\tilde{\mathcal{G}}_{\xi}, \tilde{\mathcal{G}}_{\zeta}, \tilde{\mathcal{G}}_{\varpi})$ is a CfNFI of $\mathfrak{A}$, then*

$$\begin{aligned} \text{(i).} \quad & \psi((f \between g) \between h; \mathfrak{D}_{\xi}) \cap \psi(h; \mathfrak{D}_{\xi}) \subseteq \psi(f \between (g \between (g \between f)); \mathfrak{D}_{\xi}), \\ & \psi((f \between g) \between h; \mathfrak{D}_{\zeta}) \cap \psi(h; \mathfrak{D}_{\zeta}) \subseteq \psi(f \between (g \between (g \between f)); \mathfrak{D}_{\zeta}), \\ & \psi((f \between g) \between h; \mathfrak{D}_{\varpi}) \cap \psi(h; \mathfrak{D}_{\varpi}) \subseteq \psi(f \between (g \between (g \between f)); \mathfrak{D}_{\varpi}), \\ \text{(ii).} \quad & \psi(f \between (g \between (g \between f)); \mathfrak{D}_{\xi}) \subseteq \xi((f \between g) \between h; \mathfrak{D}_{\xi}), \\ & \psi(f \between (g \between (g \between f)); \mathfrak{D}_{\zeta}) \subseteq \zeta((f \between g) \between h; \mathfrak{D}_{\zeta}), \\ & \psi(f \between (g \between (g \between f)); \mathfrak{D}_{\varpi}) \subseteq \varpi((f \between g) \between h; \mathfrak{D}_{\varpi}), \end{aligned}$$

for all $f, g, h \in \mathfrak{A}$.

Proof. Let $\mathcal{R}_{\xi} \in \psi((f \between g) \between h; \mathfrak{D}_{\xi}) \cap \psi(f; \mathfrak{D}_{\xi})$, $\mathcal{R}_{\zeta} \in \psi((f \between g) \between h; \mathfrak{D}_{\zeta}) \cap \psi(f; \mathfrak{D}_{\zeta})$, and $\mathcal{R}_{\varpi} \in \psi((f \between g) \between h; \mathfrak{D}_{\varpi}) \cap \psi(f; \mathfrak{D}_{\varpi})$ for all $f, g, h \in \mathfrak{A}$. Then

$$(f \between g) \between h \in \mathfrak{D}_{\xi}(\mathcal{R}_{\xi}) \text{ and } f \in \mathfrak{D}_{\xi}(\mathcal{R}_{\xi}),$$

$$(f \between g) \between h \in \mathfrak{D}_\zeta(\mathcal{R}_\zeta) \text{ and } f \in \mathfrak{D}_\zeta(\mathcal{R}_\zeta),$$
$$(f \between g) \between h \in \mathfrak{D}_\varpi(\mathcal{R}_\varpi) \text{ and } f \in \mathfrak{D}_\varpi(\mathcal{R}_\varpi).$$

Since $\mathfrak{D}_\xi(\mathcal{R}_\xi)$, $\mathfrak{D}_\zeta(\mathcal{R}_\zeta)$, and $\mathfrak{D}_\varpi(\mathcal{R}_\varpi)$ are "commutative ideals (CI)" of $\mathfrak{A}$ the condition (I_3) implies that $f \between (g \between (g \between f)) \in \mathfrak{D}_\xi(\mathcal{R}_\xi) \cap \mathfrak{D}_\zeta(\mathcal{R}_\zeta) \cap \mathfrak{D}_\varpi(\mathcal{R}_\varpi)$ and so that $\mathcal{R}_\xi \in \psi(f \between (g \between (g \between f)); \mathfrak{D}_\xi)$, $\mathcal{R}_\zeta \in \psi(f \between (g \between (g \between f)); \mathfrak{D}_\zeta)$, and $\mathcal{R}_\varpi \in \psi(f \between (g \between (g \between f)); \mathfrak{D}_\varpi)$. Hence, (i) is true. Now, let $\mathcal{R}_\xi \in \psi(f \between (g \between (g \between f)); \mathfrak{D}_\xi)$, $\mathcal{R}_\zeta \in \psi(f \between (g \between (g \between f)); \mathfrak{D}_\zeta)$, and $\mathcal{R}_\varpi \in \psi(f \between (g \between (g \between f)); \mathfrak{D}_\varpi)$ for all $f, g, h \in \mathfrak{A}$. Then, $f \between (g \between (g \between f)) \in \mathfrak{D}_\xi(\mathcal{R}_\xi) \cap \mathfrak{D}_\zeta(\mathcal{R}_\zeta) \cap \mathfrak{D}_\varpi(\mathcal{R}_\varpi)$. Note that

$$\begin{aligned}
&((f \between g) \between h) \between (f \between (g \between (g \between f)))\\
&\quad = ((f \between g) \between (f \between (g \between (g \between f)))) \between h\\
&\quad \le ((g \between (g \between f)) \between g) \between h\\
&\quad = ((g \between g) \between (g \between f)) \between h\\
&\quad = (0 \between (g \between f)) \between h = 0 \between h = 0,
\end{aligned}$$

which yields $((f \between g) \between h) \between (f \between (g \between (g \between f))) = 0 \in \mathfrak{D}_\xi(\mathcal{R}_\xi) \cap \mathfrak{D}_\zeta(\mathcal{R}_\zeta) \cap \mathfrak{D}_\varpi(\mathcal{R}_\varpi)$. Since $\mathfrak{D}_\xi(\mathcal{R}_\xi)$, $\mathfrak{D}_\zeta(\mathcal{R}_\zeta)$, and $\mathfrak{D}_\varpi(\mathcal{R}_\varpi)$ are "commutative ideals (CI)" and therefore ideals of $\mathfrak{A}$, it follows that $(f \between g) \between h \in \mathfrak{D}_\xi(\mathcal{R}_\xi) \cap \mathfrak{D}_\zeta(\mathcal{R}_\zeta) \cap \mathfrak{D}_\varpi(\mathcal{R}_\varpi)$. Hence, $\mathcal{R}_\xi \in \psi((f \between g) \between h; \mathfrak{D}_\xi)$, $\mathcal{R}_\zeta \in \psi((f \between g) \between h; \mathfrak{D}_\zeta)$, and $\mathcal{R}_\varpi \in \psi((f \between g) \between h; \mathfrak{D}_\varpi)$. Therefore, (ii) is valid. Let $(\psi, \mathfrak{E}, \mathfrak{D})$ be a probability-space, let $\mathfrak{F}(\mathfrak{A}) := \{f \mid f : \psi \to \mathfrak{A} \text{ is a mapping}\}$. Let $\odot$ be a binary operation on $\mathfrak{F}(\mathfrak{A})$, defined as: $((f \odot i)(\beta) = f(\beta) \between i(\beta))$ for all $\beta \in \psi$, and for all $f, i \in \mathfrak{F}(\mathfrak{A})$. Then, $(\mathfrak{F}(\mathfrak{A}); \odot, \pi)$ is a BCI/BCK-A (see [11]) where π is specified as follows: $\pi : \psi \mapsto \mathfrak{A}, \beta \mapsto 0$. Given a subset $\mathcal{C}$ of $\mathfrak{A}$, and $i_\xi, i_\zeta, i_\kappa, i_\varpi \in \mathfrak{F}(\mathfrak{A})$, consider the following:

$$\begin{aligned}
&\mathcal{C}_\xi^i := \{\mathcal{R}_\xi \in \psi \mid i_\xi(\mathcal{R}_\xi) \in \mathcal{C}\},\\
&\mathcal{C}_\zeta^i := \{\mathcal{R}_\zeta \in \psi \mid i_\zeta(\mathcal{R}_\zeta) \in \mathcal{C}\},\\
&\mathcal{C}_\xi^i := \{\mathcal{R}_\varpi \in \psi \mid i_\varpi(\mathcal{R}_\varpi) \in \mathcal{C}\} \quad \text{and}
\end{aligned}$$

$$\begin{aligned}
\mathfrak{D}_\xi &: \psi \to \mathcal{W}(\mathfrak{F}(\mathfrak{A})), \quad \mathcal{R}_\xi \mapsto \{i_\xi \in \mathfrak{F}(\mathfrak{A}) \mid i_\xi(\mathcal{R}_\xi) \in \mathcal{C}\},\\
\mathfrak{D}_\zeta &: \psi \to \mathcal{W}(\mathfrak{F}(\mathfrak{A})), \quad \mathcal{R}_\zeta \mapsto \{i_\zeta \in \mathfrak{F}(\mathfrak{A}) \mid i_\zeta(\mathcal{R}_\zeta) \in \mathcal{C}\},\\
\mathfrak{D}_\varpi &: \psi \to \mathcal{W}(\mathfrak{F}(\mathfrak{A})), \quad \mathcal{R}_\varpi \mapsto \{i_\varpi \in \mathfrak{F}(\mathfrak{A}) \mid i_\varpi(\mathcal{R}_\varpi) \in \mathcal{C}\}.
\end{aligned}$$

Then, $\mathcal{C}_\xi^i, \mathcal{C}_\zeta^i, \mathcal{C}_\varpi^i \in \mathfrak{E}$ (see [12]). □

Theorem 1. *If $\mathcal{N}$ is a "commutative ideal (CI)" of $\mathfrak{A}$, then*

$$\begin{aligned}
\mathfrak{D}_\xi(\mathcal{R}_\xi) &= \{i_\xi \in \mathfrak{F}(\mathfrak{A}) \mid i_\xi(\mathcal{R}_\xi) \in \mathcal{N}\},\\
\mathfrak{D}_\zeta(\mathcal{R}_\zeta) &= \{i_\zeta \in \mathfrak{F}(\mathfrak{A}) \mid i_\zeta(\mathcal{R}_\zeta) \in \mathcal{N}\} \quad \textit{and}\\
\mathfrak{D}_\varpi(\mathcal{R}_\varpi) &= \{i_\varpi \in \mathfrak{F}(\mathfrak{A}) \mid i_\varpi(\mathcal{R}_\varpi) \in \mathcal{N}\}
\end{aligned}$$

are "commutative ideals (CI)" of $\mathfrak{F}(\mathfrak{A})$.

Proof. Suppose that $\mathcal{N}$ is a "commutative ideal (CI)" of $\mathfrak{A}$. Since $\pi(\mathcal{R}_\xi) = 0 \in \mathcal{N}$, $\pi(\mathcal{R}_\zeta) = 0 \in \mathcal{N}$, and $\pi(\mathcal{R}_\varpi) = 0 \in \mathcal{N}$ for all $\mathcal{R}_\xi, \mathcal{R}_\zeta, \mathcal{R}_\varpi \in \psi$, we have $\pi \in \mathfrak{D}_\xi(\mathcal{R}_\xi)$, $\pi \in \mathfrak{D}_\zeta(\mathcal{R}_\zeta)$, and $\pi \in \mathfrak{D}_\varpi(\mathcal{R}_\varpi)$. Let $\mathfrak{f}_\xi, i_\xi, \mathfrak{M}_\xi \in \mathfrak{F}(\mathfrak{A})$ be such that $(\mathfrak{f}_\xi \odot i_\xi) \odot \mathfrak{M}_\xi \in \mathfrak{D}_\xi(\mathcal{R}_\xi)$, $\mathfrak{M}_\xi \in \mathfrak{D}_\xi(\mathcal{R}_\xi)$. Then, $(\mathfrak{f}_\xi(\mathcal{R}_\xi) \between i_\xi(\mathcal{R}_\xi)) \between \mathfrak{M}_\xi(\mathcal{R}_\xi) = ((\mathfrak{f}_\xi \odot i_\xi) \odot \mathfrak{M}_\xi)(\mathcal{R}_\xi) \in \mathcal{N}$, and $\mathfrak{M}_\xi(\mathcal{R}_\xi) \in \mathcal{N}$. Since $\mathcal{N}$ is a "commutative ideal (CI)" of $\mathfrak{A}$, it follows from (I_3) that $(\mathfrak{f}_\xi \odot (i_\xi \odot (i_\xi \odot \mathfrak{f}_\xi)))(\mathcal{R}_\xi) = \mathfrak{f}_\xi(\mathcal{R}_\xi) \between (i_\xi(\mathcal{R}_\xi) \between (i_\xi(\mathcal{R}_\xi) \between \mathfrak{f}_\xi(\mathcal{R}_\xi))) \in \mathcal{N}$, that is, $\mathfrak{f}_\xi \odot (i_\xi \odot (i_\xi \odot \mathfrak{f}_\xi)) \in \mathfrak{D}_\xi(\mathcal{R}_\xi)$. Hence, $\mathfrak{D}_\xi(\mathcal{R}_\xi)$ is a "commutative ideal (CI)" of $\mathfrak{F}(\mathfrak{A})$. Now, Let $\mathfrak{f}_\zeta, i_\zeta, \mathfrak{M}_\zeta \in \mathfrak{F}(\mathfrak{A})$ be such that $(\mathfrak{f}_\zeta \odot i_\zeta) \odot \mathfrak{M}_\zeta \in \mathfrak{D}_\zeta(\mathcal{R}_\zeta)$, $\mathfrak{M}_\zeta \in \mathfrak{D}_\zeta(\mathcal{R}_\zeta)$. Then, $(\mathfrak{f}_\zeta(\mathcal{R}_\zeta) \between i_\zeta(\mathcal{R}_\zeta)) \between \mathfrak{M}_\zeta(\mathcal{R}_\zeta) = ((\mathfrak{f}_\zeta \odot i_\zeta) \odot \mathfrak{M}_\zeta)(\mathcal{R}_\zeta) \in \mathcal{N}$, and $\mathfrak{M}_\zeta(\mathcal{R}_\zeta) \in \mathcal{N}$. Since $\mathcal{N}$ is a "commutative ideal (CI)" of $\mathfrak{A}$, it follows from (I_3) that $(\mathfrak{f}_\zeta \odot (i_\zeta \odot (i_\zeta \odot \mathfrak{f}_\zeta)))(\mathcal{R}_\zeta) = \mathfrak{f}_\zeta(\mathcal{R}_\zeta) \between (i_\zeta(\mathcal{R}_\zeta) \between (i_\zeta(\mathcal{R}_\zeta) \between \mathfrak{f}_\zeta(\mathcal{R}_\zeta))) \in \mathcal{N}$, that is, $\mathfrak{f}_\zeta \odot (i_\zeta \odot (i_\zeta \odot \mathfrak{f}_\zeta)) \in \mathfrak{D}_\zeta(\mathcal{R}_\zeta)$. Hence, $\mathfrak{D}_\zeta(\mathcal{R}_\zeta)$ is a "commutative ideal (CI)" of $\mathfrak{F}(\mathfrak{A})$. Now, let $\mathfrak{f}_\varpi, i_\varpi, \mathfrak{M}_\varpi \in \mathfrak{F}(\mathfrak{A})$ be such that $(\mathfrak{f}_\varpi \odot i_\varpi) \odot \mathfrak{M}_\varpi \in \mathfrak{D}_\varpi(\mathcal{R}_\varpi)$, and $\mathfrak{M}_\varpi \in \mathfrak{D}_\varpi(\mathcal{R}_\varpi)$. Then, $(\mathfrak{f}_\varpi(\mathcal{R}_\varpi) \between i_\varpi(\mathcal{R}_\varpi)) \between \mathfrak{M}_\varpi(\mathcal{R}_\varpi) = ((\mathfrak{f}_\varpi \odot i_\varpi) \odot \mathfrak{M}_\varpi)(\mathcal{R}_\varpi) \in \mathcal{N}$, and $\mathfrak{M}_\varpi(\mathcal{R}_\varpi) \in \mathcal{N}$. Since $\mathcal{N}$ is a "commutative ideal (CI)" of $\mathfrak{A}$, it follows from (I_3) that $(\mathfrak{f}_\varpi \odot (i_\varpi \odot (i_\varpi \odot \mathfrak{f}_\varpi)))(\mathcal{R}_\varpi) = \mathfrak{f}_\varpi(\mathcal{R}_\varpi) \between (i_\varpi(\mathcal{R}_\varpi) \between (i_\varpi(\mathcal{R}_\varpi) \between \mathfrak{f}_\varpi(\mathcal{R}_\varpi))) \in \mathcal{N}$, that is, $\mathfrak{f}_\varpi \odot (i_\varpi \odot (i_\varpi \odot \mathfrak{f}_\varpi)) \in \mathfrak{D}_\varpi(\mathcal{R}_\varpi)$. Hence, $\mathfrak{D}_\varpi(\mathcal{R}_\varpi)$ is a "commutative ideal (CI)" of $\mathfrak{F}(\mathfrak{A})$. The theorem proof is completed. □

Theorem 2. *Let* $(\psi, \mathfrak{E}, \mathfrak{Q}) = ([0,1], \mathfrak{E}, n)$ *be a probability-space. Then, every CNFI is a CfNFI.*

Proof. Let $\tilde{\mathcal{G}} := (\tilde{\mathcal{G}}_\xi, \tilde{\mathcal{G}}_\zeta, \tilde{\mathcal{G}}_\varpi)$ be a CNFI of $\mathfrak{A}$. Then, $\mathcal{U}_\epsilon(\tilde{\mathcal{G}}_\xi, \xi)$, $\mathcal{U}_\epsilon(\tilde{\mathcal{G}}_\zeta, \zeta)$, and $K_\epsilon(\tilde{\mathcal{G}}_\varpi, \varpi)$, $\mathcal{L}_\epsilon(\tilde{\mathcal{G}}_\varpi, \varpi)$ are "commutative ideals (CI)" of $\mathfrak{A}$, for all $\xi \in (0,1]$, $\zeta \in (0,1]$, and $\varpi \in [0,1)$. Hence, a pair $\mathfrak{D} := (\mathfrak{D}_\xi, \mathfrak{D}_\zeta, \mathfrak{D}_\varpi)$ in which

$$
\begin{aligned}
\mathfrak{D}_\xi &: [0,1] \to \mathcal{W}(\mathfrak{A}), \quad \xi \mapsto \mathcal{U}_\epsilon(\tilde{\mathcal{G}}; \xi),\\
\mathfrak{D}_\zeta &: [0,1] \to \mathcal{W}(\mathfrak{A}), \quad \zeta \mapsto \mathcal{U}_\epsilon(\tilde{\mathcal{G}}; \zeta),\\
\mathfrak{D}_\varpi &: [0,1] \to \mathcal{W}(\mathfrak{A}), \quad \varpi \mapsto \mathcal{L}_\epsilon(\tilde{\mathcal{G}}; \varpi),
\end{aligned}
$$

are an NF-cut-cloud of $\tilde{\mathcal{G}} := (\tilde{\mathcal{G}}_\xi, \tilde{\mathcal{G}}_\zeta, \tilde{\mathcal{G}}_\varpi)$, and so $\tilde{\mathcal{G}} := (\tilde{\mathcal{G}}_\xi, \tilde{\mathcal{G}}_\zeta, \tilde{\mathcal{G}}_\varpi)$ is a CfNFI of $\mathfrak{A}$. □

It is important to note that the converse of Theorem 2 does not hold, a fact demonstrated by the following example.

Example 2. Define the set $\mathfrak{A} = \{0, \mathfrak{k}_1, \mathfrak{k}_2, \mathfrak{k}_3, \mathfrak{k}_4\}$ and equip it with "$\between$", a binary operation, whose operation table is shown in Table 2.

Table 2: BCK-algebra.

$\eth$	**0**	$\mathfrak{k}_1$	$\mathfrak{k}_2$	$\mathfrak{k}_3$	$\mathfrak{k}_4$
0	0	0	0	0	0
$\mathfrak{k}_1$	$\mathfrak{k}_1$	0	$\mathfrak{k}_1$	0	$\mathfrak{k}_1$
$\mathfrak{k}_2$	$\mathfrak{k}_2$	$\mathfrak{k}_2$	0	0	$\mathfrak{k}_2$
$\mathfrak{k}_3$	$\mathfrak{k}_3$	$\mathfrak{k}_2$	$\mathfrak{k}_1$	0	$\mathfrak{k}_3$
$\mathfrak{k}_4$	$\mathfrak{k}_4$	$\mathfrak{k}_4$	$\mathfrak{k}_4$	$\mathfrak{k}_4$	0

Then, $(\mathfrak{A}; \eth, 0)$ is a BCK-A [14]. Consider $(\psi, \mathfrak{E}, \mathfrak{Q}) = ([0,1], \mathfrak{E}, n)$, and let $\mathfrak{O} := (\mathfrak{O}_\xi, \mathfrak{O}_\zeta, \mathfrak{O}_\varpi)$ be an NFRS on $\mathfrak{A}$ that is given as follows:

$$\mathfrak{O}_\xi : [0,1] \to \mathcal{W}(\mathfrak{A}), \quad f \mapsto \begin{cases} \{0, \mathfrak{k}_1\} & \text{if } f \in [0, 0.3), \\ \{0, \mathfrak{k}_2\} & \text{if } f \in [0.3, 0.65), \\ \{0, \mathfrak{k}_2, \mathfrak{k}_4\} & \text{if } f \in [0.65, 0.85), \\ \{0, \mathfrak{k}_1, \mathfrak{k}_2, \mathfrak{k}_3\} & \text{if } f \in [0.85, 1], \end{cases}$$

$$\mathfrak{O}_\zeta : [0,1] \to \mathcal{W}(\mathfrak{A}), \quad f \mapsto \begin{cases} \{0, \mathfrak{k}_1\} & \text{if } f \in [0, 0.1), \\ \{0, \mathfrak{k}_2\} & \text{if } f \in [0.1, 0.45), \\ \{0, \mathfrak{k}_2, \mathfrak{k}_4\} & \text{if } f \in [0.45, 0.55), \\ \{0, \mathfrak{k}_1, \mathfrak{k}_2, \mathfrak{k}_3\} & \text{if } f \in [0.55, 1] \end{cases} \quad \text{and}$$

$$\mathfrak{O}_\varpi : [0,1] \to \mathcal{W}(\mathfrak{A}), \quad f \mapsto \begin{cases} \{0\} & \text{if } f \in (0.98, 1], \\ \{0, \mathfrak{k}_2\} & \text{if } f \in (0.87, 0.98], \\ \{0, \mathfrak{k}_4\} & \text{if } f \in (0.69, 0.87], \\ \{0, \mathfrak{k}_2, \mathfrak{k}_4\} & \text{if } f \in (0.44, 0.69], \\ \mathfrak{A} & \text{if } f \in [0, 0.44]. \end{cases}$$

Then, $\mathfrak{O}_\xi(\mathfrak{s})$, $\mathfrak{O}_\zeta(\mathfrak{s})$, and $\mathfrak{O}_\varpi(\mathfrak{s})$ are "commutative ideals (CIs)" of $\mathfrak{A}$ for all $\mathfrak{s} \in [0,1]$. Hence, the NFfSh $\tilde{\mathcal{G}} := (\tilde{\mathcal{G}}_\xi, \tilde{\mathcal{G}}_\zeta, \tilde{\mathcal{G}}_\varpi)$ of $\mathfrak{O} := (\mathfrak{O}_\xi, \mathfrak{O}_\zeta, \mathfrak{O}_\varpi)$ is a CfNFI of $\mathfrak{A}$, and it is given as follows:

$$\tilde{\mathcal{G}}_\xi(f) = \begin{cases} 1 & \text{if } f = 0, \\ 0.55 & \text{if } f = \mathfrak{k}_1, \\ 0.9 & \text{if } f = \mathfrak{k}_2, \\ 0.35 & \text{if } f = \mathfrak{k}_3, \\ 0.3 & \text{if } f = \mathfrak{k}_4, \end{cases} \quad \tilde{\mathcal{G}}_\zeta(f) = \begin{cases} 1 & \text{if } f = 0, \\ 0.35 & \text{if } f = \mathfrak{k}_1, \\ 0.9 & \text{if } f = \mathfrak{k}_2, \\ 0.35 & \text{if } f = \mathfrak{k}_3, \\ 0.3 & \text{if } f = \mathfrak{k}_4 \end{cases} \quad \text{and}$$

$$\widetilde{\mathcal{G}}_{\varpi}(f) = \begin{cases} 0 & \text{if } f = 0, \\ 0.77 & \text{if } f \in \{\mathfrak{k}_1, \mathfrak{k}_3\}, \\ 0.41 & \text{if } f = \mathfrak{k}_2, \\ 0.34 & \text{if } f = \mathfrak{k}_4. \end{cases}$$

However, $\widetilde{\mathcal{G}} := (\widetilde{\mathcal{G}}_\xi, \widetilde{\mathcal{G}}_\zeta, \widetilde{\mathcal{G}}_\varpi)$ is not a CNFI of $\mathfrak{A}$, since $(\mathfrak{k}_3 \between \mathfrak{k}_4) \between \mathfrak{k}_2 \in \mathcal{U}_\epsilon(\widetilde{\mathcal{G}}, 0.5)$, and $\mathfrak{k}_2 \in \mathcal{U}_\epsilon(\widetilde{\mathcal{G}}, 0.7)$, but $\mathfrak{k}_3 \between (\mathfrak{k}_4(\mathfrak{k}_4 \between \mathfrak{k}_3)) = \mathfrak{k}_3 \notin \mathcal{U}_\epsilon(\widetilde{\mathcal{G}}, 0.5)$.

We provide relations between a fNFI and a CfNFI.

Theorem 3. *Consider a probability-space $(\psi, \mathfrak{E}, \mathfrak{Q})$ and let $\widetilde{\mathcal{G}} := (\widetilde{\mathcal{G}}_\xi, \widetilde{\mathcal{G}}_\zeta, \widetilde{\mathcal{G}}_\varpi)$ be the NFfSh of an NFRS $\mathfrak{D} := (\mathfrak{D}_\xi, \mathfrak{D}_\zeta, \mathfrak{D}_\varpi)$ on $\mathfrak{A}$. If $\widetilde{\mathcal{G}} := (\widetilde{\mathcal{G}}_\xi, \widetilde{\mathcal{G}}_\zeta, \widetilde{\mathcal{G}}_\varpi)$ is a CfNFI of $\mathfrak{A}$, then it is a fNFI of $\mathfrak{A}$.*

Proof. Let $\widetilde{\mathcal{G}} := (\widetilde{\mathcal{G}}_\xi, \widetilde{\mathcal{G}}_\zeta, \widetilde{\mathcal{G}}_\varpi)$ be a CfNFI of $\mathfrak{A}$. Then, $\mathfrak{D}_\xi(\mathcal{R}_\xi)$, $\mathfrak{D}_\zeta(\mathcal{R}_\zeta)$, and $\mathfrak{D}_\varpi(\mathcal{R}_\varpi)$ are "commutative ideals (CI)" of $\mathfrak{A}$ for all $\mathcal{R}_\xi, \mathcal{R}_\zeta, \mathcal{R}_\varpi \in \psi$. Thus, $\mathfrak{D}_\xi(\mathcal{R}_\xi)$, $\mathfrak{D}_\zeta(\mathcal{R}_\zeta)$, and $\mathfrak{D}_\varpi(\mathcal{R}_\varpi)$ are ideals of $\mathfrak{A}$ for all $\mathcal{R}_\xi, \mathcal{R}_\zeta, \mathcal{R}_\varpi \in \psi$. Therefore, $\widetilde{\mathcal{G}} := (\widetilde{\mathcal{G}}_\xi, \widetilde{\mathcal{G}}_\zeta, \widetilde{\mathcal{G}}_\varpi)$ is a fNFI of $\mathfrak{A}$. □

Interestingly, the converse statement of Theorem 3 lacks validity, as illustrated by the forthcoming counterexample.

Example 3. Define the set $\mathfrak{A} = \{0, \mathfrak{k}_1, \mathfrak{k}_2, \mathfrak{k}_3, \mathfrak{k}_4\}$ and equip it with "$\between$" a binary operation, whose operation table is shown in Table 3.

Table 3: BCK-algebra.

$\between$	**0**	$\mathfrak{k}_1$	$\mathfrak{k}_2$	$\mathfrak{k}_3$	$\mathfrak{k}_4$
0	0	0	0	0	0
$\mathfrak{k}_1$	$\mathfrak{k}_1$	0	0	$\mathfrak{k}_1$	0
$\mathfrak{k}_2$	$\mathfrak{k}_2$	$\mathfrak{k}_1$	0	$\mathfrak{k}_2$	0
$\mathfrak{k}_3$	$\mathfrak{k}_3$	$\mathfrak{k}_3$	$\mathfrak{k}_3$	0	$\mathfrak{k}_3$
$\mathfrak{k}_4$	$\mathfrak{k}_4$	$\mathfrak{k}_4$	$\mathfrak{k}_4$	$\mathfrak{k}_4$	0

Then, $(\mathfrak{A}; \between, 0)$ is a BCK-A [14]. Let us consider $(\psi, \mathfrak{E}, \mathfrak{Q}) = ([0,1], \mathfrak{E}, n)$, and let $\mathfrak{D} := (\mathfrak{D}_\xi, \mathfrak{D}_\zeta, \mathfrak{D}_\varpi)$ on $\mathfrak{A}$ be an NFRS on $\mathfrak{A}$ that is given by

$$\mathfrak{D}_\xi : [0,1] \to \mathcal{W}(\mathfrak{A}), \quad f \mapsto \begin{cases} \{0, \mathfrak{k}_3\} & \text{if } f \in [0, 0.37), \\ \{0, \mathfrak{k}_1, \mathfrak{k}_2, \mathfrak{k}_3\} & \text{if } f \in [0.37, 0.76), \\ \{0, \mathfrak{k}_1, \mathfrak{k}_2, \mathfrak{k}_4\} & \text{if } f \in [0.76, 1], \end{cases}$$

$$\mathfrak{D}_\zeta : [0,1] \to \mathcal{W}(\mathfrak{A}), \quad f \mapsto \begin{cases} \{0, \mathfrak{k}_3\} & \text{if } f \in [0, 0.17), \\ \{0, \mathfrak{k}_1, \mathfrak{k}_2, \mathfrak{k}_3\} & \text{if } f \in [0.17, 0.56), \\ \{0, \mathfrak{k}_1, \mathfrak{k}_2, \mathfrak{k}_4\} & \text{if } f \in [0.56, 1] \end{cases} \quad \text{and}$$

$$\mathfrak{O}_{\varpi} : [0,1] \to \mathcal{W}(\mathfrak{A}), \quad f \mapsto \begin{cases} \{0\} & \text{if } f \in (0.94, 1], \\ \{0, \mathfrak{k}_3\} & \text{if } f \in (0.86, 0.94], \\ \{0, \mathfrak{k}_1, \mathfrak{k}_2, \mathfrak{k}_4\} & \text{if } f \in (0.68, 0.86], \\ \mathfrak{A} & \text{if } f \in [0, 0.68]. \end{cases}$$

Then, $\mathfrak{O}_{\xi}(\mathfrak{s})$, $\mathfrak{O}_{\zeta}(\mathfrak{s})$, and $\mathfrak{O}_{\varpi}(\mathfrak{s})$ are ideals of $\mathfrak{A}$ for all $\mathfrak{s} \in [0,1]$. Hence, the NFfSh $\tilde{\mathcal{G}} := (\tilde{\mathcal{G}}_{\xi}, \tilde{\mathcal{G}}_{\zeta}, \tilde{\mathcal{G}}_{\varpi})$ of $\mathfrak{O} := (\mathfrak{O}_{\xi}, \mathfrak{O}_{\zeta}, \mathfrak{O}_{\varpi})$ is a fNFI of $\mathfrak{A}$. However, it is not a CfNFI of $\mathfrak{A}$ because if $\xi \in [0, 0.37)$, $\zeta \in [0, 0.17)$, and $\varpi \in (0.86, 0.94]$, then, $\mathfrak{O}_{\xi}(\xi) = \{0, \mathfrak{k}_3\}$, $\mathfrak{O}_{\zeta}(\zeta) = \{0, \mathfrak{k}_3\}$, and $\mathfrak{O}_{\varpi}(\varpi) = \{0, \mathfrak{k}_3\}$ are not "commutative ideals (CI)" of $\mathfrak{A}$, respectively.

Every ideal in a "commutative BCK-algebra (C-BCK-A)" is commutative, yielding the theorem below.

Theorem 4. *Consider a probability-space $(\psi, \mathfrak{E}, \mathfrak{Q})$, and let $\tilde{\mathcal{G}} := (\tilde{\mathcal{G}}_{\xi}, \tilde{\mathcal{G}}_{\zeta}, \tilde{\mathcal{G}}_{\varpi})$ be an NFfSh of an NFRS $\mathfrak{O} := (\mathfrak{O}_{\xi}, \mathfrak{O}_{\zeta}, \mathfrak{O}_{\varpi})$ on a "commutative BCK-algebra (C-BCK-A)." If $\tilde{\mathcal{G}} := (\tilde{\mathcal{G}}_{\xi}, \tilde{\mathcal{G}}_{\zeta}, \tilde{\mathcal{G}}_{\varpi})$ is a fNFI of $\mathfrak{A}$, then it is a CfNFI of $\mathfrak{A}$.*

Corollary 1. *Let a probability-space $(\psi, \mathfrak{E}, \mathfrak{Q})$ and a BCK-A $\mathfrak{A}$ fulfil one of the following conditions:*
(i) $f \le g \Rightarrow f \le g \between (g \between f)$;
(ii) $f \le g \Rightarrow f = g \between (g \between f)$;
(iii) $f \between (f \between g) = g \between (g \between (f \between (f \between g)))$;
(iv) $f, g \le h, h \between g \le h \between f \Rightarrow f \le g$;
(v) $f \le h, h \between g \le h \between f \Rightarrow f \le g$, *let $\tilde{\mathcal{G}} := (\tilde{\mathcal{G}}_{\xi}, \tilde{\mathcal{G}}_{\zeta}, \tilde{\mathcal{G}}_{\varpi})$ be an NFfSh of an NFRS $\mathfrak{O} := (\mathfrak{O}_{\xi}, \mathfrak{O}_{\zeta}, \mathfrak{O}_{\varpi})$ on $\mathfrak{A}$. If $\tilde{\mathcal{G}} := (\tilde{\mathcal{G}}_{\xi}, \tilde{\mathcal{G}}_{\zeta}, \tilde{\mathcal{G}}_{\varpi})$ is a fNFI of $\mathfrak{A}$, then it is a CfNFI of $\mathfrak{A}$.*

4 Conclusion

This study introduces the concept of CfNFI in BCK-algebras, establishing key relationships and properties. Every commutative neutrosophic fuzzy ideal is shown to be a CfNFI, but the converse is demonstrated to be false via a counterexample. Conditions for a falling neutrosophic fuzzy ideal to be commutative are identified. By introducing CfNFI, this work extends fuzzy logic applications to BCK-algebras, bridging existing literature gaps and paving the way for further research in algebraic structures and uncertainty handling.

Potential future directions include exploring applications of CfNFI in decision-making processes, fuzzy systems, and cryptography. Further investigation into the properties of CfNFI in other algebraic structures like Bd-algebras or MV-algebras, PU, BCH-algebras could also yield valuable insights. Extending these concepts to other fuzzy frameworks like Pythagorean fuzzy sets or spherical fuzzy sets might broaden their applicability.

Bibliography

[1] Atanassov, K. T. (1986). Intuitionistic fuzzy sets. *Fuzzy Sets and Systems*, *20*, 87–96.

[2] Bhuvaneswari, D., Satyanarayana, B., & Naik, K. A. (2026). Fuzzy sets as equivalence classes of random sets. In S. Singh, S. Kaur, E. Güler, B. Bhushan & S. Malik (Eds.), *Analysis, Algebra, and Associated Topics* (pp. 165–183). Springer Nature Switzerland AG.

[3] Borzooei, R. A., Rezaei, G. R., Kologhani, M. A., & Jun, Y. B. (2021). Falling shadow theory with applications in hoops. *Bulletin of the Section of Logic*, *50*(3), 337–353.

[4] Borzooei, R. A., Xin, X. L., & Jun, Y. B. (2023). Intuitionistic falling shadows applied to commutative ideals in BCK-algebras. *Journal of Algebraic Systems*, *10*(2), 285–297.

[5] Goodman, I. R. (1982). Fuzzy sets as equivalence classes of random sets. In R. Yager (Ed.), *Recent Developments in Fuzzy Sets and Possibility Theory* (pp. 327–343). Pergamon.

[6] Huang, Y. (2006) *BCI-Algebra*. Science Press.

[7] Jun, Y. B., & Roh, E. H. (1994). Fuzzy commutative ideals of BCK-algebras. *Fuzzy Sets and Systems*, *64*, 401–405.

[8] Jun, Y. B., & Kim, K. H. (2000). Intuitionistic fuzzy ideals of BCK-algebras. *International Journal of Mathematics and Mathematical Sciences*, *24*(12), 839–849.

[9] Jun, Y. B., Lee, D. S., & Park, C. H. (2008). Intuitionistic fuzzy commutative ideals of BCK-algebras. *Journal of the Korea Society of Mathematical Education Series B: Pure and Applied Mathematics*, *15*(1), 73–84.

[10] Jun, Y. B., & Kang, M. S. (2011). Falling fuzzy commutative ideals of BCK-algebras. *Journal of the Korea Society of Mathematical Education Series B: Pure and Applied Mathematics*, *18*(1), 41–50.

[11] Jun, Y. B., & Park, C. H. (2012). Falling shadows applied to subalgebras and ideals of BCK/BCI-algebras. *Honam Mathematical Journal*, *34*(2), 135–144.

[12] Jun, Y. B., Song, S. Z., & Lee, K. J. (2018). Intuitionistic falling shadow theory with application in BCK/BCI-algebras. *Mathematics*, *6*(8), 138.

[13] Meng, J. (1991). Commutative ideals in BCK-algebras. *Pure and Applied Mathematics*, *9*, 49–53.

[14] Meng, J., & Jun, Y. B. (1994). *BCK-Algebras*. Kyungmoonsa Co.

[15] Naik, K. A., Satyanarayana, B., & Bhuvaneswari, D. On interval-valued neutrosophic fuzzy (implicative and commutative) ideals of BCK-algebras. Manuscript accepted for Springer Proceedings.

[16] Naik, K. A., Bhuvaneswari, D., Madhavi, U. B., Krishna, B. T., & Satyanarayana, B. (2026). On neutrosophic fuzzy (implicative, commutative, positive implicative) hyper BCK-ideals of BCK-algebras. *Neutrosophic Sets and Systems*, *95*, 341–358.

[17] Naik, K. A., Satyanarayana, B., Ramesh, D., & Krishna, B. T. (2026). Interval-valued neutrosophic fuzzy n-fold positive implicative ideals of BCK-algebras: properties and characterizations. *Neutrosophic Sets and Systems*, *95*, 341–358.

[18] Tan, S. K., Wang, P. Z., & Lee, E. S. (1993). Fuzzy set operations based on the theory of falling shadows. *Journal of Mathematical Analysis and Applications*, *174*, 242–255.

[19] Tan, S. K., Wang, P. Z., & Zhang, X. Z. (1993). Fuzzy inference relation based on the theory of falling shadows. *Fuzzy Sets and Systems*, *53*, 179–188.

[20] Wang, P. Z., & Sanchez, E. (1982). Treating a fuzzy subset as a projectable random set. In M. M. Gupta & E. Sanchez (Eds.), *Fuzzy Information and Decision* (pp. 212–219). Pergamon.

[21] Wang, P. Z. (1985). *Fuzzy Sets and Falling Shadows of Random Sets*. Beijing Normal University Press.

Gaurav Singh, Suraj Roy, and Apu Kumar Saha

LO-AHA: an improved artificial hummingbird algorithm based on a Laplacian operator for global optimization

Abstract: The Artificial Hummingbird Algorithm (AHA) is a recently developed bioinspired optimization technique to handle complex optimization problems. However, the AHA exhibits an imbalance between exploration and usage capabilities; sometimes prematurely convergence with poor precision. Hence, in this chapter, an improved version of AHA using the Laplacian operator, named LO-AHA is proposed to address the inherent lacunae of AHA. The proposed LO-AHA has been utilized to solve 15 classical benchmark functions and the IEEE CEC 2019 suite to examine its effectiveness compared with other state-of-the-art algorithms. The analysis is demonstrated via Friedman's rank test for the mean values of the proposed algorithm and other state-of-the-art algorithms for benchmark functions as well as IEEE CEC 2019 and it is found that LO-AHA is better in competition with other comparing algorithms. It has been found that the proposed algorithm performs better and similar results in 94 % occasions while 6 % are inferior. The Friedman's rank test also confirms our claim, where the proposed algorithms has obtained the rank 1. The results illustrate that the proposed approach is a reliable and efficient method for addressing challenging optimization problems.

Keywords: Artificial hummingbird algorithm, Laplacian operator, IEEE CEC 2019 functions, benchmark functions, Friedman rank test

1 Introduction

The algorithms for optimization draw motivation from different aspects of nature, including biological evolution, the social and collective behaviors of living organisms, physical or chemical processes, and human habits and beliefs. Mathematical, medical, engineering, military, and nearly every field faces global optimization challenges. The method of determining the most optimal possible values for a given technique's parameters consisting of all the potential values in order to optimize its output is known as optimization. In an optimization problem, particular objective functions are either maximized or minimized, for the purpose of generating mathematical models with the complex system (evaluating the solution's adequacy). Furthermore, despite the development of a mathematical model, it faces numerous problems such as complex nonlinear

Gaurav Singh, Apu Kumar Saha, Department of Mathematics, National institute of Technology Agartala, Tripura, India, e-mail: gauravsingh.nita22@gmail.com
Suraj Roy, Department of Mathematics, Tripura University, Agartala, Tripura, India

https://doi.org/10.1515/9783112234525-008

constraints, a large search space, and high computing costs. The abundance of research conducted in the past few decades on population-based algorithms for optimization is indicative of the significance of this field of study. All kinds of optimization issues in academia and business are resolved by these methods. These algorithms' ease of use and simplicity are making them increasingly popular among users across several fields. Metaheuristic algorithms' greatest strength is that, in contrast to other algorithms, they do not require any preconditions in order to tackle optimization issues. In order to simplify and make the solution more accessible, metaheuristic algorithms (MHAs) mainly take input and output into account. MHAs facilitate multiagent exploration in a limited search space. Many concepts are incorporated into MHA, enabling operators to scan larger areas and make money on noteworthy locations. Such algorithms are divided into four important families: evolutionary-based techniques such as differential evolution (DE) [1], swarm-based techniques such as particle swarm optimization (PSO) [2], physical science-based algorithms such as the gravitational search algorithm (GSA) [4], and human behavior-based like teaching-learning-based optimization (TLBO) [5]. Recently, several other highly recognized algorithms have also been developed such as cuckoo search (CS) [6], and artificial bee colony (ABC) [7].

Enhancing the effectiveness of swarm-based method requires balancing both exploration and extraction. Exploration is a global search capability used in swarm-based techniques to break out of local optima and look in new locations using randomization. Exploitation speeds up the convergence of these techniques, allowing for better solutions near the present optimal solution. However, finding the exact solution of any complex nonlinear optimization problem is a difficult task.

Over the decades, numerous modifications have been made to various existing algorithms to deal with the limitations. A modified SCA variation, MAMSCA [8], deals with situations where existing algorithms become stuck in adjacent optima and uneven exploitation, also balanced global and local searches. The modified adaptive ant colony optimization algorithm (MAACO) [9], an improved variation of ACO, suggests unevenly distributing initial pheromone concentration to avoid blind searching. CNPSO [10], a modification of PSO, a modified Genetic Algorithm approach for classification of abnormal magnetic resonance brain tumor images [11], the improved TLBO with Laplacian operator [12] algorithm to keep exploration and exploitation in balance are some of the modified algorithms of traditional base algorithms. A fuzzy gain-based dynamic ant colony optimization (FGDACO) [13] is a useful method for planning collision-free and smooth pathways with viable path lengths and minimum time. A Bayesian inference-based learning automaton with a multiobjective brainstorming [14] method is used to identify the optimal solutions for combinatorial optimization problems, such as truck routing, the quantum-inspired ant colony optimization (Qi-ACO) [15] is used to solve a sustainable four-dimensional traveling salesman problem (4DTSP).

In the same order, the Artificial Hummingbird Algorithm (AHA) [16], a swarm intelligence-based algorithm inspired from the foraging behavior of hummingbirds, was recently proposed by Zhao et al. in 2022. The AHA mimics three specific flight skills as well as three efficient foraging tactics in habitats that are natural. In AHA, the territorial foraging approach supports exploitation while the migration foraging approach ensures exploration of the search area, while the guided foraging approach prioritizes exploration initially and exploitation afterwards. The algorithm differs from other algorithms in that it uses a unique memory update technique. Each hummingbird uses a visit table to determine their preferred food source based on when they last visited each one. The AHA outperforms various other algorithms with regard to computational effectiveness and precision for tackling engineering optimization challenges, as evidenced in the literature. Some of the modifications of AHA named as LCAHA, a hybrid artificial hummingbird algorithm [17] and the Quantum Artificial Hummingbird Algorithm [18] have been suggested in the last few years. AHA requires adaptation to tackle problems in other domains, making it challenging to attain optimal results. Some problems have a unimodal search space while some have multimodal search spaces, which can impact search performance. In this sense, AHA can result in inferior solutions. Several techniques have been offered in the literature to increase MHA performance. One technique of mutation, the interaction between one hummingbird to another exchange the information that results in obtaining more effective exploration and exploitation in the search area.

On the improvement of the Artificial Hummingbird Algorithm, some of the latest studies include a multistrategy boosted hybrid artificial hummingbird algorithm (LC-AHA) [3], a multiobjective AHA (MOAHA) [19] for solving multiobjective complex optimization problems. All these the studies in the literature show the shortfall of the balance between the exploration and exploitation in the existing AHA and suggests the need to overcome such problems and perform the algorithm effectively.

In this chapter, we propose an improved variant of AHA named LO-AHA using the Laplacian mutation operator (LO). The contributions of the chapter are described as follows:
1. The proposed algorithm incorporates a Laplacian operator during the guided foraging phase that helps in improving the exploration of the search space and hence the convergence of the solution.
2. Laplace crossover boosts the efficiency of searches by improving the algorithm's local exploitation capabilities, allowing them to escape local optima as well as locate superior solutions. It accomplishes this by producing offspring with greater step sizes during the early phases of the search, boosting their chances of jumping out of local extrema.
3. To test the efficacy of the proposed method it has been contrasted with ten state-of-the-art algorithms to solve forty benchmark functions consisted of various unimodal and multimodal functions.
4. To further investigate the performance of the proposed algorithm it has been applied on the IEEE CEC 2019 function suite and the outcomes are contrasted with eight basic algorithms.
5. The statistical performance of the algorithm is tested using Friedman's rank test.

2 Related work

This section presents a basic discussion regarding the Laplacian operator and the Artificial Hummingbird Algorithm.

2.1 Laplacian crossover operator

Crossover, mutation, and selection are the three selection operators used in genetic algorithms [20]. The operator of crossovers in GA takes responsibility for exploring the search area. This operator generates a new solution through the combination of two parent values. This characteristic is thought to produce a superior solution using their parents' best qualities. Deep and Thakur (2007) introduce a real-coded GA operator called Laplace crossover. In the Laplacian operator, instead of one offspring, two are produced, each of which is expected to inherit the parents' characteristics. It is thus a parent-centered strategy. The Laplacian operator generates two offspring (y_1 and y_2) from parents (p_1 and p_2). Following the Laplace distribution to create a random number α, for which we first generate a uniform random number $u \in [0,1]$, we then calculate the value of α using the following equation:

$$\alpha = \begin{cases} a - b\ln(u), & u \leq 0.5, \\ a + b\ln(u), & u > 0.5, \end{cases} \tag{1}$$

where $a \in \mathbb{R}$ is termed the location parameter, and $b > 0$ is known as the scale parameter. Taking $\gamma = 0.5$, the two new offspring are produced as

$$y_1 = p_1 + \alpha(p_1 - p_2) \tag{2a}$$

$$y_2 = p_2 + \alpha(p_1 - p_2) \tag{2b}$$

$$y = \gamma y_1 + (1 - \gamma)y_2. \tag{3}$$

2.2 Artificial hummingbird algorithm (AHA)

The AHA is a revolutionary population-based SI methodology. AHA begins with creating randomized possible solutions. The AHA algorithm has three mathematical representations that imitate hummingbirds' foraging behaviors: guided foraging, territorial foraging, and migrating foraging.

In guided foraging, birds use their predetermined knowledge to find food sources. Hummingbirds save their knowledge in their memory or visit tables as notations. If multiple food sources possess a similar visit level, the bird will choose the one having the greatest nectar-refilling efficiency. The three distinct kinds of flying skills used at the time of foraging are: axial flight, omnidirectional flight, and diagonal flight.

In territorial foraging, hummingbirds will look for better possibilities in nearby areas (territory) rather than returning to previous ones.

A random number is generated between [0,1] to select the foraging behavior that satisfies for guided foraging the value of random number must be less than 0.5, otherwise territorial foraging takes place.

In migrating foraging, a hummingbird regularly visits a food-deficient area, it typically migrates to an additional distanced food source to obtain nourishment. If a certain number of iterations crosses the fixed migration parameter, then the hummingbirds that locate at the food source having the nectar-refilling efficiency as worst will move to a randomly selected fresh food source within the search area.

3 Proposed method

In this chapter, AHA incorporates the Laplacian operator named as LOAHA, to enhance the search space exploration. The Laplacian operator is implemented in the AHA's guided foraging, which enhances the search efficiency by improving the local exploration skill. The LOAHA avoids the local guiding, which helps to deal with structured local optima points in the search space, and accomplishes this by boosting offspring's step sizes during the early phases of the search, improving their chances of leaping out of local extrema.

The initializing step begins by generating its population. Let N hummingbirds be put on N food sources, that will initialize the hummingbird's locations (p) randomly by using equation (4) as follows:

$$p_i = L_b + r * (U_b - L_b), \quad i = 1, 2, \ldots N, \tag{4}$$

where L_b and U_b are the upper and lower bounds for a dim-dimensional problem respectively, $r \in [0,1]$ a random vector, p_i refers to the position of the ith food source assigned to the ith hummingbird. In the guided phase, one hummingbird interacts with others to exchange their information. The equation in AHA for mathematics is designed to simulate guided foraging actions and identify potential food sources as

$$v^{(i)}(t+1) = p_{i,\text{tar}}(t) + a * D_q * (p_i(t) - p_{i,\text{tar}}(t)), \quad q \in [A, D, O], \tag{5}$$

where for the ith food source $p_i(t)$ refers to the position at time t, $p_{i,\text{tar}}(t)$ refers to the position of targeted food sources that are planned to be visited by the ith hummingbird and a guided factor $a \sim \mathbb{N}(0,1)$ that satisfies the normal distribution with mean of zero and standard deviation of one.

In LO-AHA, equation (5) is updated by using the information exchanged in the guided phase to this mutation of birds mathematically expressed in the form of the Laplacian operator, in which each hummingbird shares their information of visits with all the others. With the help of equation (6a) and equation (6b) two new offsprings v_1^i and v_2^i are generated for a fixed ith hummingbird from the parents' position p_i and p_j (where j varies among all the hummingbirds other than the ith) then the new position is updated performing the mutation using equation (7) that helps to better explore the search space

$$v_1^i = p_{i,\text{tar}}(t) + \alpha * D_q * (p_i(t) - p_j(t)), \quad q \in [A, D, O] \tag{6a}$$

$$v_2^i = p_{i,\text{tar}}(t) + \alpha * D_q * (p_j(t) - p_i(t)), \quad q \in [A, D, O] \tag{6b}$$

$$v^{(i)}(t+1) = \gamma v_1^i + (1-\gamma) v_2^i, \tag{7}$$

where $v_1^{(i)}$, $v_2^{(i)}$ are two new positions calculated after the interaction of the ith and the jth hummingbird, p_i and p_j are the present positions of the ith and the jth hummingbird, α is a random number, created by using the Laplace distribution equation (1), which is used as the guided factor and taking $\gamma = 0.5$ to avoid complexity.

After that, the equation for updating the position of the ith food source, whenever a hummingbird discovers a closer spot to its intended food supply is

$$p_i(t+1) = \begin{cases} p_i(t), & f(p_i(t)) \le f(v^{(i)}(t+1)), \\ v^{(i)}(t+1), & f(p_i(t)) > f(v^{(i)}(t+1)), \end{cases} \tag{8}$$

where the objective function for the fitness value is indicated by $f(\cdot)$. According to equation (8), if the potential food source has a higher nectar-refilling capacity than that of the present one, the hummingbird will switch to the candidate based on equation (7) to feed. After guided foraging, the mathematical formula for replicating hummingbirds' territorial foraging behavior and potential food source is as follows:

$$v^{(i)}(t+1) = p_i(t) + b * D_q * p_i(t), \quad q \in [A, D, O], \tag{9}$$

where a territorial factor $b \sim \mathbb{N}(0,1)$ that satisfies the normal distribution with mean of zero and standard deviation of one and D_q is one of the flying modes. The location of the hummingbird during iteration $p_i(t+1)$ is obtained as equation (8). Then, the migration foraging behavior of a hummingbird through the place of the food source with the nectar-refilling efficiency as worst to a fresh one randomized can be described as follows:

$$p_{\text{worst}}(t+1) = L_b + r * (U_b - L_b), \quad \text{when } M = t, \tag{10}$$

where p_{worst} is the food source with nectar-refilling efficiency as worst among the population, t refers to the present iteration, and M is the migration factor that is generally M and $2N$ is the number of hummingbirds or population size. Algorithm 1 is the pseudocode of the LO-AHA algorithm that gives the brief overview for the MATLAB code to observe the experimental results and conclude.

Algorithm 1: Pseudo-Code Of LO-AHA ALGORITHM.

Input: *Set the values of N, dim, L_b, U_b, f, Maximum_iteration.*
Initialization:
For $i = 1$ *to* N
Set the initial values via $p_i = L_b + r(U_b + L_b)$;
For $j = 1$ *to* N
Create the initialization of the visit table;
If $i \neq j$
Then $VT_{i,j} = 1$;
Else $VT_{i,j} = null$;
End If
End For
End For
While $t < maximum\_iteration$
For $i = 1$ *to* N
$r = rand$
If $r < \frac{1}{3}$
Then *execute for diagonal flight;*
Else If $r > \frac{2}{3}$
Then *execute for omnidirectional flight;*
Else *execute for axial flight;*
End If
execute Eq. 1 to calculate $\propto$;
For $j = 1$ *to* N
If $rand < 0.5$
execute Eq. 3 to perform the guided foraging based on Laplacian operator;
If $f\left(v^{(i)}(t+1)\right) < f(z_i(t))$
Then $z_i(t+1) = v^{(i)}(t+1)$;
If $j \neq i$
$VT_{i,j} = VT_{i,j} + 1$;
Else
$VT_{i,tar} = 0$; $VT_{i,j} = \max_{k \in N \text{ and } k \neq j} VT_{i,j} + 1$;
End If
Else *Visit table updated as;*
If $j \neq i$
$VT_{i,j} = VT_{i,j} + 1$; $VT_{i,tar} = 0$
End If
End If
Else *execute Eq. 9 for territorial foraging;*
If $f\left(v^{(i)}(t+1)\right) < f(z_i(t))$
Then $z_i(t+1) = v^{(i)}(t+1)$;
If $j \neq i$
$VT_{i,j} = VT_{i,j} + 1$;
Else $VT_{j,i} = \max_{k \in N \text{ and } k \neq j} VT_{j,k} + 1$;
End If
Else
If $j \neq i$
$VT_{i,j} = VT_{i,j} + 1$;
End If
End If
End If
End For
End For
If *mod (t,2N) == 0*
Then *execute Eq. 10 for migration foraging*
For $j = 1$ *to* N
If $j \neq worst$
$VT_{worst,j} = VT_{worst,j} + 1$;
Else $VT_{j,worst} = \max_{k \in N \text{ and } k \neq j} VT_{j,k} + 1$;
End If
End For
End If
End While

4 Results and discussion

This section shows the test results in a way that is easy to comprehend. The improved algorithm with LO is compared with the AHA and some basic recently introduced metaheuristic algorithms, namely the whale optimization algorithm (WOA), BOA, SCA, the arithmetic optimization algorithm (AOA) [21], GWO, AVOA, BWOA, HHO, and the white shark optimizer (WSO) [22]. All basic techniques were tested on benchmark functions as well as the IEEE CEC 2019 suites for comparison.

4.1 Benchmark functions used for experimental analysis

The reference functions from the published literature were used to evaluate the method's performance. Table 7 in appendix presents the function expression, where the first column is 'Name', which shows the name of the function; the second one 'C' refers to the characteristics; the next column 'Function' has the expressions of the named functions; 'Dim' is the dimension of the function; the column named as 'Range' contacts the value of the upper and lower bounds, and the next column 'f_{opt}' represents the optimal value. These test functions are generally used for minimization. There are four types of functions covered by this test set, which is rather large: unimodal, multimodal, nonseparable, and separable. Multimodal functions contain many local extrema, while unimodal functions only have one. Finding the local optima of algorithms is always made simple by their multimodal feature. The distinction between the nonseparable characteristic and the separable characteristic is that the former may be separated as a product of functions for each variable, whereas the latter cannot due to the interrelationships between the variables. The global optimum is typically hard to find due to the nonseparable property. Table 1 reports the experimental results of LO-AHA and other algorithms for benchmark functions to assess the performance of the proposed work. As in Table 2, the comparison to the improved variant is much better for the benchmark functions.

Friedman's rank test for benchmark functions has been also computed in Table 3 and shows that the LO-AHA performs as rank first. While for the functions F_{13} and F_{14} the minimum value of the improved one is worse.

Statistical analysis is done to compare the performance of LO-AHA with various algorithms. The experimental results for LO-AHA and other algorithms for the 15 benchmark functions are shown in Table 1. F_1 to F_6 unimodal functions (Stepint, Quartic, Easom, Matyas, Trid10, Schwefel2.22) are utilized to analyze the convergence rate of the proposed algorithm and F_7 to F_{15} multimodal functions (Branin, Rastrigin, Michalewicz10, Bohachevsky3, Kowalik, Shekel7, Perm, PowerSum, Hartman6) are used to assess the capability of the proposed algorithm to escape the local optima stagnation (for the mathematical expressions, refer to Table 7). For comparison to LO-AHA with other basic algorithms for these 15 benchmark functions refer to Table 2, in which

Table 1: Experimental results of LO-AHA and other algorithms for the benchmark functions defined.

Function	Index	LO-AHA	AHA	WOA	WSO	AOVA	BOA	BWOA	HHO	SCA	AOA	GWO
F1	Mean	-4.50×10^{0}	7.07×10^{0}	-5.00×10^{0}	-5.00×10^{0}	-5.00×10^{0}	-1.54×10^{2}	-5.00×10^{0}	-5.00×10^{0}	-5.00×10^{0}	-5.00×10^{0}	-5.00×10^{0}
	SD	1.53×10^{0}	2.52×10^{0}	0.00×10^{0}	0.00×10^{0}	0.00×10^{0}	5.55×10^{1}	0.00×10^{0}	0.00×10^{0}	0.00×10^{0}	0.00×10^{0}	0.00×10^{0}
F2	Mean	1.66×10^{5}	1.67×10^{5}	7.86×10^{5}	2.86×10^{2}	7.15×10^{4}	5.33×10^{1}	4.37×10^{-5}	5.25×10^{-5}	2.25×10^{-2}	1.43×10^{-3}	1.50×10^{-3}
	SD	3.24×10^{5}	1.91×10^{5}	6.23×10^{5}	2.27×10^{2}	7.61×10^{4}	3.35×10^{1}	4.62×10^{-5}	5.42×10^{-5}	1.81×10^{-2}	1.69×10^{-3}	8.94×10^{-4}
F3	Mean	-1.00×10^{0}	-6.67×10^{-2}	-1.00×10^{0}	-1.57×10^{-1}	-1.00×10^{0}	-1.59×10^{-2}	-1.00×10^{0}	-1.00×10^{0}	-9.99×10^{-1}	-1.00×10^{0}	-1.00×10^{0}
	SD	3.64×10^{-6}	2.54×10^{-1}	0.00×10^{0}	3.33×10^{-1}	2.26×10^{-11}	0.00×10^{0}	0.00×10^{0}	1.38×10^{-7}	7.33×10^{-4}	2.63×10^{-8}	4.46×10^{-7}
F4	Mean	0.00×10^{0}	0.00×10^{0}	1.13×10^{-282}	0.00×10^{0}	1.09×10^{-71}	5.42×10^{-2}	0.00×10^{0}	1.67×10^{-254}	1.10×10^{-123}	0.00×10^{0}	3.33×10^{-136}
	SD	0.00×10^{0}	0.00×10^{0}	0.00×10^{0}	0.00×10^{0}	5.40×10^{-71}	8.24×10^{-2}	0.00×10^{0}	0.00×10^{0}	5.90×10^{-123}	0.00×10^{0}	1.83×10^{-135}
F5	Mean	-2.10×10^{2}	-7.97×10^{1}	-2.10×10^{2}	5.19×10^{0}	-2.09×10^{2}	4.33×10^{3}	-5.56×10^{6}	-2.10×10^{2}	-5.20×10^{1}	-2.10×10^{2}	-1.76×10^{2}
	SD	8.96×10^{-2}	8.59×10^{0}	4.18×10^{-6}	0.00×10^{0}	8.20×10^{-1}	2.11×10^{3}	4.62×10^{1}	1.74×10^{-1}	6.61×10^{1}	1.01×10^{-2}	4.83×10^{1}
F6	Mean	0.00×10^{0}	0.00×10^{0}	2.44×10^{-150}	0.00×10^{0}	5.07×10^{-24}	2.55×10^{6}	0.00×10^{0}	2.40×10^{-102}	4.12×10^{-6}	1.44×10^{-110}	7.01×10^{-20}
	SD	0.00×10^{0}	0.00×10^{0}	1.33×10^{-149}	0.00×10^{0}	2.19×10^{-23}	1.09×10^{7}	0.00×10^{0}	7.44×10^{-102}	8.13×10^{-5}	3.74×10^{-110}	6.09×10^{-20}
F7	Mean	3.98×10^{-1}	4.05×10^{-1}	3.98×10^{-1}	6.28×10^{-1}	3.98×10^{-1}	7.32×10^{-1}	3.98×10^{-1}	3.98×10^{-1}	3.99×10^{-1}	3.98×10^{-1}	3.98×10^{-1}
	SD	3.13×10^{-6}	7.61×10^{-3}	0.00×10^{0}	2.76×10^{-1}	2.26×10^{-10}	4.42×10^{-1}	0.00×10^{0}	1.41×10^{-7}	1.11×10^{-3}	6.75×10^{-7}	3.04×10^{-6}
F8	Mean	0.00×10^{0}	0.00×10^{0}	0.00×10^{0}	0.00×10^{0}	0.00×10^{0}	3.55×10^{1}	0.00×10^{0}	0.00×10^{0}	7.80×10^{0}	5.68×10^{-15}	1.55×10^{0}
	SD	0.00×10^{0}	0.00×10^{0}	0.00×10^{0}	0.00×10^{0}	0.00×10^{0}	3.13×10^{1}	0.00×10^{0}	0.00×10^{0}	1.20×10^{1}	2.29×10^{-14}	2.33×10^{0}
F9	Mean	-9.65×10^{0}	-3.93×10^{0}	-9.66×10^{0}	-3.95×10^{0}	-7.00×10^{0}	-3.46×10^{0}	-6.17×10^{0}	-6.21×10^{0}	-4.21×10^{0}	-6.38×10^{0}	-7.51×10^{0}
	SD	8.20×10^{-3}	3.85×10^{-1}	7.62×10^{-3}	6.76×10^{-1}	1.31×10^{0}	5.14×10^{-1}	8.58×10^{-1}	5.61×10^{-1}	4.71×10^{-1}	9.35×10^{-1}	9.76×10^{-1}
F10	Mean	0.00×10^{0}	0.00×10^{0}	0.00×10^{0}	0.00×10^{0}	0.00×10^{0}	1.21×10^{2}	0.00×10^{0}	0.00×10^{0}	0.00×10^{0}	2.41×10^{-17}	0.00×10^{0}
	SD	0.00×10^{0}	0.00×10^{0}	0.00×10^{0}	0.00×10^{0}	0.00×10^{0}	1.69×10^{2}	0.00×10^{0}	0.00×10^{0}	0.00×10^{0}	4.76×10^{-17}	0.00×10^{0}

Table 1 (continued)

Function	Index	LO-AHA	AHA	WOA	WSO	AOVA	BOA	BWOA	HHO	SCA	AOA	GWO
F11	Mean	5.01×10^{-4}	8.78×10^{-3}	3.07×10^{-4}	3.08×10^{-2}	6.62×10^{-4}	3.33×10^{-2}	3.43×10^{-3}	3.24×10^{-4}	9.45×10^{-4}	6.34×10^{-4}	4.40×10^{-3}
	SD	6.06×10^{-4}	2.21×10^{-2}	1.94×10^{-19}	2.17×10^{-2}	2.70×10^{-4}	2.52×10^{-2}	1.20×10^{-2}	1.32×10^{-5}	4.06×10^{-4}	3.19×10^{-4}	8.12×10^{-3}
F12	Mean	-1.04×10^{1}	-5.03×10^{0}	-1.04×10^{1}	-4.88×10^{0}	-1.04×10^{1}	-1.61×10^{0}	-7.29×10^{0}	-5.09×10^{0}	-4.89×10^{0}	-8.04×10^{0}	-1.04×10^{1}
	SD	1.53×10^{-3}	1.75×10^{0}	1.14×10^{-15}	3.27×10^{0}	8.32×10^{-7}	1.23×10^{0}	2.68×10^{0}	2.48×10^{-4}	1.69×10^{0}	3.21×10^{0}	7.06×10^{-4}
F13	Mean	3.10×10^{0}	3.19×10^{1}	5.38×10^{-2}	1.83×10^{2}	4.15×10^{0}	4.81×10^{1}	4.99×10^{-1}	6.24×10^{-1}	4.99×10^{-1}	4.03×10^{0}	7.76×10^{-1}
	SD	5.05×10^{0}	9.88×10^{1}	1.18×10^{-1}	2.94×10^{2}	9.14×10^{0}	5.70×10^{1}	1.98×10^{0}	5.14×10^{-1}	4.31×10^{-1}	5.20×10^{0}	7.88×10^{-1}
F14	Mean	7.66×10^{-2}	1.15×10^{0}	1.34×10^{-4}	1.09×10^{1}	9.38×10^{-1}	1.06×10^{1}	1.48×10^{-3}	7.89×10^{-2}	3.42×10^{0}	8.36×10^{-1}	1.26×10^{-1}
	SD	9.00×10^{-4}	1.01×10^{0}	1.51×10^{-4}	1.11×10^{1}	3.59×10^{1}	1.71×10^{1}	6.25×10^{-3}	1.01×10^{-1}	6.76×10^{0}	8.45×10^{-1}	2.30×10^{-1}
F15	Mean	-3.27×10^{0}	-3.12×10^{0}	-3.31×10^{0}	-2.47×10^{0}	-3.27×10^{0}	-2.23×10^{0}	-3.17×10^{0}	-3.19×10^{0}	-2.94×10^{0}	-3.22×10^{0}	-3.26×10^{0}
	SD	5.99×10^{-2}	6.77×10^{-2}	4.11×10^{-2}	4.40×10^{-1}	7.08×10^{-2}	4.40×10^{-2}	1.52×10^{-2}	6.41×10^{-2}	2.55×10^{-1}	1.69×10^{-1}	8.32×10^{-2}

Table 2: Comparison of LO-AHA with other algorithms based on superiority analysis.

Category	AHA	WOA	WSO	AOVA	SCA	BOA	GWO	BWOA	AOA	HHO
Superior	10	7	11	8	10	12	13	9	12	11
Equal	5	5	4	7	4	3	2	4	2	3
Inferior	0	3	0	0	1	0	0	2	1	1

Table 3: Friedman mean rank and final ranking of the compared algorithms.

Index	LO-AHA	WOA	AHA	AOA	AVOA	BOA	BWOA	GWO	HHO	SCA	WSAO
Mean Rank	3.34	5.68	3.79	7.05	5.08	10.53	5.19	5.55	4.38	7.70	7.74
Rank	1	7	2	8	4	11	5	6	3	9	10

the proposed algorithm is superior in 10, equal in 5 functions in comparison to AHA; superior in 7, equal in 5 functions in comparison to WOA; superior in 8, equal in 7 functions in comparison to AOVA; superior in 10, equal in 4 functions in comparison to SCA; superior in 12, equal in 3 functions in comparison to BOA; superior in 13, equal in 2 functions in comparison to GWO; superior in 9, equal in 4 functions in comparison to BWOA; superior in 12, equal in 2 functions in comparison to AOA, and superior in 11, equal in 3 functions in comparison to HHO.

4.2 Friedman's test

For identifying variations in approaches (methods) used in several efforts (functions), Friedman's test is a nonparametric method. Given that the test is nonparametric, it does not presume that the distribution of our data is normal. When an ANOVA test's basic premise—that is, when data do not originate from a normal population—is broken, it is utilized in its place. When it contains more than three treatments (methods), the "Paired samples Wilcoxon signed-rank test" is extended and employed in this test. Both tests are the same when there are two treatments (strategies).

4.3 Convergence analysis

The rate at which an algorithm generates its solution is measured using a convergence graph. Different algorithms' convergence graphs are examined in order to identify the algorithm that converges more quickly. This chapter's techniques all generate convergence graphs with a fixed population size of 50 and a maximum iteration of 100. We plotted the optimal fitness values determined by the algorithms in relation to the iteration. Figure 1 shows the convergence graphs of LO-AHA with basic algorithms using benchmark functions given for randomly selected six benchmark functions. Figure 1(a)

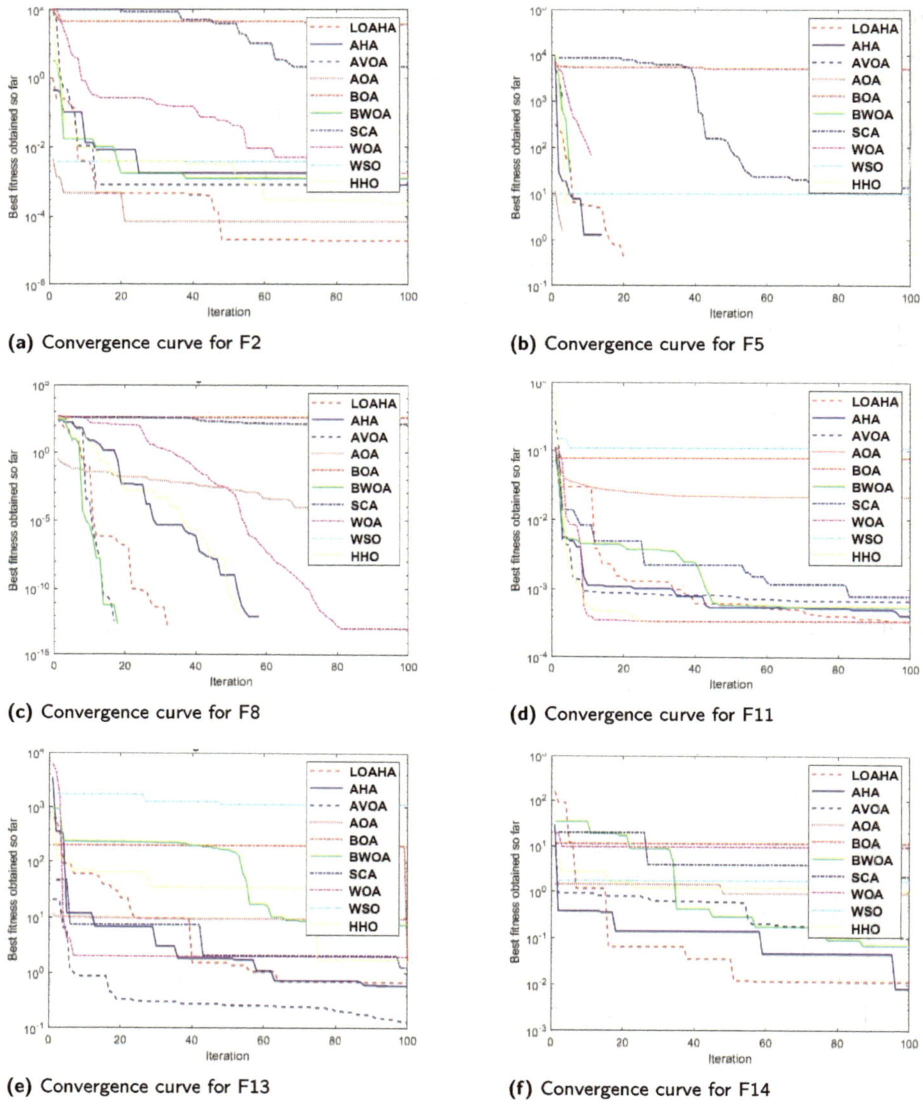

(a) Convergence curve for F2

(b) Convergence curve for F5

(c) Convergence curve for F8

(d) Convergence curve for F11

(e) Convergence curve for F13

(f) Convergence curve for F14

Figure 1: Convergence graphs of LO-AHA with basic algorithms using benchmark functions.

represents the convergence curve for the Quartic function as F_2; similarly, Figure 1(b) for the Trid10 function as F_5, Figure 1(c) for the Rastrigin function as F_8, Figure 1(d) for the Kowalik function F11, Figure 1(e) for the Perm function F_{13}, and Figure 1(f) for the PowerSum function as F_{14}.

Table 3 gives the rankings in Friedman's rank test for benchmark functions. LO-AHA ranked as 1 with a mean rank of 3.34 while WOA ranked as 7 with a mean rank of 5.68,

AHA ranked as 2 with a mean rank of 3.79, AOA ranked as 8 with a mean rank of 7.05, AVAO ranked as 4 with a mean rank of 5.08, BOA ranked as 11 with a mean rank of 10.53, BWOA ranked as 5 with a mean rank of 5.19, GWO ranked as 6 with a mean rank of 5.55, HHO ranked as 3 with a mean rank of 4.38, SCA ranked as 9 with a mean rank of 7.70, and WSAO ranked as 10 with a mean rank of 7.74. This concludes the performance of LO-AHA, which is the best among the above algorithms and has very good comparison analysis in ranking.

The convergence curve for function 5 represents the convergence of the LO-AHA algorithm with population size of 50 and maximum iterations of 100, compared with AHA, AVOA, AOA, BOA, BWOA, SCA, WOA, WSO, and HHO using different colors to plot the graphs. Similarly, for the other curves.

Thus, the randomly chosen six functions results the compression as better among the various algorithms presented for experiment.

4.4 IEEE CEC 2019 functions

To assess the effectiveness of the suggested LO-AHA, the benchmark suite for the CEC 2019 100-digit challenge is used. The objective of this benchmark suite, which consists of 10 functions, is to calculate the minimum value of each function to a 10-digit precision level without being constrained by time. The authors of [23] provide a brief explanation of these 10 functions that has been used to compare the proposed algorithm with other algorithms.

Comparison of LO-AHA with another 8 basic algorithms for IEEE CEC 2019 functions and the experimental results are recoded in Table 4 by performing functions in MATLAB code to calculate minimum (min), standard deviation (SD), and average (mean).

Quantitative analysis for IEEE CEC 2019 functions is shown in Table 5, which gives the comparison as the proposed algorithm as superior for 10 functions with SCA, 9 functions with MFO, 9 functions with WSO, 9 functions with LHHO, 9 functions with BWOA, 10 functions with LCA, 8 functions with AO, and 6 functions with BBO. Also, there is an equal comparison with various base algorithms. This discussion results as much better and overtakes the previous existing algorithms.

Table 4 shows the IEEE CEC 2019 results of the improved algorithm with SCA, MFO, AOA, and some others. As in Table 6, the ranking of LO-AHA is decided by Friedman's rank test for IEEE CEC 2019. LO-AHA ranked as 1 with a mean rank of 1.8, while SCA ranked as 7 with a mean rank of 6.2, MFO ranked as 5 with a mean rank of 4.7, WSO ranked as 8 with a mean rank of 7.6, LHHO ranked as 3 with a mean rank of 3.5, BWOA ranked as 6 with a mean rank of 6.05, LCA ranked as 9 with a mean rank of 8.3, AO ranked as 2 with a mean rank of 3.1, and BBO ranked as 4 with a mean rank of 3.75. Hence, the comparison results calculated here give better performance for the improved algorithm.

Table 4: Statistical results (mean and SD) of different algorithms on benchmark functions.

Algorithm	Index	F1	F2	F3	F4	F5	F6	F7	F8	F9	F10
LO-AHA	Mean	1.0000×10^{0}	4.2285×10^{0}	1.0031×10^{0}	9.9656×10^{0}	1.1241×10^{0}	2.0431×10^{0}	3.6033×10^{2}	2.8142×10^{0}	1.1565×10^{0}	2.0990×10^{1}
	SD	1.0000×10^{0}	4.8283×10^{0}	2.0592×10^{0}	2.8970×10^{1}	1.7516×10^{0}	4.9008×10^{0}	7.1193×10^{2}	3.9330×10^{0}	1.3193×10^{0}	2.1015×10^{1}
WSO	Mean	1.0000×10^{0}	5.0000×10^{0}	7.7972×10^{0}	6.0699×10^{1}	3.7133×10^{1}	1.0086×10^{1}	1.6638×10^{3}	4.7696×10^{0}	2.1663×10^{0}	2.1419×10^{1}
	SD	1.0000×10^{0}	5.0000×10^{0}	1.1231×10^{1}	1.1121×10^{2}	1.2967×10^{2}	1.3064×10^{1}	2.3446×10^{3}	5.2735×10^{0}	4.3253×10^{0}	2.1793×10^{1}
BWOA	Mean	1.0000×10^{0}	4.7809×10^{0}	3.7012×10^{0}	2.0564×10^{1}	5.5852×10^{0}	6.2977×10^{0}	8.7464×10^{2}	4.0720×10^{0}	1.1420×10^{0}	2.1080×10^{1}
	SD	1.0000×10^{0}	4.9848×10^{0}	5.6084×10^{0}	6.1706×10^{1}	3.0508×10^{1}	9.3038×10^{0}	1.6378×10^{3}	4.6420×10^{0}	1.9449×10^{0}	2.1361×10^{1}
SCA	Mean	1.9800×10^{6}	3.1400×10^{3}	8.8100×10^{0}	4.5200×10^{1}	8.1600×10^{0}	7.4300×10^{0}	1.5000×10^{3}	4.3900×10^{0}	1.5500×10^{0}	2.1400×10^{1}
	SD	2.4100×10^{6}	1.7500×10^{3}	1.3600×10^{0}	5.6800×10^{0}	2.0600×10^{0}	1.0700×10^{0}	2.4200×10^{2}	2.2600×10^{-1}	1.5300×10^{-1}	6.4500×10^{-2}
MFO	Mean	1.0800×10^{7}	2.0300×10^{3}	6.7200×10^{0}	2.7900×10^{1}	2.0900×10^{0}	5.3600×10^{0}	9.2100×10^{2}	4.5900×10^{0}	1.3400×10^{0}	2.1200×10^{1}
	SD	1.8600×10^{7}	2.4500×10^{3}	2.1500×10^{0}	7.1500×10^{0}	2.7000×10^{0}	1.8400×10^{0}	3.8000×10^{2}	3.6500×10^{-1}	1.5900×10^{-1}	1.5300×10^{-1}
LHHO	Mean	1.0000×10^{0}	4.4088×10^{0}	1.4098×10^{0}	1.7309×10^{1}	1.4538×10^{0}	2.7822×10^{0}	5.3012×10^{2}	4.0115×10^{0}	1.1131×10^{0}	2.0981×10^{1}
	SD	1.0000×10^{0}	4.9308×10^{0}	2.8604×10^{0}	3.4975×10^{1}	2.0067×10^{0}	6.7175×10^{0}	1.0532×10^{3}	4.4730×10^{0}	1.4208×10^{0}	2.1031×10^{1}
LCA	Mean	6.1400×10^{4}	1.5900×10^{2}	1.1900×10^{1}	1.1600×10^{2}	1.2500×10^{2}	1.3300×10^{1}	2.6000×10^{3}	5.2200×10^{0}	4.6800×10^{0}	2.1700×10^{1}
	SD	7.0300×10^{4}	1.2100×10^{2}	7.4500×10^{-1}	1.9200×10^{1}	3.5300×10^{1}	1.4300×10^{0}	2.2100×10^{2}	1.6700×10^{-1}	6.3400×10^{-1}	1.3100×10^{-1}
AOA	Mean	1.0000×10^{0}	5.0000×10^{0}	5.5800×10^{0}	3.0600×10^{1}	1.7600×10^{0}	4.9800×10^{0}	9.9600×10^{2}	4.3300×10^{0}	1.4400×10^{0}	1.8400×10^{1}
	SD	0.0000×10^{0}	0.0000×10^{0}	1.3400×10^{0}	1.1000×10^{1}	1.3800×10^{-1}	1.0100×10^{0}	3.2800×10^{2}	3.0500×10^{-1}	1.5800×10^{-1}	6.4700×10^{0}
BBO	Mean	1.0000×10^{0}	4.9700×10^{0}	1.9200×10^{0}	5.0900×10^{1}	2.4700×10^{1}	7.4300×10^{0}	1.0900×10^{3}	4.5400×10^{0}	1.3000×10^{0}	2.1000×10^{1}
	SD	0.0000×10^{0}	1.4300×10^{-1}	4.4200×10^{-1}	7.6700×10^{0}	1.0800×10^{1}	9.3000×10^{-1}	2.0100×10^{2}	2.6800×10^{-2}	8.2600×10^{-2}	1.9900×10^{-1}

Table 5: Performance comparison of algorithms based on superiority, equality, and inferiority.

Criteria	SCA	MFO	WSO	LHHO	BWOA	LCA	AO	BBO
Superior	10	9	9	9	9	10	8	6
Equal	0	0	1	1	1	0	1	1
Inferior	0	1	0	0	0	0	1	3

Table 6: Mean rank and final ranking of the compared algorithms.

Index	LO-AHA	SCA	MFO	WSO	LHHO	BWOA	LCA	AO	BBO
Mean Rank	1.8	6.2	4.7	7.6	3.5	6.05	8.3	3.1	3.75
Rank	1	7	5	8	3	6	9	2	4

The discussion ends with the calculation of various function values on the benchmark and performing IEEE CEC 2019. The improved variant shows greater accuracy toward the optimal values.

5 Conclusion

In order to create a resilient, robust, and effective metaheuristic algorithm, the current study proposes an improved algorithm called LO-AHA, which expands on the AHA approach. In LO-AHA, the modified mutualism phase incorporated with the Laplacian Crossover Operator was employed to deepen the intensification process close to the prospective solutions. The guided phase of AHA updated with the crossover operator for the better search space exploration and also for exploitation. The algorithm's diversity and intensity were better balanced by using the self-adaptive parameter setting. The IEEE CEC 2019 benchmark suite and forty single objective classical benchmark functions are solved by the suggested LO-AHA. The results are contrasted with a few advanced metaheuristics currently in use for performance analysis. The evaluation's findings show that the suggested LO-AHA works better than the algorithms that were compared. The Friedman rank test is utilized to ascertain whether the recommended approach is statistically superior. The convergence graphs show that the convergence rate is faster for LO-AHA in comparison to other base algorithms with best outcomes.

The LO-AHA technique is promoted as a superior approach that may be used to handle challenging engineering optimization problems by all of the findings, remarks, and analyses. For future work, the proposed algorithm can be further modified by introducing more novelty of operators that will improve the exploration capability and then it can be applied to solve various critical real-world engineering problems such as car side impact, welded beam design, pressure vessel design, and spacecraft trajectory design problems. LOAHA can also be extended to multimodal techniques, hybrid operators, and be combined with more relatable techniques, which deal with complex problems and can be applied to the multiobjective issues to maintain the balance between exploration and exploitation. It will also serve as the gateway to the problems that are generally structured in their local extrema part.

Appendix. Benchmark functions

Table 7: Benchmark functions used for the experimental results [16].

Name	C	Function	Dim	Range	f_{opt}	Type
F_1	US	$F_1(\mathbf{x}) = 25 + \sum_{i=1}^{n} \lfloor x_i \rfloor$	5	[−5.12, 5.12]	0	Stepint
F_2	US	$F_2(\mathbf{x}) = \sum_{i=1}^{n} i x_i^4 + \text{rand}[0, 1)$	30	[−1.28, 1.28]	0	Quartic
F_3	UN	$F_3(\mathbf{x}) = -\cos(x_1)\cos(x_2)\exp[-(x_1-\pi)^2 - (x_2-\pi)^2]$	2	[−100, 100]	−1	Easom
F_4	UN	$F_4(\mathbf{x}) = 0.26(x_1^2 + x_2^2) - 0.48 x_1 x_2$	2	[−10, 10]	0	Matyas
F_5	UN	$F_5(\mathbf{x}) = \sum_{i=1}^{n}(x_i - 1)^2 + \sum_{i=2}^{n} x_i x_{i-1}$	10	$[-D^2, D^2]$	−210	Trid10
F_6	UN	$F_6(\mathbf{x}) = \sum_{i=1}^{n} \lvert x_i \rvert + \prod_{i=1}^{n} \lvert x_i \rvert$	30	[−10, 10]	0	Schwefel 2.22
F_7	MS	$F_7(\mathbf{x}) = (x_2 - \frac{5.1}{4\pi^2} x_1^2 + \frac{5}{\pi} x_1 - 6)^2 + 10(1 - \frac{1}{8\pi}) \cos x_1 + 10$	2	[−5.10] × [0.15]	0.398	Branin
F_8	MS	$F_8(\mathbf{x}) = -\sum_{i=1}^{n} [x_i^2 - 10\cos(2\pi x_i) + 10]$	30	[−5.12, 5.12]	0	Rastrigin
F_9	MS	$F_9(\mathbf{x}) = -\sum_{i=1}^{n} \sin(x_i)[\sin(\frac{i x_i^2}{\pi})]^{20}$	10	$[0, \pi]$	−9.6602	Michalewicz 10
F_{10}	MN	$F_{10}(\mathbf{x}) = x_1^2 + 2x_2^2 - 0.3\cos(3\pi x_1 + 4\pi x_2) + 0.3$	2	[−100, 100]	0	Bohachevsky 3
F_{11}	MN	$F_{11}(\mathbf{x}) = \sum_{i=1}^{11} \lvert a_i - \frac{x_1(b_i^2 + b_i x_2)}{b_i^2 + b_i x_3 + x_4} \rvert$	4	[−5, 5]	0.00031	Kowalik
F_{12}	MN	$F_{12}(\mathbf{x}) = -\sum_{i=1}^{7} [(\mathbf{x} - \mathbf{a}_i)(\mathbf{x} - \mathbf{a}_i)^T + c_i]^{-1}$	4	[0, 10]	−10.4028	Shekel 7
F_{13}	MN	$F_{13}(\mathbf{x}) = \sum_{k=1}^{n} (\sum_{i=1}^{n} (i^k + \beta)[(\frac{x_i}{i})^k - 1])^2$	4	$[-D, D]$	0	Perm
F_{14}	MN	$F_{14}(\mathbf{x}) = \sum_{k=1}^{n} (\sum_{i=1}^{n} x_i^k - b_k)^2$	4	$[0, D]$	0	PowerSum
F_{15}	MN	$F_{15}(\mathbf{x}) = -\sum_{i=1}^{4} \exp[-\sum_{j=1}^{6} a_{ij}(x_j - p_{ij})^2]$	6	[0, 1]	−3.32	Hartman 6

Bibliography

[1] Storn, R., & Price, K. (1997). Differential evolution—a simple and efficient heuristic for global optimization over continuous spaces. *Journal of Global Optimization, 11*(4), 341–359.

[2] Eberhart, R., & Kennedy, J. (1995). A new optimizer using particle swarm theory. In *Proceedings of the Sixth International Symposium on Micro Machine and Human Science (MHS'95)* (pp. 39–43). IEEE.

[3] Hu, G., Zhong, J., Zhao, C., Wei, G., & Chang, C.-T. (2023). LCAHA: a hybrid artificial hummingbird algorithm with multi-strategy for engineering applications. *Computer Methods in Applied Mechanics and Engineering, 415*, 116238.

[4] Rashedi, E., Nezamabadi-Pour, H., & Saryazdi, S. (2009). GSA: a gravitational search algorithm. *Information Sciences*, *179*(13), 2232–2248.

[5] Rao, R. V., Savsani, V. J., & Vakharia, D. P. (2012). Teaching–learning-based optimization: an optimization method for continuous non-linear large scale problems. *Information Sciences*, *183*(1), 1–15.

[6] Yang, X. S., & Deb, S. (2009). Cuckoo search via Lévy flights. In *Proceedings of the World Congress on Nature & Biologically Inspired Computing (NaBIC)* (pp. 210–214). IEEE.

[7] Karaboga, D., & Akay, B. (2009). A comparative study of artificial bee colony algorithm. *Applied Mathematics and Computation*, *214*(1), 108–132.

[8] Saha, A. K. (2022). Multi-population-based adaptive sine cosine algorithm with modified mutualism strategy for global optimization. *Knowledge-Based Systems*, *251*, 109326.

[9] Wu, L., Huang, X., Cui, J., Liu, C., & Xiao, W. (2023). Modified adaptive ant colony optimization algorithm for mobile robot path planning. *Expert Systems with Applications*, *215*, 119410.

[10] Wang, C., & Song, W. (2019). A modified particle swarm optimization algorithm based on velocity updating mechanism. *Ain Shams Engineering Journal*, *10*(4), 847–866.

[11] Hemanth, D. J., & Anitha, J. (2019). Modified genetic algorithm approaches for classification of abnormal MRI brain tumour images. *Applied Soft Computing*, *75*, 21–28.

[12] Garg, V., Deep, K., & Bansal, S. (2023). Improved teaching learning algorithm with Laplacian operator for nonlinear optimization. *Engineering Applications of Artificial Intelligence*, *124*, 106549.

[13] Sangeetha, V., Krishankumar, R., Ravichandran, K. S., Cavallaro, F., Kar, S., Pamucar, D., & Mardani, A. (2021). A fuzzy gain-based dynamic ant colony optimization for path planning in dynamic environments. *Symmetry*, *13*, 280.

[14] Tunga, H., Kar, S., & Giri, D. (2024). Multi-objective brainstorm optimization algorithm based on Bayesian inference learning automaton. *Applied Soft Computing*, *162*, 111755.

[15] Das, M., Roy, A., Maity, S., & Kar, S. (2023). A quantum-inspired ant colony optimization for solving a sustainable four-dimensional traveling salesman problem under type-2 fuzzy variable. *Advanced Engineering Informatics*, *55*, 101816.

[16] Zhao, W., Wang, L., & Mirjalili, S. (2022). Artificial hummingbird algorithm: a new bio-inspired optimizer with its engineering applications. *Computer Methods in Applied Mechanics and Engineering*, *388*, 114194.

[17] Hu, G., Zhong, J., Zhao, C., Wei, G., & Chang, C.-T. (2023). LCAHA: a hybrid artificial hummingbird algorithm with multi-strategy for engineering applications. *Computer Methods in Applied Mechanics and Engineering*, *415*, 116238.

[18] Abd Elaziz, M., Dahou, A., Al-Betar, M. A., El-Sappagh, S., Oliva, D., & Aseeri, A. O. (2023). Quantum artificial hummingbird algorithm for feature selection of social IoT. *IEEE Access*, *11*, 66257–66278.

[19] Zhao, W., Zhang, Z., Mirjalili, S., Wang, L., Khodadadi, N., & Mirjalili, S. M. (2022). An effective multi-objective artificial hummingbird algorithm. *Computer Methods in Applied Mechanics and Engineering*, *398*, 115223.

[20] Deep, K., & Thakur, M. (2007). A new mutation operator for real coded genetic algorithms. *Applied Mathematics and Computation*, *193*(1), 211–230.

[21] Abualigah, L., Diabat, A., Mirjalili, S., Abd Elaziz, M., & Gandomi, A. H. (2021). The arithmetic optimization algorithm. *Computer Methods in Applied Mechanics and Engineering*, *376*, 113609.

[22] Braik, M., Hammouri, A., Atwan, J., Al-Betar, M. A., & Awadallah, M. A. (2022). White shark optimizer: a novel bio-inspired meta-heuristic algorithm. *Knowledge-Based Systems*, *243*, 108457.

[23] Lu, J., Zhou, X., Ma, Y., Wang, M., Wan, J., & Wang, W. (2019). A novel artificial bee colony algorithm for CEC 2019 benchmark problems. In *Proceedings of the IEEE Congress on Evolutionary Computation (CEC)* (pp. 387–394).

Navneet Rana, Rakesh Kumar, and Abhijit Sarkar

Bifurcation analysis of a planktonic system with harvesting of infected zooplankton

Abstract: The phytoplankton and zooplankton system is a cornerstone of aquatic ecosystems. Their interactions and roles in nutrient cycling, carbon sequestration, and food webs dynamics illustrate their profound impact on both marine and terrestrial environments. The phytoplankton are infected directly by external toxins such as industrial waste, agricultural pesticides, and heavy metals. The zooplankton are influenced by the ingestion of infected phytoplankton. In this study, we investigate the two-species interplay, with the assumption that the zooplankton population is further divided into two types – susceptible zooplankton and infected zooplankton. It is also postulated that the growth rate of phytoplankton and intraspecific competition rates of phytoplankton are equal. This study introduces environmental toxins and harvesting of infected zooplankton as a novel strategy to manage disease outbreaks and sustain stability in the planktonic system. The existence of various equilibrium points, and their stability analysis has been explored. The system exhibits different types of bifurcation such as transcritical bifurcation and Hopf bifurcation. The numerical simulations are performed using MATLAB to confirm the analytical findings.

Keywords: Phytoplankton, susceptible zooplankton, infected zooplankton, stability analysis, transcritical bifurcation, environmental toxins

MSC 2020: 92D25, 92D30, 34D20, 34C23, 37G15, 92D40

1 Introduction

Phytoplankton and zooplankton create a foundation for marine food webs. Phytoplankton are at the bottom of the aquatic trophic chain. Due to the presence of chlorophyll pigment in the cells, phytoplankton grow autotrophically in aquatic environments.

Acknowledgement: The present work was completed in the Numerical analysis lab of Shaheed Bhagat Singh State University, Ferozepur, Punjab. We are grateful for the assistance given by the staff to finish this study.

Navneet Rana, Department of Applied Sciences and Humanities, Shaheed Bhagat Singh State University, Ferozepur, 152004, Punjab, India; and PG Department of Mathematics, Guru Nanak College, Sri Muktsar Sahib, 152026, Punjab, India, e-mail: rana.navneet2@gmail.com
Rakesh Kumar, Department of Applied Sciences and Humanities, Shaheed Bhagat Singh State University, Ferozepur, 152004, Punjab, India, e-mail: rakesh@sbsstc.ac.in
Abhijit Sarkar, Department of Mathematics, SRM Institute of Science and Technology, SRM Nagar, Kattankulathur, 603203, Tamil Nadu, India, e-mail: abhijit2012sarkar@gmail.com

https://doi.org/10.1515/9783112234525-009

These unicellular organisms serve as energy sources, with energy flowing via food webs to higher trophic levels. Any changes in these main producers will have an impact on the entire food chain structure. Environmental toxins can have severe and far-reaching impacts on both phytoplankton and zooplankton, disrupting aquatic ecosystems and the larger food web [1]. Phytoplankton are often the first to absorb environmental toxins. Zooplankton that consume contaminated phytoplankton ingest these toxins [2, 3]. Factors like pollution, changes in water temperature, and low oxygen levels can weaken zooplankton and make them more prone to infection. Some bacteria and viruses can infect zooplankton, either directly harming them or causing them to spread the pathogens within aquatic food webs. A decrease in healthy zooplankton populations can disrupt the availability of food for higher tropic levels, ultimately affecting fish, birds, and marine mammals. Understanding the dynamics between susceptible and infected zooplankton is essential for comprehending disease spread in aquatic ecosystems, as well as its potential ripple effects through various trophic levels. Several mathematical models have been proposed by researchers to investigate the underlying dynamics of planktonic systems [4–7]. Kumar et al. [8] have analyzed how toxins released by phytoplankton influence the stability and interaction patterns in the phytoplankton–zooplankton model. Panja et al. [9] showed that the stability of a plankton–fishery system is affected by environmental toxins, phytoplankton-released toxins, and fish harvesting. Mandal et al. [10] explored the role of surrounding pollutants in shaping the phytoplankton–zooplankton system, with the assumption that zooplankton survive not only on phytoplankton but also on alternative food resources. Mandal et al. [11] analyzed the contaminated environmental impacts of refuge patterns of phytoplankton species. Harvesting of species from the aquatic ecosystem is important for the sea food industry [12, 13]. Harvesting infected zooplankton typically involves collecting zooplankton specimens from water bodies where they may naturally carry infections, or it may involve experimental infection within controlled environments for research purposes. By studying infected zooplankton, researchers can gain valuable insights into the broader effects of infections on aquatic ecosystems, the spread of aquatic pathogens, and the resilience of food webs under changing environmental conditions [14, 15]. The study of delay models is essential as it incorporates time lags in biological processes, providing a realistic framework to analyze stability, oscillations, and long-term dynamics in population systems [16–21].

In this chapter, we have extended the work of [22, 23] by assuming that the growth rate and intraspecific rate of phytoplankton are equal. Secondly, we have considered the harvesting of infected zooplankton species only. Our main goal is to reveal insights into the system dynamics based on the following facts:

1. To explore how the harvesting of infected zooplankton helps to stabilize the system.
2. To examine the occurrence of transcritical and Hopf bifurcations.
3. To carry out a sensitivity analysis for inspecting the positive and negative impact of parameters.
4. To study the role of other important parameters on the system.

The organization of the study is outlined as follows: the next section describes the structure of the mathematical model that includes the phytoplankton, susceptible zooplankton, and infected zooplankton. The mathematical analysis of the model has been performed in the third section, which includes positivity, boundedness, the existence of equilibrium points, stability analysis, and bifurcation analysis. In the fourth section, numerical simulation is performed to verify the analytical results. The chapter ends with a discussion.

2 The mathematical model

The mathematical structure of the system consists of two populations – phytoplankton and zooplankton. The zooplankton population is further classified into two categories as susceptible zooplankton and infected zooplankton. Let $x(t)$ be the phytoplankton density, $y_1(t)$ be the susceptible zooplankton density, and $y_2(t)$ be the infected zooplankton density at any time t. The structure is constructed on the basis of the following assumptions:

1. It is assumed that r is the growth rate of prey. The intraspecific competition rate is also assumed to be r. This implies that the phytoplankton population has reached a stable equilibrium. At this point, the phytoplankton population neither increases nor decreases significantly.
2. It is assumed that both forms of zooplankton consume phytoplankton following a Holling type-II functional response.
3. It is assumed that environmental toxins directly impact the phytoplankton, and zooplankton that feed on infected phytoplankton are impacted indirectly.
4. It is assumed that infected zooplankton are harvested at the rate E.

On the basis of these postulates, the mathematical model is presented as

$$\begin{cases} \frac{dx}{dt} = r(1-x)x - \frac{\alpha x y_1}{K_1+x} - \frac{\beta x y_2}{K_2+x} - \mu_1 x^3, \\ \frac{dy_1}{dt} = \frac{\alpha_1 x y_1}{K_1+x} - \delta_1 y_1 - \mu_2 {y_1}^2, \\ \frac{dy_2}{dt} = \frac{\beta_1 x y_2}{K_2+x} - (\delta_2 + E) y_2 - \mu_3 {y_2}^2. \end{cases} \tag{1}$$

The initial conditions are defined as: $x(0) > 0, y_1(0) > 0$, and $y_2(0) > 0$. The explanation of the parameters involved in model (1), along with their assumed values are given in Table 1. The term $\mu_1 x^3$ in the first equation represents the infection of external toxic substances on phytoplankton species. The infection in susceptible and infected zooplankton species is represented by terms $\mu_2 {y_1}^2$ and $\mu_3 {y_2}^2$, respectively.

Table 1: The explanation of parameters.

Parameter	Explanation	Values
r	Phytoplankton growth rate	1.5
a	Capture rate by susceptible zooplankton	1.61
β	Capture rate by infected zooplankton	0.81
K_1	Half-saturation constant of susceptible zooplankton	2
K_2	Half-saturation constant of infected zooplankton	0.4
a_1	Conversion rate of susceptible zooplankton	1.53
β_1	Conversion rate of infected zooplankton	0.2
δ_1	Mortality rate of susceptible zooplankton	0.02
δ_2	Mortality rate of infected zooplankton	0.008
E	Harvesting effort of infected zooplankton	0.01
μ_1	Environmental toxin effect on phytoplankton	0.3
μ_2	Environmental toxin effect on susceptible zooplankton	0.015
μ_3	Environmental toxin effect on infected zooplankton	0.012

3 Analysis of the model (1)

The present section deals with some important dynamical properties of the system such as the nonnegativity and boundedness of solutions. The presence of feasible equilibrium points, and their stability is explored using a proper analytical technique.

3.1 Positivity and boundedness

The first equation of the model (1) describes the expression as

$$\frac{dx}{x} = \left(r - rx - \frac{ay_1}{K_1 + x} - \frac{\beta y_2}{K_2 + x} - \mu_1 x^2\right)dt.$$

On integrating this mathematical expression over the interval $[0, t]$, we obtain

$$x(t) = x(0)\exp\left\{\int_0^t \left(r - rx - \frac{ay_1}{K_1 + x} - \frac{\beta y_2}{K_2 + x} - \mu_1 x^2\right)dt\right\} > 0.$$

In a similar way, the second and third equations of the model (1) can be solved to obtain $y_1(t) > 0$ and $y_2(t) > 0$.

Next, we carry out the analysis for the boundedness of the solution. For this, the first equation of the system (1) can be restructured as $\frac{dx}{dt} \leq rx(1 - x)$, which gives $\limsup_{t\to\infty} x(t) \leq 1$. Further, consider the function

$$N(t) = x(t) + \frac{a}{a_1}y_1(t) + \frac{\beta}{\beta_1}y_2(t),$$

to obtain the result $\frac{dN}{dt} + kN \leq 2r - \mu_1$, where $k = \min\{r, \frac{a\delta_1}{a_1}, \frac{\beta(\delta_2+E)}{\beta_1}\}$. Applying the result of the theorem of differential inequalities [24], we obtain

$$0 < N(t) \leq \frac{N(x(0), y_1(0), y_2(0))}{e^{kt}} + \frac{2r - \mu_1}{k}.$$

As $t \to \infty$, we obtain

$$0 \leq N(t) \leq \frac{2r - \mu_1}{k}.$$

Thus, the system admits all ultimately bounded solutions.

3.2 Equilibrium points

The possible equilibrium states of the model (1) are described as:

1. Trivial equilibrium point, $V_0(0, 0, 0)$: In this state, no population exists in the system.
2. Predator-free equilibrium point, $V_1(x_1, 0, 0)$: This state exists when only the prey population is present and they interact with each other. The phytoplankton density $x(t)$ satisfies the relation

$$\mu_1 x^2 + rx - r = 0.$$

   The values of x are given as: $\frac{-r \pm \sqrt{r^2+4\mu_1 r}}{2\mu_1}$. The phytoplankton density x is always positive in $\frac{-r+\sqrt{r^2+4\mu_1 r}}{2\mu_1}$. This indicates that the prey population alone can sustain itself in the environment.
3. Infected zooplankton-free equilibrium point, $V_2(x', y_1', 0)$: This state accounts for the interplay between phytoplankton and susceptible zooplankton in the system. The value of y_1' is given by the relation $y_1' = \frac{1}{\mu_2}(\frac{a_1 x'}{K_1+x'} - \delta_1)$, and x' satisfies the equation

$$A_1 x'^4 + A_2 x'^3 + A_3 x'^2 + A_4 x' + A_5 = 0, \tag{2}$$

   where

$$\begin{cases} A_1 = \mu_1 \mu_2 > 0, \\ A_2 = r\mu_2 + 2\mu_1\mu_2 K_1 > 0, \\ A_3 = \mu_1\mu_2 {K_1}^2 - r\mu_2 + 2r\mu_2 K_1, \\ A_4 = a a_1 - a\delta_1 + \mu_2 r {K_1}^2 - 2r\mu_2 K_1, \\ A_5 = -(a\delta_1 K_1 + r\mu_2 {K_1}^2) < 0. \end{cases}$$

   Equation (2) has one positive root if $A_3 > 0$ and $A_4 > 0$, or $A_3 < 0$ and $A_4 < 0$, or $A_3 > 0$ and $A_4 < 0$, or three positive roots if $A_3 < 0$ and $A_4 > 0$. The infected zooplankton-free equilibrium biologically means a disease-free steady state of the ecosystem, where only healthy zooplankton survive along with phytoplankton.

4. Susceptible zooplankton-free equilibrium point, $V_3(x'', 0, y_2'')$: This state gives the interplay between phytoplankton and infected zooplankton in the system. The susceptible zooplankton density y_2'' is given by the expression

$$y_2'' = \frac{1}{\mu_3}\left(\frac{\beta_1 x''}{K_2 + x''} - (\delta_2 + E)\right)$$

and x'' satisfies the equation

$$B_1 x''^4 + B_2 x''^3 + B_3 x''^2 + B_4 x'' + B_5 = 0, \tag{3}$$

where

$$\begin{cases} B_1 = \mu_1\mu_3 > 0, \\ B_2 = r\mu_3 + 2\mu_1\mu_3 K_2 > 0, \\ B_3 = \mu_1\mu_3 {K_2}^2 - r\mu_3 + 2r\mu_3 K_2, \\ B_4 = \beta\beta_1 - \beta\delta_2 - \beta E + \mu_3 r {K_2}^2 - 2r\mu_3 K_2, \\ B_5 = -(\beta\delta_2 K_2 + \beta K_2 E + r\mu_3 {K_2}^2) < 0. \end{cases}$$

Equation (3) has one positive root if $B_3 > 0$ and $B_4 > 0$, or $B_3 > 0$ and $B_4 < 0$, or $B_3 < 0$ and $B_4 < 0$, or three positive roots if $B_3 < 0$ and $B_4 > 0$. The susceptible zooplankton-free equilibrium biologically represents a disease-dominated state, where infection eradicated the healthy zooplankton population, and only infected individuals persist along with phytoplankton.

5. Coexisting equilibrium point, $V_4(x^*, {y_1}^*, {y_2}^*)$: This steady state involves all the three populations interacting with each other. Here, $(x^*, {y_1}^*, {y_2}^*)$ are solutions of the equations

$$\begin{cases} rx - rx^* - \frac{\alpha {y_1}^*}{K_1 + x^*} - \frac{\beta {y_2}^*}{K_2 + x^*} - \mu_1 {x^*}^2 = 0, \\ \frac{\alpha_1 x^*}{K_1 + x^*} - \delta_1 - \mu_2 {y_1}^* = 0, \\ \frac{\beta_1 x^*}{K_2 + x^*} - (\delta_2 + E) - \mu_3 {y_2}^* = 0. \end{cases}$$

On simplifying these equations, the values of ${y_1}^*$ and ${y_2}^*$ are given as $\frac{1}{\mu_2}(\frac{\alpha_1 x^*}{K_1 + x^*} - \delta_1)$, and $\frac{1}{\mu_3}(\frac{\beta_1 x^*}{K_2 + x^*} - (\delta_2 + E))$, respectively. The value of x^* satisfies the expression

$$r - rx^* - \frac{\alpha}{\mu_2}\left[\frac{\alpha_1 x^* - \delta_1(K_1 + x^*)}{(K_1 + x^*)^2}\right] - \frac{\beta}{\mu_3}\left[\frac{\beta_1 x^* - (\delta_2 + E)(K_2 + x^*)}{(K_2 + x^*)^2}\right] - \mu_1 {x^*}^2 = 0.$$

This is the most biologically meaningful equilibrium because it represents a sustainable ecosystem where none of the populations are wiped out.

3.3 Stability analysis

The stability analysis reveals whether an equilibrium state is biologically feasible, indicating if populations will persist, collapse, or coexist under small ecological disturbances. To carry out the stability analysis, we first linearize the system around each steady state using the Jacobian matrix. The sign of the eigenvalues of the Jacobian matrix helps us to determine whether the steady state is stable or unstable.

1. At the trivial equilibrium state V_0, the Jacobian matrix has eigenvalues as r, $-\delta_1$ and $-(\delta_2 + E)$. The steady-state V_0 is unstable as it is a saddle point.
2. The eigenvalues of the Jacobian matrix at the predator-free steady state V_1 are $r - 2rx_1 - 3\mu_1 {x_1}^2$, $\frac{\alpha_1 x_1}{K_1+x_1} - \delta_1$, and $\frac{\beta_1 x_1}{K_2+x_1} - (\delta_2 + E)$. The steady state V_1 is LAS if $r < 2rx_1 + 3\mu_1 {x_1}^2$, $\frac{\alpha_1 x_1}{K_1+x_1} < \delta_1$, and $\frac{\beta_1 x_1}{K_2+x_1} < (\delta_2 + E)$. Biologically, the stability of the predator-free equilibrium indicates that the predator's death rate exceeds its growth rate, preventing its survival and allowing only the prey population to persist.
3. At the infected zooplankton-free equilibrium point, $V_2(x', y_1', 0)$, the Jacobian matrix takes the form

$$\mathbf{J}_{\mathbf{V}_2} = \begin{bmatrix} A_{11} & A_{12} & A_{13} \\ A_{21} & A_{22} & 0 \\ 0 & 0 & A_{33} \end{bmatrix},$$

where

$$\begin{cases} A_{11} = r - 2rx' - \frac{\alpha K_1 y_1'}{(K_1+x')^2} - 3\mu_1 x'^2, \\ A_{12} = -\frac{\alpha x'}{K_1+x'}, \\ A_{13} = -\frac{\beta x'}{K_2+x'}, \\ A_{21} = \frac{\alpha_1 K_1 y_1'}{(K_1+x')^2}, \\ A_{22} = -\mu_2 y_1', \\ A_{33} = \frac{\beta_1 x'}{K_2+x'} - (\delta_2 + E). \end{cases}$$

At J_{V_2}, the characteristic equation is calculated as

$$(A_{33} - \lambda)(\lambda^2 - \gamma_1 \lambda + \gamma_2) = 0, \tag{4}$$

where $\gamma_1 = A_{11} + A_{22}$ and $\gamma_2 = A_{11}A_{22} - A_{12}A_{21}$. Among the eigenvalues, one is A_{33} and the remaining two are roots of the equation $\lambda^2 - \gamma_1 \lambda + \gamma_2 = 0$.
Let the quadratic equation have roots of λ_1 and λ_2. Then, $\lambda_1 + \lambda_2 = \gamma_1 = A_{11} + A_{22}$ and $\lambda_1 \lambda_2 = \gamma_2 = A_{11}A_{22} - A_{12}A_{21}$. By the result of the criterion of Routh–Hurwitz [25], equation (4) has roots, whose real part is negative if and only if $A_{33} < 0$, $A_{11} + A_{22} < 0$, and $A_{11}A_{22} - A_{12}A_{21} > 0$.

Thus, for the equilibrium state $V_2(x', y_1', 0)$ to be LAS, the conditions are

$$\begin{cases} \frac{\beta_1 x'}{K_2+x'} < (\delta_2 + E), \\ r < 2rx' + \frac{aK_1 y_1'}{(K_1+x')^2} + 3\mu_1 x'^2 + \max\{\mu_2 y_1', \frac{a\alpha_1 K_1 x'}{\mu_2 (K_1+x')^3}\}. \end{cases}$$

These represent the stability conditions for the infected zooplankton-free equilibrium point V_2. The stability of the infected zooplankton-free equilibrium signifies that the disease cannot invade the system, and the healthy zooplankton population along with phytoplankton persists without infection.

4. At the susceptible zooplankton-free equilibrium point $V_3(x'', 0, y_2'', 0)$, the Jacobian matrix takes the form

$$\mathbf{J}_{\mathbf{V}_3} = \begin{bmatrix} B_{11} & B_{12} & B_{13} \\ 0 & B_{22} & 0 \\ B_{31} & 0 & B_{33} \end{bmatrix},$$

where

$$\begin{cases} B_{11} = r - 2rx'' - \frac{\beta K_2 y_2''}{(K_2+x'')^2} - 3\mu_1 x''^2, \\ B_{12} = -\frac{ax''}{K_1+x''}, \\ B_{13} = -\frac{\beta x''}{K_2+x''}, \\ B_{22} = \frac{\alpha_1 x''}{K_1+x''} - \delta_1, \\ B_{31} = \frac{\beta_1 K_2 y_2''}{(K_2+x'')^2}, \\ B_{33} = -\mu_3 y_2''. \end{cases}$$

At J_{V_3}, the characteristic equation is calculated as

$$(B_{22} - \lambda)(\lambda^2 - \eta_1 \lambda + \eta_2) = 0, \tag{5}$$

where $\eta_1 = B_{11} + B_{33}$ and $\gamma_2 = B_{11}B_{33} - B_{13}B_{31}$. Among the eigenvalues, one is B_{22} and the remaining two are roots of the equation $\lambda^2 - \eta_1\lambda + \eta_2 = 0$.

Let λ_1 and λ_2 represent the roots of the quadratic term. Then, $\lambda_1 + \lambda_2 = \eta_1 = B_{11} + B_{33}$ and $\lambda_1\lambda_2 = \eta_2 = B_{11}B_{33} - B_{13}B_{31}$. By the result of the criterion of Routh–Hurwitz [25], equation (5) has roots, whose real part is negative if and only if $B_{22} < 0$, $B_{11} + B_{33} < 0$ and $B_{11}B_{33} - B_{13}B_{31} > 0$.

Thus, for the steady state $V_3(x'', 0, y_2'')$ to be LAS, the conditions are

$$\begin{cases} \frac{\alpha_1 x''}{K_1+x''} < \delta_1, \\ r < 2rx'' + \frac{\beta K_2 y_2''}{(K_2+x'')^2} + 3\mu_1 x''^2 + \max\{\mu_3 y_2'', \frac{\beta\beta_1 K_2 x''}{\mu_3 (K_2+x'')^3}\}. \end{cases}$$

These are the stability conditions for the susceptible zooplankton-free equilibrium state V_3. The stability of the susceptible zooplankton-free equilibrium implies that infection dominates the zooplankton population, preventing the recovery of healthy individuals and allowing only infected zooplankton to persist.

5. At the coexisting equilibrium point, V_4, the Jacobian matrix has the value

$$\mathbf{J}_{\mathbf{V}_4} = \begin{bmatrix} C_{11} & C_{12} & C_{13} \\ C_{21} & C_{22} & 0 \\ C_{31} & 0 & C_{33} \end{bmatrix},$$

where

$$\begin{cases} C_{11} = r - 2rx^* - \frac{\alpha K_1 y_1{}^*}{(K_1+x^*)^2} - \frac{\beta K_2 y_2{}^*}{(K_2+x^*)^2} - 3\mu_1 x^{*2}, \\ C_{12} = -\frac{\alpha x^*}{K_1+x^*}, \\ C_{13} = -\frac{\beta x^*}{K_2+x^*}, \\ C_{21} = \frac{\alpha_1 K_1 y_1{}^*}{(K_1+x^*)^2} \\ C_{22} = -\mu_2 y_1{}^*, \\ C_{31} = \frac{\beta_1 K_2 y_2{}^*}{(K_2+x^*)^2}, \\ C_{33} = -\mu_3 y_2{}^*. \end{cases}$$

The characteristic equation at J_{V_4} is computed as

$$\lambda^3 + \Omega_1\lambda^2 + \Omega_2\lambda + \Omega_3 = 0, \tag{6}$$

where the coefficients are noted as

$$\Omega_1 = -(C_{11} + C_{22} + C_{33}),$$
$$\Omega_2 = C_{11}C_{22} + C_{11}C_{33} + C_{22}C_{33} - C_{12}C_{21} - C_{13}C_{31},$$
$$\Omega_3 = C_{11}C_{22}C_{33} - C_{12}C_{21}C_{33} - C_{13}C_{31}C_{22}.$$

The steady point $V_4(x^*, y_1{}^*, y_2{}^*)$ is LAS if $\Omega_1\Omega_2 > \Omega_3$. If $\Omega_1\Omega_2 = \Omega_3$, a loss of stability in the system gives rise to a Hopf bifurcation. The stability of the coexisting equilibrium indicates a sustainable ecosystem where all species persist together, maintaining ecological balance despite small disturbances.

3.4 Bifurcation analysis

In the present section, we perform the analysis to investigate the occurrence of bifurcation in the system.

3.4.1 Transcritical bifurcation

Let us consider the representation of the system as $Z = (x, y_1, y_2)$, $Z' = H(Z)$, where $H = (H_1, H_2, H_3)^T$. Then

$$\mathbf{DH} = \begin{bmatrix} G_{11} & G_{12} & G_{13} \\ G_{21} & G_{22} & 0 \\ G_{31} & 0 & G_{33} \end{bmatrix},$$

where

$$\begin{cases} G_{11} = r - 2rx - \frac{\alpha K_1 y_1}{(K_1+x)^2} - \frac{\beta K_2 y_2}{(K_2+x)^2} - 3\mu_1 x^2, \\ G_{12} = -\frac{\alpha x}{K_1+x}, \\ G_{13} = -\frac{\beta x}{K_2+x}, \\ G_{21} = \frac{\alpha_1 K_1 y_1}{(K_1+x)^2}, \\ G_{22} = \frac{\alpha_1 x}{K_1+x} - \delta_1 - 2\mu_2 y_1, \\ G_{31} = \frac{\beta_1 K_2 y_2}{(K_2+x)^2}, \\ G_{33} = \frac{\beta_1 x}{K_2+x} - (\delta_2 + E) - 2\mu_3 y_2. \end{cases}$$

1. About steady point $V_0(0,0,0)$: The eigenvalues are r, $-\delta_1$, and $-(\delta_2 + E)$. No eigenvalue becomes zero. Hence, transcritical bifurcation and saddle node bifurcation do not exist about V_0.
2. About steady point $V_1(x_1, 0, 0)$: The eigenvalues are $r - 2rx_1 - 3\mu_1 {x_1}^2$, $\frac{\alpha_1 x_1}{K_1+x_1} - \delta_1$, and $\frac{\beta_1 x_1}{K_2+x_1} - (\delta_2 + E)$. If $\frac{\beta_1 x_1}{K_2+x_1} - (\delta_2 + E) = 0$, then one eigenvalue becomes zero. Two eigenvectors corresponding to J_{V_1} and $J_{V_1}^T$ are calculated as $e_1 = (1, 0, \frac{e}{f})$, where $e = r - 2rx_1 - 3\mu_1 {x_1}^2$ and $f = \frac{\beta x_1}{K_2+x_1}$, and $e_2 = (0, 0, 1)$, respectively. Then
   (i) ${e_2}^T[H_E(V_1, E = E^*)] = 0$;
   (ii) ${e_2}^T[DH_E(V_1, E = E^*)e_1] = \frac{-e}{f} \neq 0$;
   (iii) ${e_2}^T[D^2H(V_1, E = E^*)(e_1, e_1)] = \frac{2\beta_1 K_2}{(K_2+x_1)^2}\frac{e}{f} - 2\mu_3(\frac{e^2}{f^2}) \neq 0$.
   By the result of Sotomayor's bifurcation theorem [26], the transcritical bifurcation exists around V_1 at $E = E^*$. The value of E is calculated as $E^* = \frac{\beta_1 x_1}{K_2+x_1} - \delta_2$. Numerically, it is verified that $E^* = 0.1282$.
3. About steady point $V_2(x', y_1', 0)$: One of the roots of equation (3), say $A_{33} = 0$. Then, zero becomes the eigenvalue of J_{V_2}. The two eigenvectors corresponding to J_{V_2} and $J_{V_2}^T$ are $f_1 = (1, g, h)$, where $g = \frac{-g_1}{g_2}$, $g_1 = \frac{\alpha_1 K_1 y_1'}{(K_1+x')^2}$, and $g_2 = \frac{\alpha_1 x'}{K_1+x'} - \delta_1 - 2\mu_2 y_1'$, $h = \frac{1}{g_3}(g_4 g - g_5)$, where $g_3 = -\frac{\beta x'}{K_2+x'}$, $g_4 = \frac{\alpha x'}{K_1+x'}$, $g_5 = r - 2rx' - \frac{\alpha K_1 y_1'}{(K_1+x')^2} - 3\mu_1 x'^2$, and $f_2 = (0, 0, 1)$, respectively. Then
   (i) ${f_2}^T[H_E(V_2, E = E^*)] = 0$;

(ii) $f_2{}^T[DH_E(V_2, E = E^*)f_1] = -h \neq 0$;
(iii) $f_2{}^T[D^2H(V_2, E = E^*)(f_1, f_1)] = -2\mu_3 h^2 + \frac{2\beta_1 K_2}{(K_2+x')^2} h \neq 0$.
Thus, transcritical bifurcation appears around V_2 by the result of Sotomayor's bifurcation theorem [26]. The value of E is calculated as $E^* = \frac{\beta_1 x'}{K_2+x'} - \delta_2$.

3.4.2 Hopf bifurcation

If $\Omega_1\Omega_2 - \Omega_3 = 0$, then the characteristic equation (6) becomes

$$(\lambda^2 + \Omega_2)(\lambda + \Omega_1) = 0. \tag{7}$$

The roots of (7) are calculated as $\Omega_1, \iota\sqrt{\Omega_2}, -\iota\sqrt{\Omega_2}$, which assures the emergence of a Hopf bifurcation within the system. Further, assume the general form of the complex root $\lambda_1 = u + \iota v$, of equation (6). Substituting this value of the root in equation (6), separating the real and imaginary parts, and after simplification, we obtain

$$-8u^3 - 8u^2\Omega_1 - 2u\Omega_2 + 2x\Omega_1^2 - \Omega_1\Omega_2 + \Omega_3 = 0.$$

Differentiating this result with respect to parameter E, and using the result that $u(E^*) = 0$, we obtain

$$\Omega_1'\Omega_2 + \Omega_1\Omega_2' - \Omega_3' \neq 0 \quad \text{at } E = E^*, \tag{8}$$

which is the transversality condition for the occurrence of a Hopf bifurcation.

4 Numerical simulation

In this section, we are going to perform numerical simulation to verify the analytical results. The set of parameters required for numerical simulation is given in Table 1. We have taken the initial density of three species as $x(t) = 0.1, y_1(t) = 0.6$, and $y_2(t) = 0.3$. The system shows unstable behavior at $E = 0.001$, which is clearly visible in Figure 1. Further, for $E = 0.008$, the solution becomes periodic (see Figure 2). It is observed that at $E = 0.00905$, the Hopf bifurcation appears in the system, and the solution starts approaching a stable trajectory from an unstable path. This behavior is noted in Figure 3. Moreover, Figure 4 exhibits that the system converges to a stable solution around $V_4(0.0498, 1.098, 0.31127)$ at $E = 0.01$. Biologically, the harvesting of infected zooplankton reduces disease prevalence in the system, prevents further spread of infection, and promotes the stability of zooplankton and phytoplankton populations, thereby maintaining the overall balance of the aquatic ecosystem.

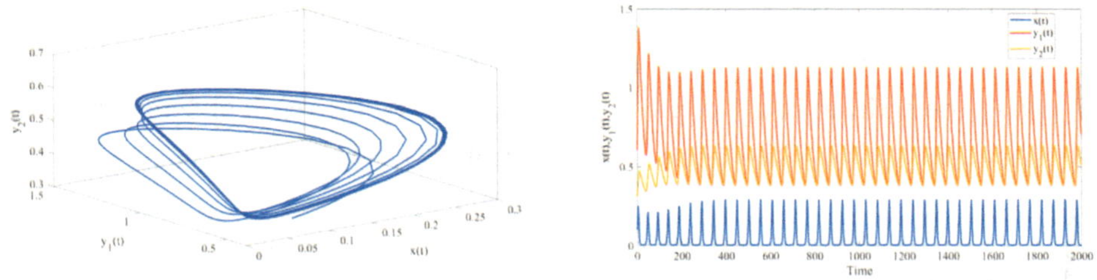

Figure 1: Phase portrait and time-series curve showing unstable behavior at $E = 0.001$.

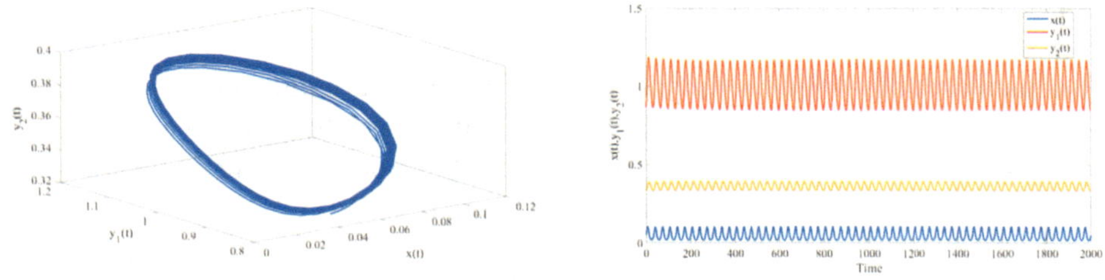

Figure 2: Periodic solution arises in the system at $E = 0.008$.

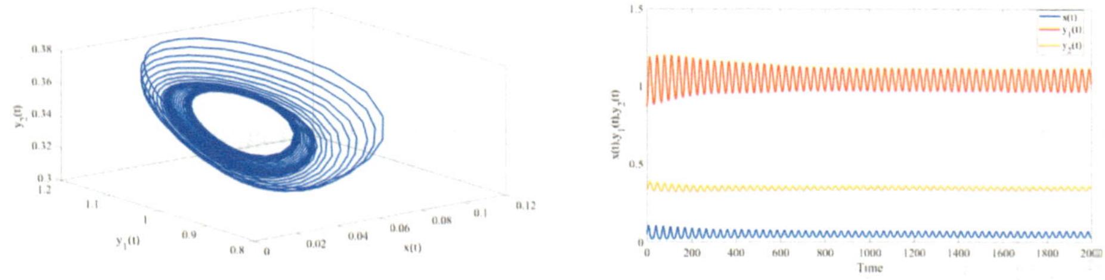

Figure 3: Bifurcation starts appearing in the system at $E = 0.00905$.

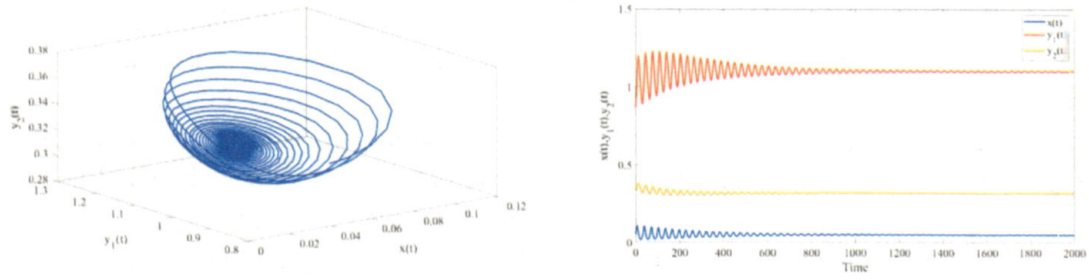

Figure 4: Phase portrait and time-series curve showing convergence of the system to a stable solution at $E = 0.01$.

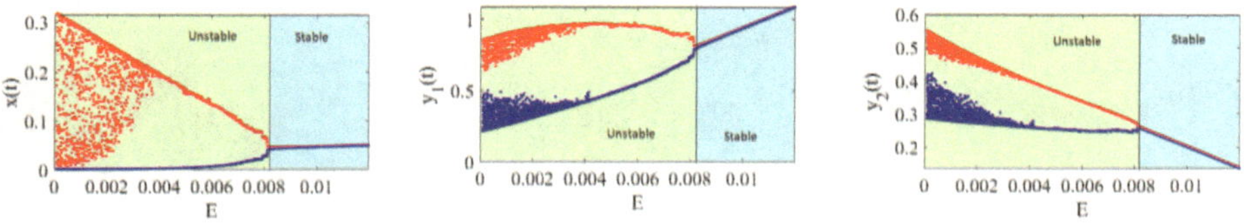

Figure 5: Bifurcation diagram of the system in relation to the rate of harvesting efforts on infected zooplankton *E*.

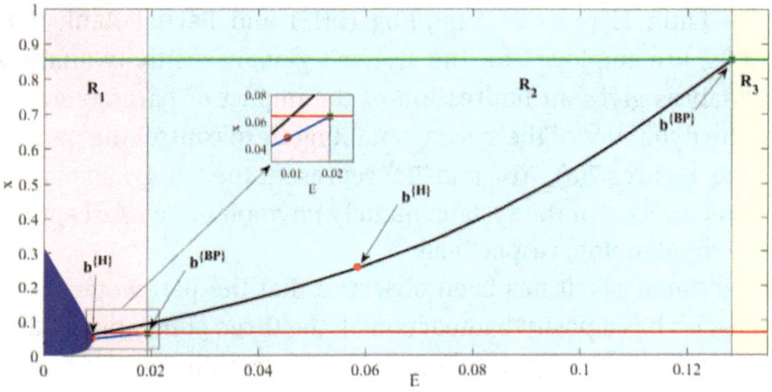

Figure 6: One-parameter bifurcation w. r. t. *E*.

In Figure 5, the bifurcation diagrams are plotted in connection with the rate of harvesting efforts on infected zooplankton E. It is observed from Figure 5 that as the harvesting rate increases beyond a certain threshold, the system undergoes a transition from instability to stability. Biologically, this means that harvesting suppresses the persistence of infection, reduces disease pressure, and stabilizes the ecosystem by maintaining a healthy balance among interacting populations.

One-parameter bifurcation in connection with the rate of harvesting efforts on infected zooplankton E: MATCONT, a package in MATLAB is used to plot the one-parameter bifurcation diagram, which is portrayed in Figure 6. The Hopf point occurs at $E = 0.00905$ (denoted by the red dot), where the Hopf bifurcation appears in the system. The unstable patterns changes to stable responses at this point. The branch point takes place at $E = 0.01935$, denoted by the green square in the diagram. At this point, transcritical bifurcation appears in the system, where the coexisting steady point V_4 (shown by the blue line) loses its stability, and the infected zooplankton-free steady point V_2 (shown by the red line) gains stability. Further, if we consider the susceptible zooplankton-free steady point (shown by the black line), again a Hopf point occurs at $E = 0.05858$, and a branch point occurs at $E = 0.1282$. At this branch point, V_1 and V_3 exchange their stability. It is observed from Figure 6 that the coexisting steady state remains stable from $E = 0.00905$ to $E = 0.011934$.

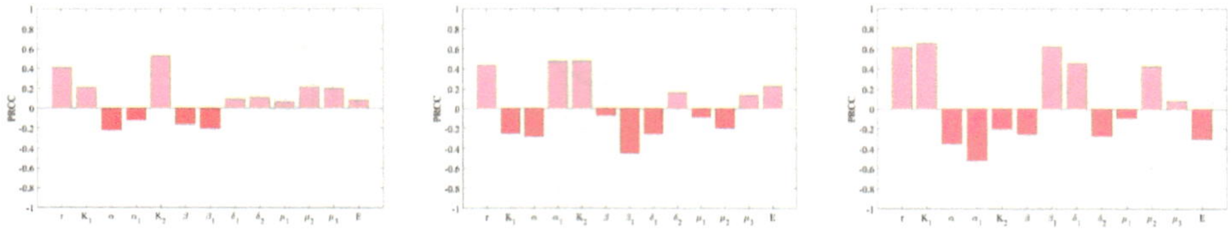

Figure 7: Sensitivity analysis based on the PRCC values of (a) Phytoplankton, (b) Susceptible zooplankton, and (c) Infected zooplankton.

Sensitivity analysis: The Latin Hypercube Sampling (LHS) and Partial Rank Correlation Coefficients (PRCCs) are employed for the system's global sensitivity analysis [27, 28]. The sensitivity analysis gives an impression of the impact of parameters involved in the system, on the dynamics of the system, enabling us to control the parameters for ecological balance. Figures 7(a), 7(b), and 7(c) represent the bar graphs based on the PRCC values for three species in the system, namely phytoplankton, susceptible zooplankton, and infected zooplankton, respectively.

Based on the sensitivity analysis, it has been observed that the parameter r, the growth rate of the prey species has a positive impact on all the three populations. Also, it is noted that the consumption rate of phytoplankton by susceptible zooplankton α has a negative impact on all the species. Hence, we can extend our analysis further by focusing on these two parameters. The bifurcation diagrams are plotted in connection with the rate of growth of phytoplankton r, and consumption rate α of susceptible zooplankton. It is observed from Figure 8(a) that stability is maintained at small values of r, while instability arises at larger values. Biologically, the higher growth of phytoplankton fuels a rapid zooplankton increase, which enhances infection spread and leads to instability in the system. Figure 8(b) depicts that the system is unstable for small values of α, and attains stability for its higher values. Biologically, a higher consumption rate of phytoplankton by susceptible zooplankton can stabilize the system by regulating excess phytoplankton growth, controlling resource-driven outbreaks, and maintaining balanced interactions among the three species.

One-parameter bifurcation:

(a) The one-parameter bifurcation plot is presented in connection to parameter r, the population growth rate. In Figure 9(a), the blue line is used to denote the coexisting steady state, and the red line is for the infected zooplankton-free steady state. The red dot represents the Hopf point, and the green square represents the branch point. The transcritical bifurcation appears at $r = 0.5307$, and we proceed to region R_2 from R_1. The Hopf bifurcation occurs at $r = 1.8473$, where the system loses its stability around V_4. One more Hopf point appears on the blue line (coexisting steady point) at $r = 8.3550$.

(b) Figure 9(b) portrays the one-parameter bifurcation in connection with the consumption rate α of phytoplankton by susceptible zooplankton. The Hopf point (red dot) takes place at $\alpha = 1.3986$. At this point, the coexisting equilibrium state (blue line) gains stability. Further, the branch point (green square) appears at $\alpha = 4.5532$. At this point,

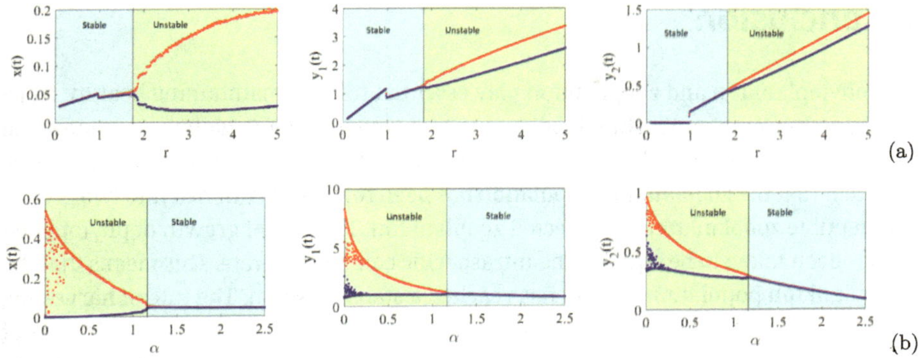

Figure 8: Bifurcation diagrams of the system with respect to (a) the growth rate of phytoplankton r, and (b) the consumption rate α of phytoplankton by susceptible zooplankton.

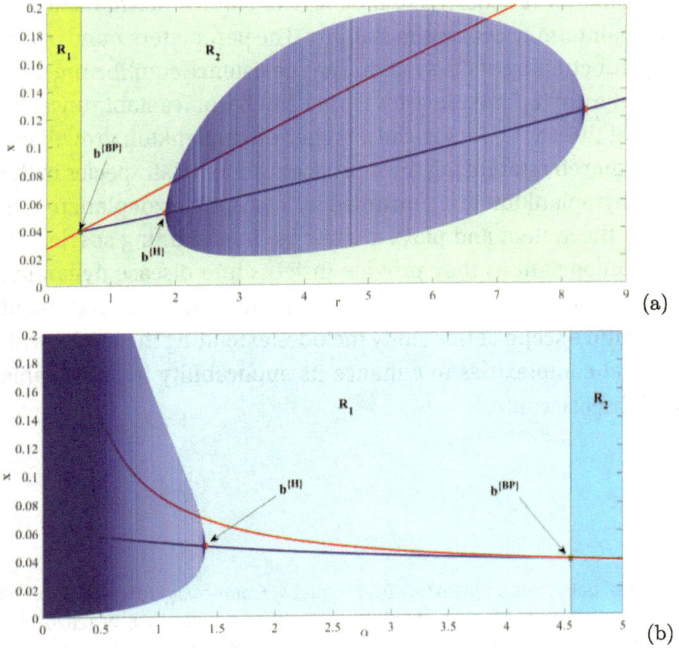

Figure 9: One-parameter bifurcation in connection to (a) the growth rate of prey species r, and (b) the consumption rate of phytoplankton by susceptible zooplankton α.

the steady state V_4 loses its stability, and the steady state V_2 (red line) gains stability. It is clear from the figure that the coexisting steady point V_4 remains stable for $\alpha = 1.3986$ to $\alpha = 4.5532$. It has been observed that for very small and high values of α, the system becomes unstable. In region R_2, the coexisting steady state does not exist.

5 Conclusion

Both phytoplankton and zooplankton play essential roles in maintaining healthy, functioning ecosystems, supporting biodiversity and contributing to global environmental processes. In the present chapter, we have considered a two-species system, with the assumption that the zooplankton population has been further partitioned into two groups – susceptible zooplankton and infected zooplankton. The rate of growth of phytoplankton has been taken to be equal to the intraspecific competition rate. This means that the phytoplankton population density has reached a stable position. The rate of harvesting efforts on infected zooplankton E plays a crucial role in maintaining the coexistence of the three species within the system. The coexisting steady state is unstable for lower values of E, and then becomes stable with the increase in the value of E. For higher values of E, the coexisting equilibrium point exchanges its stability with the infected zooplankton-free equilibrium point. This result has biological significance too, as the harvesting of infected zooplankton functions like a biological control mechanism balancing disease spread and maintaining ecosystem stability. The parameters r and α also contribute towards the well functioning of the system. The coexistence equilibrium does not exist for lower values of the phytoplankton growth rate, but becomes stable once the growth rate exceeds a critical threshold. Biologically, higher phytoplankton growth ensures sufficient food supply, thereby stabilizing the mutual existence of all species in the system. The rate at which phytoplankton are consumed by susceptible zooplankton, α, regulates energy transfer in the system and plays a vital role in sustaining species coexistence. Such models are important as they provide insights into disease dynamics, resource utilization, and ecological balance, thereby guiding sustainable management of aquatic ecosystems. The future scope of this study includes extending the model with environmental and ecological complexities to enhance its applicability in sustainable fisheries management and disease control.

Bibliography

[1] Walsh, G. (1978). Toxic effects of pollutants on plankton. In *Principles of Ecotoxicology* (pp. 257–274).

[2] Häder, D.-P., Banaszak, A. T., Villafañe, V. E., Narvarte, M. A., González, R. A., & Helbling, E. W. (2020). Anthropogenic pollution of aquatic ecosystems: emerging problems with global implications. *Science of the Total Environment*, *713*, 136586.

[3] Sprules, W. G. (1975). Factors affecting the structure of limnetic crustacean zooplankton communities in central Ontario lakes. *Verhandlungen der Internationalen Vereinigung für theoretische und angewandte Limnologie*, *19*(1), 635–643.

[4] Kumar, R., & Rana, N. (2022). Mathematical analysis of effect of nutrients on plankton model with time delay. In *International Joint Conference on Advances in Computational Intelligence* (pp. 689–703). Springer.

[5] Rana, N., Kumar, R., & Sarkar, A. (2024). Role of fear factor in controlling chaos of a plankton–fish system with prey refuge and an additional food. *The European Physical Journal Plus*, *139*(1), 1–24.

[6] Rana, N., Kumar, R., Sarkar, A., & Mondal, B. (2025). Unraveling dynamics of bursting, transient, and tipping behavior in toxic plankton–fish system with fear and zooplankton refuge. *Journal of Computational Science*, *85*, 102527.

[7] Rana, N., Kumar, R., & Sarkar, A. (2025). Bi-stability behaviors of plankton–fish model taking phytoplankton as a basic nutritional source for fish. *The European Physical Journal Plus*, *140*(9). https://doi.org/10.1140/epjp/s13360-025-06779-y.

[8] Kumar, R., Rana, N., & Kumar, V. Bifurcation behaviors of a phytoplankton–zooplankton model with toxicity. *Computational Analysis of Mathematical Systems and Wireless Networks, 149*.

[9] Panja, P., Mondal, S. K., & Jana, D. K. (2017). Effects of toxicants on phytoplankton–zooplankton–fish dynamics and harvesting. *Chaos, Solitons & Fractals*, *104*, 389–399.

[10] Mandal, A., Tiwari, P. K., Samanta, S., Venturino, E., & Pal, S. (2020). A nonautonomous model for the effect of environmental toxins on plankton dynamics. *Nonlinear Dynamics*, *99*, 3373–3405.

[11] Mandal, A., Biswas, S., Tiwari, P. K., & Pal, S. (2024). Seasonal refuge patterns of phytoplankton trigger irregular bloom events in a contaminated environment. *Scientific Reports*, *14*(1), 25248.

[12] Kumar, R., & Rana, N. (2023). Mathematical modelling and reliability in harvesting of seafood for food-processing industries. In *Industrial Reliability and Safety Engineering* (pp. 17–34). CRC Press.

[13] Pei, Y., Lv, Y., & Li, C. (2012). Evolutionary consequences of harvesting for a two-zooplankton one-phytoplankton system. *Applied Mathematical Modelling*, *36*(4), 1752–1765.

[14] Cáceres, C. E., Tessier, A. J., Duffy, M. A., & Hall, S. R. (2014). Disease in freshwater zooplankton: what have we learned and where are we going? *Journal of Plankton Research*, *36*(2), 326–333.

[15] Roberts, A. J., & Suttle, C. A. (2023). Pathogens and passengers: roles for crustacean zooplankton viruses in the global ocean. *Microorganisms*, *11*(4), 1054.

[16] Kumar, R., Sharma, A. K., & Agnihotri, K. (2017). Dynamics of an innovation diffusion model with time delay. *East Asian Journal on Applied Mathematics*, *7*(3), 455–481.

[17] Kumar, R., Sharma, A. K., & Agnihotri, K. (2018). Stability and bifurcation analysis of a delayed innovation diffusion model. *Acta Mathematica Scientia*, *38*(2), 709–732.

[18] Kumar, R., Sharma, A. K., & Agnihotri, K. (2018). Delay-induced Hopf bifurcation in a nonlinear innovation diffusion model. *East Asian Journal on Applied Mathematics*, *8*(3), 422–446.

[19] Kumar, P., et al. (2023). Investigating the impact of toxicity on plant growth dynamics through the zero of a fifth-degree exponential polynomial. *Chaos, Solitons & Fractals*, *171*, 113457.

[20] Kumar, P., & Dipesh (2022). Effect of time delay on dynamic of plant competition under allelopathy. *Mathematical Methods in the Applied Sciences*, *45*(16), 9308–9321.

[21] Chen, Q., Dipesh, K. P., & Baskonus, H. M. (2024). Stability analysis of GDP–national debt dynamics using delay differential equations. *Fractals*, 2540059.

[22] Agnihotri, K., & Kaur, H. (2021). Optimal control of harvesting effort in a phytoplankton–zooplankton model with infected zooplankton. *Mathematics and Computers in Simulation*, *190*, 946–964.

[23] Chakraborty, K., & Das, K. (2015). Modeling and analysis of a two-zooplankton one-phytoplankton system in the presence of toxicity. *Applied Mathematical Modelling*, *39*(3–4), 1241–1265.

[24] Birkhoff, G., & Rota, G.-C. (1989). *Ordinary Differential Equations*, Boston, MA: Ginn.

[25] David, G., Schaeffer, C., & John, W. (2018). *Ordinary Differential Equations: Basics and Beyond*. Springer.

[26] Perko, L. (2013). *Differential Equations and Dynamical Systems* (Vol. 7). Springer.

[27] Mondal, B., Sarkar, A., Santra, S. S., Majumder, D., & Muhammad, T. (2023). Sensitivity of parameters and impact of white noise on a predator–prey model. *The European Physical Journal Plus*, *138*(12), 1–13.

[28] Sarkar, A., Tiwari, P. K., & Pal, S. (2021). Effect of additional food on predator–prey interactions with water-level fluctuation. *Journal of Biological Systems*, *29*(4), 995–1022.

Ritu Ahuja, Rakesh Kumar, and Kulbhushan Agnihotri

Modeling the fear effect and prey refuge in a predator–prey system with harvesting

Abstract: This study proposes a predator–prey model that investigates how fear indirectly affects prey reproduction and birth rates, while also considering the influence of prey refuge. The model is constructed using a Holling type-II functional response framework and incorporates a linear harvesting approach for predators. The conditions for species extinction and uniform persistence, as well as all biologically viable equilibrium points, are identified. A comprehensive stability analysis is conducted to determine both local and global stability conditions, shedding light on population stabilization over time. The study also includes comprehensive bifurcation analysis, examining transcritical and Hopf bifurcations to understand critical changes in system behavior. The findings suggest that increasing prey refuge can dampen oscillatory dynamics, steering the system towards equilibrium. Additionally, stability transitions occur when the fear parameter surpasses a critical threshold. The modulation of a system by a harvesting parameter is also discussed by one-parameter bifurcation. Numerical simulations in MATLAB are employed to validate the analytical findings and to explore how key parameters impact the system's dynamics.

Keywords: Refuge, functional response, bifurcations, stability

MSC 2020: 92D25, 34D20, 34C23, 37G15, 92D40

1 Introduction

Prey–predator interaction is a fundamental area of research in mathematical biology. The intricate interactions among prey and predators shape the overall system dynamics and permanence of the system. Earlier, the only way predators were thought to influence the ecosystem was through direct killing of prey. However, over recent times, vari-

Acknowledgement: The authors would like to express sincere gratitude to the Department of Applied Sciences and Humanities, Shaheed Bhagat Singh State University, Ferozepur, Punjab, India, for their invaluable support and resources provided to carry out this research. We also thank the faculty and staff for their assistance and encouragement throughout the work.

Ritu Ahuja, Department of Applied Sciences and Humanities, Shaheed Bhagat Singh State University, Ferozepur, 152004, Punjab, India; and Khalsa College for Women, Ludhiana, 141001, Punjab, India, e-mail: rituahuja.kcw@gmail.com
Rakesh Kumar, Department of Applied Sciences and Humanities, Shaheed Bhagat Singh State University, Ferozepur, 152004, Punjab, India, e-mail: rakesh@sbsstc.ac.in
Kulbhushan Agnihotri, Department of Mathematics, Panjab University, Chandigarh, India

https://doi.org/10.1515/9783112234525-010

ous researchers have shown in their work [1, 8, 15–17], that predators' influence on prey is far-reaching, resulting in multiple changes in the behavior, psychology, and physiology of prey species, finally changing the population composition of the system. When prey and predators interact, the prey feels threatened by the risk of predation. The perception triggers feelings of fearfulness or distress in the prey animal. Historically, it was believed that the stress or fear caused by predators was necessarily short-lived, acute, and had minimal impact on demographic processes. However, this perspective began to shift when numerous experiments were conducted. These studies aimed to investigate whether the fear induced by predators could significantly affect prey populations, even in scenarios where direct predation was intentionally prevented. Frightened prey forages less, which leads to starvation and less reproduction [2, 10]. Creel et al. [4] observed in their work that the predation risk of wolves leads to a change in reproductive physiology in elks. Hua et al. [11] in their study noted that if they control the vocal signals of predators, the bluebirds undergoing reproduction cycles can adjust their breeding strategies as per those signals. Eggers et al. [8] investigated the perceived risk of nest predation in Siberian jays (Perisoreus infaustus) by playing recordings of predator calls near their nesting sites. These sound signals simulated a heightened risk of predation. As a result, the jays chose safer nesting spots with more cover and laid fewer eggs. Research has shown that exposure to predators or their cues can negatively affect prey species' reproductive success and long-term survival. However, these investigations did not measure the demographic impact of predator-induced fear on prey populations. A groundbreaking field experiment conducted by Zanette et al. [29] in 2011 addressed this gap by actively preventing direct predation to assess the effect of predator risk or fear on prey demographics. Their study revealed that the mere perception of predation risk, manipulated through predator call playbacks, resulted in about a 40 percent decrease in the annual production of song sparrow chicks. This finding highlighted the significant impact of 'intimidation' alone on prey populations. The fear response in prey animals can lead to the development of certain antipredator behaviors as a means of avoiding predators. These may include secreting toxins, congregating in groups, feigning death, imitating other species, moving in herds and blending into their surroundings (camouflage). An effective antipredator response is often achieved through the use of refuges. Prey animals often protect themselves from being eaten by predators by hiding in a secure location or moving to a new habitat that makes them harder to find or access. These are safe areas or conditions that allow prey to avoid predation, such as physical structures like underwater vegetation or rocks, long grass, burrows or environmental conditions like low-light zones, a temporary change of habitat or merely reducing foraging activity that allows prey to avoid confrontation with predators [6, 21]. Employing refuges allows prey populations to remain partially concealed from predators. A fraction of the prey is consistently defended from predators. By providing safe zones, refuges reduce predation rates and influence prey survival and population dynamics. However, the longer a prey stays in the refuge, the fewer chances it has to feed and mate. Refuges have a significant influence on the coexistence of species within the system [25]. Incorporating the fear element and refuge into ecological models is vital for capturing a more accurate and practical view of these interactions [7, 30]. Krivan V [13] explored how a behavioral refuge, created either by the predator's optimal foraging or the prey's adaptive antipredator behavior, impacts the Gause predator–prey model. The findings show that both mechanisms support the coexistence of predators and prey, either in a stable state or along limit cycles. Wang et al. [26] were the first to introduce the fear effect into a mathematical model and observed a decline in prey reproduction due to fear. They observed that fear with higher values leads the system towards stability, by destroying periodic solutions. Kar [12] in their study showed that increased refuge in a prey–predator model can lead to a population burst. Their model was extended by incorporation of a fear factor and a stabilizing effect of fear on the system was observed. Hassel [9] depicted that a system exhibiting divergent oscillations shifts towards asymptotically stable equilibrium with addition of a large refuge to system. Wang and Zou [27] formulated a stage-structured predator–prey interplay by incorporating the effect of fear of predator population on prey population. They considered a delay differential system with maturation delay and divided the prey species into juvenile and mature stages. Naji and Majeed [18] explored a stage-structured prey–predator system involving two prey species (immature and mature) and one predator, incorporating prey refuge. They found that the inclusion of the refuge factor contributes to stabilizing the system. Raw et al. [23] investigated a tritrophic predator–prey model consisting of one prey and two predators, where both group defence mechanisms and immigration within the top predator population were considered. They employed a Monod–Haldane-type functional response to represent interactions between species. Chinnathambi and Rihan [3] analyzed the dynamics of a fractional-order delayed prey–predator system with a Monod–Haldane response function. Das and Samanta [5] examined the impact of fear on prey in a stochastic prey–predator model when additional food is provided to the predator. Pandey et al. [19] investigated a three-species food chain model where the growth of the middle predator is diminished due to the fear of the top predator and the growth rate of the prey is similarly reduced by fear of the middle predator. The interactions among the species were modeled using a Holling type-II functional response. Kumar and Dubey [14] developed a model that incorporates the fear effect and prey refuge in a predator–prey system with a gestation time delay using a Crowley–Martin-type functional response. Sarkar and Khajanchi [24] studied a prey–predator system with the cost of fear into prey reproduction using a Holling type-II functional response. Their study suggests that prey species are more inclined to exhibit antipredator defences when the rate of predation becomes larger. Yadav et al. [28] analyzed a spatial and associated nonspatial fear-induced prey–predator system taking into account group defence and prey refuge using a Holling type-IV functional response.

Knowledge gap and significance

The exploitation of ecological resources including the harvesting of species, is a common practice across various sectors such as fisheries, forestry, and agriculture. Harvesting can be stabilizing, destabilizing, or can even induce oscillations in populations. Many researchers have integrated fear effects and prey refuge into predator–prey models using various types of functional responses. Additionally, the impact of harvesting on predator–prey dynamics has been widely studied. However, models that combine fear effects, prey refuge, and harvesting have not yet been explored. Motivated by the work done by Yadav et al. [28], Kumar and Dubey [14], and Sarkar and Khajanchi [24], in the present chapter, we examine a predator–prey interaction system influenced by fear effects, incorporating prey refuge and linear harvesting of the predator population, with the following study objectives:
- To examine the impact of the fear effect when predators are being harvested.
- To determine the conditions under which prey and predator populations may become extinct or persist.
- To assess how various levels of harvesting effort affect the system.
- To examine the modulation of system dynamics by a prey refuge parameter.

2 Formulation of the mathematical model

(i). The model system consists of the prey and the predator populations whose densities at any instant of time are $x(t)$ and $y(t)$, respectively. We make the following assumptions. Prey species protect themselves by taking refuge due to fear of predation. In the absence of predators, the logistic growth is assumed for the prey population; r being the birth rate of prey, μ_1 is the natural mortality rate, and μ_2 is the mortality due to intraspecific competition among them. Hence, the prey population can be modeled by the differential equation

$$\frac{dx}{dt} = (r - \mu_1)x - \mu_2 X^2.$$

(ii). Researches have shown that predator-induced fear reduces the prey's birth rate [26]. The growth term is multiplied by the monotonically decreasing term $g(K, y)$. In accordance with Wang et al. [26], we assume the effect of fear as

$$g(K, y) = \frac{1}{1 + yK},$$

where y is the predator population and K stands for the cost of fear. $g(K, y)$ satisfies the following characteristics:

(i) $g(0, y) = 1 = g(K, 0)$,

i. e., the prey species grows at their growth rate when no fear is experienced from the predator population or the predator population is absent.

(ii) $\lim_{K\to\infty} g(K,y) = 0 = \lim_{y\to\infty} g(K,y)$,

i. e., due to a very high level of predator-induced fear or the presence of extremely large number of predators, the growth of prey species decreases and tends to zero.

(iii) $\frac{\partial g}{\partial K} < 0$, $\frac{\partial g}{\partial y} < 0$,

i. e., with a rise in the fear level or predator density, the growth rate of prey species decreases.

(iii). The number of prey consumed per predator as a function of prey density is called the functional response. We assume that prey is consumed by predators according to a Holling type-II functional response at a rate a, c is the food conversion efficiency of predator. Thus, prey and predator interaction can be written as

$$\frac{dx(t)}{dt} = \frac{1}{1+Ky}rx - \mu_1 x - \mu_2 x^2 - \frac{axy}{b+x}$$
$$\frac{dy(t)}{dt} = \frac{caxy}{b+x} - (d_1 x + qE)y - d_2 y^2,$$

where b is the half-saturation constant, d_1 is the natural rate of death of predators, and d_2 is the density-dependent death rate owing to intraspecific competition among them. Additionally, predators are being harvested at a linear rate, E being the harvesting effort and q is the catchability coefficient.

(iv). In various ecosystems, prey populations often find refuges where they keep away from predators. Consequently, the entire prey population is not confronted with predators. It is assumed that a fraction mx of the prey species is completely shielded from predators by refuges, and only a portion $(1-m)x$ of the prey species is accessible to predators. With incorporation of m, the system dynamics can be written as:

$$\begin{aligned} \frac{dx(t)}{dt} &= \frac{rx}{1+Ky} - \mu_1 x - \mu_2 x^2 - \frac{a(1-m)x(t)y(t)}{x(1-m)+b}, \\ \frac{dy(t)}{dt} &= \frac{ac(1-m)x(t)y(t)}{(1-m)x+b} - (d_1 x + qE)y - d_2 y^2. \end{aligned} \tag{1}$$

System (1) will be studied with the initial conditions

$$x(0) > 0, \quad y(0) > 0. \tag{2}$$

A brief description of the parameters used and their biological relevance is given in Table 1.

Table 1: Brief description of the parameters used in the model.

Parameter	Description	Value	Dimensions/units
x	Prey population density		mass
y	Predator population density		mass
r	Intrinsic growth rate of prey population	2	1/time
K	Level of fear	0.4	mass^{-1}
m	Refuge taken by prey	0.2	dimensionless
a	Predation rate of predators	10	time^{-1} mass^{-1}
b	Half-saturation constant	5	mass
c	Rate of conversion of prey biomass to predator biomass	0.4	dimensionless
μ_1	Natural mortality rate of $x(t)$	0.03	1/time
d_1	Natural mortality rate of $y(t)$	0.5	1/time
μ_2	Death rate of prey due to intraspecific competition	0.1	mass^{-1} time^{-1}
d_2	Death rate of predators due to intraspecific competition	0.025	mass^{-1} time^{-1}
q	Catchability coefficient of $y(t)$	0.04	dimensionless
E	Harvesting effort	5	dimensionless

3 Positivity and boundedness of system

The boundedness and positive invariance of the solutions of the system ensure the ecological feasibility of the model. Ecologically, positivity refers to the survival of the population, whereas boundedness means that a population does not explode by growing indefinitely large.

Theorem 1. *All the solutions of model* (1) *corresponding to initial conditions* $x(0) > 0$, $y(t) > 0$ *remain positive in* $\mathbb{R}^2_+$, $\forall t \geq 0$.

Proof. Rewriting the system (1) as

$$\frac{dx(t)}{x} = f(x,y).x,$$
$$\frac{dy(t)}{y} = g(x,y).y,$$

where

$$f(x,y) = \frac{r}{1+Ky} - \mu_1 - \mu_2 x - \frac{a(1-m)y(t)}{(1-m)x+b};$$
$$g(x,y) = \frac{ca(1-m)x(t)}{(1-m)x+b} - (d_1 x + qE) - d_2 y.$$

Integrating the equations, yields

$$x(t) = x(0).\exp\left[\int_0^t f(x,y)ds\right] > 0$$

$$y(t) = y(0).\exp\left[\int_0^t g(x,y)ds\right] > 0.$$

Thus, all the solutions of system (1) starting with positive initial conditions remain positive for all $t \geq 0$. □

Theorem 2. *The solutions of the system* (1) *with positive initial conditions* $(x(0) > 0, y(0) > 0)$ *remain confined in the region* $U = \{(x,y) \in \mathbb{R}_+^2 : 0 \leq x(t) \leq \frac{r}{\mu_2}, 0 \leq x(t) + \frac{1}{c}y(t) \leq \frac{r^2}{\eta\mu_2}\}$, *where* $\eta = \min(\mu_1, d_1 + qE)$.

Proof. Let $(x(t), y(t))$ be any solution of the system at any time t, with positive initial conditions.

Consider the first equation of (1), we have

$$\begin{aligned}\frac{dx(t)}{dt} &\leq \frac{rx}{1+Ky} - \mu_2 x^2 \\ &< rx - \mu_2 x^2.\end{aligned}$$

By the comparison theorem, this implies

$$\lim_{t\to\infty} \operatorname{Sup} x(t) \leq \frac{r}{\mu_2}.$$

Assume a time-dependent function, $W(t) = x(t) + \frac{y(t)}{c}$, then we have

$$\begin{aligned}\frac{dW}{dt} &= \frac{dx}{dt} + \frac{1}{c}.\frac{dy}{dt} \\ &< rx - \mu_2 x^2 - \left(\mu_1 x + \frac{(d_1 + qE)}{c} y\right) \\ &\leq \frac{r^2}{\mu_2} - \eta W \\ \frac{dW}{dt} + \eta W &\leq \frac{r^2}{\mu_2},\end{aligned}$$

for $\eta = \min \mu_1, d_2 + qE$.

Applying the theorem on differential inequalities [20], we obtain

$$0 \leq W(x,y) \leq \frac{r^2}{\mu_2 \eta}(1 - e^{-\eta t}) + (x(0), y(0))e^{-\eta t},$$

which, upon letting $t \to \infty$, yields $0 \le W(x,y) \le \frac{r^2}{\mu_2 \eta}$. Hence, the solutions of the system (1) and (2) arising in $\mathbb{R}^2_+$ stay in region

$$U = \left\{(x(t), y(t) : 0 \le W(t) \le \frac{r^2}{\mu_2 \eta}\right\},$$

i. e., the solutions of the system (1) and (2) are bounded in U for all $t > 0$. □

4 Extinction and persistence

4.1 Extinction of species

Theorem 3. (i) *In the absence of predators,* $\lim_{t\to\infty} x(t) \to 0$ *if* $r < \mu_1$.
(ii) *When predators are present,* $\lim_{t\to\infty} x(t) \to 0$ *if* $\underline{y} > \frac{r(1-m)a\mu_2}{b\mu_2+(1-m)(r+\mu_2)}$, *where* $\underline{y} = \lim_{t\to\infty} \operatorname{Inf} y(t)$.

Proof. (i) First, equation (1) can be written as

$$\begin{aligned} \frac{dx}{dt} &\le rx - \mu_1 x - \mu_2 x^2 \\ &< (r - \mu_1)x \\ \Rightarrow \quad x(t) &\le x(0) \exp\left(\int_0^t (r - \mu_1) ds\right). \end{aligned}$$

Note that for $r - \mu_1 < 0$, $\exp(\int_0^t (r - \mu_1) ds) \to 0$ as $t \to \infty$.

As a result, $x(t) \to 0$ as $t \to \infty$.

Ecologically, it can be interpreted that if predators are missing, the prey population will find it difficult to survive if the birth rate is lower than the natural death rate.

(ii) By Theorem 2

$$\lim_{t\to\infty} \operatorname{Sup} x(t) \le \frac{r}{\mu_2}.$$

For $\epsilon > 0$, we have $t_1 > 0$, so that for $t > t_1$ $x(t) \le \frac{r}{\mu_2} + \epsilon$.

Let $\underline{y} = \lim_{t\to\infty} \operatorname{Inf} y(t)$. By the definition of Lim inf $y(t)$, for $\epsilon_1 > 0$, we can find $t_1 > 0$ so that for $t > t_1, y(t) > \underline{y} - \epsilon_1$.

Consider $t > t_1, t_2$ and $L = \frac{(1-m)a\mu_2}{b\mu_2+(1-m)(r+\mu_2)}$.

Choose $\epsilon_1 < \underline{y} - \frac{r}{L}$.

From the first equation of the system (1)

$$\frac{dx}{dt} \le rx - \frac{a(1-m)xy}{b + (1-m)x}$$

$$\leq rx - \frac{a(1-m)x(\underline{y}-\epsilon_1)}{b+(1-m)(\frac{r}{\mu_2}+\epsilon)}$$
$$\leq rx - \frac{a(1-m)x(\underline{y}-\epsilon_1)}{b+(1-m)(\frac{r}{\mu_2}+1)}$$
$$= -Lx\left(\underline{y}-\epsilon_1-\frac{r}{L}\right)$$
$$< 0$$
$$\Rightarrow \quad \lim_{t\to\infty} x(t) = 0 \quad \text{for } \underline{y} > \frac{r(1-m)a\mu_2}{b\mu_2+(1-m)(r+\mu_2)}.$$

This means that if the predator population remains above a threshold, the prey population eventually struggles to survive. □

Theorem 4. (i) *The density-dependent extinction in predators occurs if*

$$\underline{y} > \frac{1}{d_2}\left(\frac{ca(1-m)(r\mu_2)}{b\mu_2} - (d_1+qE)\right).$$

(ii) *The predator population goes to extinction due to the prey population if* $\underline{y} > \frac{r}{d_2}$.

Proof. (i) By Theorem 2

$$\lim_{t\to\infty} \operatorname{Sup} x(t) \leq \frac{r}{\mu_2}.$$

For $\epsilon > 0$, there exists $t_1 > 0$ such that for every $t > t_1$, $x(t) \leq \frac{r}{\mu_2} + \epsilon$.

Also, $\underline{y} = \lim_{t\to\infty} \operatorname{Inf} y(t)$. By the definition of $\operatorname{Lim\,inf} y(t)$, for $\epsilon_1 > 0$, there exists $t_1 > 0$ such that for every $t > t_1, y(t) > \underline{y} - \epsilon_1$.

Consider $t > t_1, t_2$ and $N = \frac{1}{d_2}(\frac{ca(1-m)(r\mu_2)}{b\mu_2} - (d_1+qE))$.

Choose $0 \leq \epsilon_1 < \underline{y} - N$

$$\frac{dy}{dt} \leq \frac{ca(1-m)xy}{(1-m)x+b} - (d_1+qE)y - d_2y^2$$
$$\leq \left(\frac{ca(1-m)}{b}\left(\frac{r}{\mu_2}+\epsilon\right) - (d_1+qE) - d_2y\right)y$$
$$\leq y\left(\frac{ca(1-m)}{b}\left(\frac{r}{\mu_2}+1\right) - (d_1+qE) - d_2(\underline{y}-\epsilon_1)\right)$$
$$= -d_2y(\underline{y}-\epsilon_1-N)$$
$$\leq 0$$
$$\Rightarrow \quad \lim_{t\to\infty} y(t) = 0.$$

(ii) The predator dynamics are

$$\begin{aligned}\frac{dy}{dt} &= \frac{ca(1-m)xy}{b+(1-m)x} - (d_1+qE)y - d_2y^2\\ &\le y\left(\frac{ca(1-m)}{b}\left(\frac{r}{\mu_2}+1\right) - (d_1+qE)\right)\\ &= y\left(\frac{ca(1-m)(r+\mu_2)}{b\mu_2} - (d_1+qE)\right).\end{aligned}$$

By the differential inequality theorem, this gives

$$y(t) \le y(0)\exp\left(\int_0^t \left[\frac{ca(1-m)(r+\mu_2)}{b\mu_2} - (d_1+qE)\right] ds\right).$$

Note that

$$\exp\left(\int_0^t \left[\frac{ca(1-m)(r+\mu_2)}{b\mu_2} - (d_1+qE)\right] ds\right) \to 0 \quad \text{as } t \to \infty$$

if

$$b(d_1+qE)\mu_2 > ca(1-m)(r+\mu_2).$$

Hence,

$$y(t) \to 0, \quad \text{as } t \to \infty.$$ □

4.2 Persistence of species

Theorem 5. (i) *The prey species is uniformly persistent if*

$$r > \left(1 + K\frac{r^2}{\eta\mu_2}\right)\left(\mu_1 + \frac{a(1-m)r^2}{b\eta\mu_2}\right).$$

(ii) *The predator population remains persistent provided*

$$qE \le \frac{ca(1-m)A_1}{b+(1-m)A_2} - d_1.$$

Proof. (i) By Theorem 2

$$\lim_{t\to\infty} \operatorname{Sup} x(t) \le \frac{r}{\mu_2}.$$

Hence, there exists a $T_1 > 0$ such that

$$x(t) \leq \frac{r}{\mu_2} = A_2(\text{say}) \quad \text{for all } t \geq T_1. \tag{3}$$

Also, there exists a $T_2 > 0$ such that for every $t \geq T_2, y(t) \leq \frac{r^2}{\mu_2 \eta}$.
For all $t > (t_1, t_2)$. The first equation of (1) can be written as

$$\begin{aligned}\frac{dx}{dt} &\geq \left(\frac{rx}{1+K\frac{r^2}{\eta\mu_2}} - \mu_1 x - \frac{a(1-m)xr^2}{b\eta\mu_2}\right) - \mu_2 x^2 \\ &= x\left(\frac{r}{1+K\frac{r^2}{\eta\mu_2}} - \mu_1 - \frac{a(1-m)r^2}{b\eta\mu_2}\right) - \mu_2 x^2.\end{aligned}$$

This simplifies to give

$$\lim_{t\to\infty} \operatorname{Inf} x(t) \geq \frac{1}{\mu_2}\left(\frac{r}{1+K\frac{r^2}{\eta\mu_2}} - \mu_1 - \frac{a(1-m)r^2}{b\eta\mu_2}\right) = A_1(\text{say}). \tag{4}$$

Inequalities (3) and (4) imply prey species $x(t)$ is persistent provided

$$r > \left(1+K\frac{r^2}{\eta\mu_2}\right)\left(\mu_1 + \frac{a(1-m)r^2}{b\eta\mu_2}\right).$$

(ii) The last equation of the system (1) gives

$$\begin{aligned}\frac{dy(t)}{dt} &\geq \frac{ca(1-m)A_1 y}{b+(1-m)A_2} - (d_1+qE)y - d_2 y^2 \\ &= y\left(\frac{ca(1-m)A_1 y}{b+(1-m)A_2} - (d_1+qE)\right) - d_2 y^2,\end{aligned}$$

which gives,

$$\lim_{t\to\infty} \operatorname{Inf} y(t) \geq \frac{1}{d_2}\left(\frac{ca(1-m)A_1}{b+(1-m)A_2} - (d_1+qE)\right) = D_1(\text{say}).$$

This implies that there exists $t_1 > 0$, such that for every $t \geq t_1, y(t) \geq D_1$.
Also, we have a $t_2 > 0$ such that for every $t \geq T_2, y(t) \leq \frac{r^2}{\mu_2\eta} = D_2(\text{say})$.
Taking $T = \max(t_1, t_2), D_1 \leq y(t) \leq D_2$, for all $t \geq T$.
Hence, the predator species is persistent provided $qE \leq \frac{ca(1-m)A_1}{b+(1-m)A_2} - d_1$. □

5 Existence of equilibrium points and stability analysis

The model system (1) and (2) possesses the following biologically feasible equilibrium points:

(i) $E_0(0,0)$ is the trivial steady state; which always exists.
(ii) $E_1(x_1,0)$ is the predator-free steady state, where $x_1 = \frac{r-\mu_1}{\mu_2}$.

$E_1(x_1,0)$ exists if $r > \mu_1$.

(iii) The coexisting steady state is $E_* = (x_*, y_*)$, where (x_*, y_*) is given by the solution of the equations

$$\frac{r}{1+Ky} - \mu_1 - \mu_2 x - \frac{a(1-m)y}{(1-m)x+b} = 0$$
$$\frac{acx(1-m)}{(1-m)x+b} - (d_1+qE) - d_2 y = 0. \quad (5)$$

For $E_* = (x_*, y_*)$, the predator component y_* is given by the second equation as

$$y_* = \left(\frac{acx(1-m)}{(1-m)x+b} - (d_1+qE)\right)\frac{1}{d_2}$$

and the prey component is given by the positive solution of the equation

$$d_2((1-m)x+b)(r-(\mu_1+\mu_2 x)+(qE+d_1-ac(1-m)x)K(\mu_1+\mu_2 x))$$
$$= \frac{a(1-m)}{d_2}\left(d_2 - K(d_1+qE) + \frac{Kac(1-m)x}{(1-m)x+b}\right)\left(\frac{ca(1-m)x}{(1-m)x+b} - (d_1+qE)\right). \quad (6)$$

Analytically, it is hard to find the solution of this equation. However, the shape and relative position of coexisting nullclines help in determining the possible number of coexisting equilibrium points. Figure 1 shows the existence of a unique equilibrium point for the set of parameter values listed in Table 1. Keeping other parameters the same, we vary the harvesting parameter E. As depicted by Figure 2, the prey nullcline intersects the predator nullcline once for $E = 30$ (blue curve), $E = 50$ (yellow curve), and $E = 60$ (green curve) and does not intersect for $E = 64$ (pink curve) and $E = 66$ (cyan curve). Hence, the model system (1) has a unique coexisting equilibrium point for the set of parameter values considered. The same is also observed from the one-parameter bifurcation diagram for parameter E.

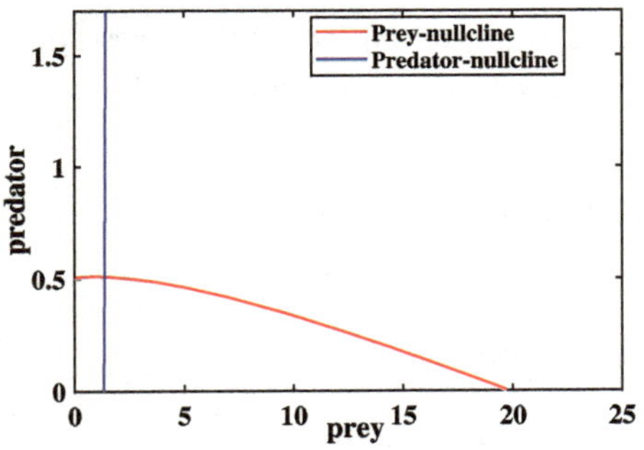

Figure 1: Intersection of nullclines showing the existence of a coexisting equilibrium point E^*.

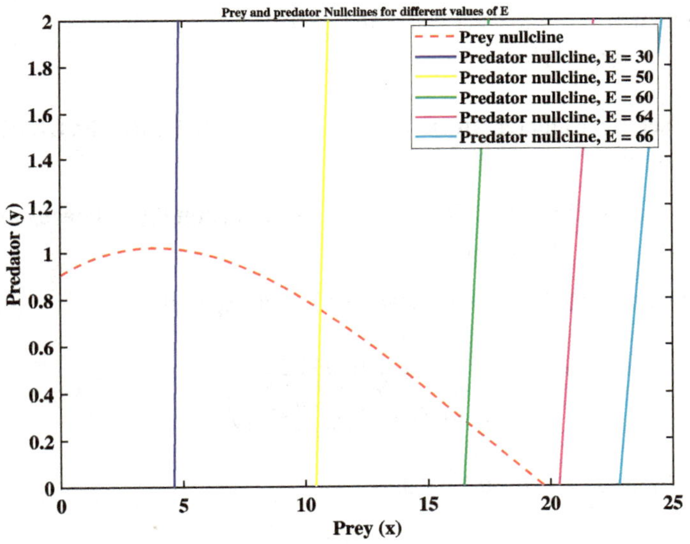

Figure 2: Prey and predator nullclines for various values of the harvesting parameter E.

5.1 Eigenvalue analysis to establish local asymptotic stability

Local stability analysis of the system (1) is done by computing the variational matrix around the respective biological feasible equilibria. The general variational matrix for system (1) at any point is

$$U(x,y) = \begin{pmatrix} \frac{r}{1+Ky} - \mu_1 - 2\mu_2 x - \frac{ba(1-m)y}{(b+(1-m)x)^2} & -\frac{rKx}{(1+Ky)^2} - \frac{a(1-m)x}{b+(1-m)x} \\ \frac{bca(1-m)y}{(b+(1-m)x)^2} & -(d_1 + qE) - 2d_2 y + \frac{ca(1-m)x}{b+(1-m)x} \end{pmatrix}.$$

Theorem 6. (i) *The trivial steady state $E_0(0,0)$ is locally asymptotic stable for $r < \mu_1$.*
(ii) *The predator-free steady state $E_1(x_1, 0)$ is locally asymptotically stable if $(d_1 + qE) > \frac{ca(1-m)(r-\mu_1)}{b\mu_2+(1-m)(r-\mu_2)}$ holds.*

Proof. (i) The Jacobian matrix evaluated for $E_0(0,0)$ is

$$U_0 = \begin{pmatrix} r-\mu_1 & 0 \\ 0 & -(d_1+qE) \end{pmatrix}.$$

The associated eigenvalues are $\lambda_1 = r - \mu_1$, $\lambda_2 = -(d_1 + qE)$.
Since all parameters are positive, λ_2 is negative.
Hence, the trivial steady point E_0 is
(a) asymptotically stable if $\lambda_1 = r - \mu_1 < 0$, i. e., $r < \mu_1$;
(b) a saddle point if $\lambda_1 = r - \mu_1 > 0$, i. e., $r > \mu_1$.

Mathematically, for the set of parameters in Table 1, $r > \mu_1$ $(2 > 0.03)$. Hence, the trivial steady state is a saddle point.

Remark. It is observed that the existence of predator-free equilibrium $E_1(x_1, 0)$ leads to the instability of the trivial steady point $E_0(0,0)$.

(ii) The variational matrix for the predator-free steady state $E_1(x_1, 0)$ is

$$U_1 = \begin{pmatrix} -(r-\mu_1) & -\left(\frac{rK(r-\mu_1)}{\mu_2} + \frac{a(1-m)(r-\mu_1)}{b\mu_2+(1-m)(r-\mu_2)}\right) \\ 0 & -(d_1+qE) + \frac{ca(1-m)(r-\mu_1)}{b\mu_2+(1-m)(r-\mu_2)} \end{pmatrix}.$$

The eigenvalues of U_1 are $\lambda_1 = -(r - \mu_1) < 0$ and

$$\lambda_2 = -(d_1 + qE) + \frac{ca(1-m)(r-\mu_1)}{b\mu_2 + (1-m)(r-\mu_2)}.$$

It can be noted that the steady state $E_1(x_1, 0)$ is
(a) asymptotically stable if $(d_1 + qE) > \frac{ca(1-m)(r-\mu_1)}{b\mu_2+(1-m)(r-\mu_2)}$;
(b) a saddle point if $(d_1 + qE) < \frac{ca(1-m)(r-\mu_1)}{b\mu_2+(1-m)(r-\mu_2)}$.

Mathematically, making use of the parameters in Table 1, $E_1(19.7, 0)$ has been calculated with its eigenvalues −19.7 and 2.34. Since one of the eigenvalues is positive, $E_1(19.7, 0)$ is a saddle point for this set of values. However, for $m = 0.95$, the eigenvalues of E_1 are −1.97, −0.05 (both negative). The condition of stability, i. e., $(d_1 + qE) > \frac{ca(1-m)(r-\mu_1)}{b\mu_2+(1-m)(r-\mu_2)}$, $(0.7 > 0.65)$ is verified. E_1 is locally asymptotically stable in this case, as is also verified from Figure 6. □

Theorem 7. *The coexisting steady state $E^* = (x^*, y^*)$ is locally asymptotically stable if*

$$\mu_2 > \frac{a(1-m)^2 y^*}{(b+(1-m)x^*)^2}.$$

Proof. The Jacobian matrix for the coexisting steady state is given by

$$U^*(x^*, y^*) = \begin{pmatrix} -\mu_2 x^* + \frac{a(1-m)^2 x^* y^*}{(b+(1-m)x^*)^2} & -(\frac{rKx^*}{(1+Ky^*)^2} + \frac{a(1-m)x^*}{b+(1-m)x^*}) \\ \frac{bca(1-m)y^*}{(b+(1-m)x^*)^2} & -d_2 y^* \end{pmatrix}.$$

The matrix has a characteristic equation given by $\lambda^2 + B_1\lambda + B_2 = 0$, where

$$B_1 = -\operatorname{trace} J^* = d_2 y^* + \left(\mu_2 x^* - \frac{a(1-m)^2 x^* y^*}{(b+(1-m)x^*)^2}\right)$$

$$B_2 = \det J^* = \left(\mu_2 x^* - \frac{a(1-m)^2 x^* y^*}{(b+(1-m)x^*)^2}\right) d_2 y^* + \frac{bca(1-m)^2 y^*}{(b+(1-m)x^*)^2} \times \left(\frac{rKx^*}{(1+Ky^*)^2} + \frac{a(1-m)x^*}{b+(1-m)x^*}\right).$$

The real part of the eigenvalues of J^* will be negative if $\operatorname{trace} J^* < 0$ and $\det J^* > 0$, i. e., If $B_1 > 0$ and $B_2 > 0$.

From the expressions above, this holds good if

$$\mu_2 > \frac{a(1-m)^2 y^*}{(b+(1-m)x^*)^2}.$$ □

Theorem 8. (i) *The coexisting equilibrium point $E^*(x^*, y^*)$ is a saddle point for $B_2 < 0$.*
(ii) *The coexisting equilibrium point $E^*(x^*, y^*)$ is unstable if $B_2 > 0$ and $B_1 < 0$.*

Proof. (i) The characteristic equation corresponding to E^* is $\lambda^2 + B_1\lambda + B_2 = 0$.

For $B_2 < 0$, both the eigenvalues of U^* are real with opposite signs.

I. e., one of the eigenvalues of U^* is positive and hence $E^*(x^*, y^*)$ is a saddle point.

(ii) Evidently, for $B_2 > 0$ and $B_1 < 0$, the eigenvalues of J^* must either be complex conjugates with positive real part or real and positive.

Hence, $E^*(x^*, y^*)$ is unstable in this case. □

5.2 Global stability

Theorem 9. *The interior equilibrium state $E^* = (x^*, y^*)$ is globally asymptotically stable if $4d_2(\mu_2 - \frac{a(1-m)^2 y^*}{b+(1-m)x^*}) > \frac{r^2K^2}{(1+K\frac{r^2}{\eta\mu_2})^2(1+Ky^*)^2}$ and $a(1-m)^2 y^* < \mu_2(b+(1-m)x^*)$ holds.*

Proof. Construct a Lyapunov function $V(x, y)$ as

$$V = \left((x - x^*) - x^* \log \frac{x}{x^*}\right) + A\left((y - y^*) - y^* \log \frac{y}{y^*}\right),$$

where A is a positive constant specified ahead.

Clearly, $V(x^*, y^*) = 0$ and $V(x, y) > 0, \forall (x, y) \neq (x^*, y^*)$, i. e., function V is a positive definite function around equilibrium point E^*.

Also, $V(X, Y)$ is radically unbounded.

Next, we show that $\frac{dV}{dt} < 0$ for all $(x, y) \neq (x^*, y^*)$. Differentiating V w. r. t. time

$$\begin{aligned}
\frac{dV}{dt} &= \frac{(x - x^*)}{x}\frac{dx}{dt} + A\frac{(y - y^*)}{y}\frac{dy}{dt} \\
&= (x - x^*)\left(r\left(\frac{1}{1 + Ky} - \frac{1}{1 + Ky^*}\right) - \mu_2(x - x^*)\right. \\
&\quad \left. - (1 - m)a\left(\frac{y}{(1 - m)x + b} - \frac{y^*}{(1 - m)x^* + b}\right)\right) \\
&\quad + A(y - y^*)\left(ca(1 - m)\left[\frac{x}{(1 - m)x + b} - \frac{x^*}{(1 - m)x^* + b}\right] - d_2(y - y^*)\right) \\
&= (x - x^*)\left(-\mu_2(x - x^*) + \frac{a(1 - m)^2 y^*(x - x^*)}{((1 - m)x + b)((1 - m)x^* + b)}\right) + A(y - y^*)^2(-d_2) \\
&\quad - (y - y^*)(x - x^*)\left(\frac{rK}{(1 + Ky)(1 + Ky^*)} + \frac{a(1 - m)}{(1 - m)x + b}\right. \\
&\quad \left. - \frac{Acab(1 - m)}{((1 - m)x + b)((1 - m)x^* + b)}\right) \\
&= (x - x^*)^2\left(-\mu_2 + \frac{a(1 - m)^2 y^*}{(b + (1 - m)x)(b + (1 - m)x^*)}\right) - A(y - y^*)^2 d_2 \\
&\quad - (y - y^*)(x - x^*)\left(\frac{rK}{(1 + Ky)(1 + Ky^*)} + \frac{a(1 - m)}{b + (1 - m)x}\right. \\
&\quad \left. - \frac{Acab(1 - m)}{(b + (1 - m)x)(b + (1 - m)x^*)}\right).
\end{aligned}$$

Choosing $A = \frac{b + (1-m)x^*}{bc}$, we obtain

$$\frac{dV}{dt} \leq -(x - x^*)^2\left(\mu_2 - \frac{a(1 - m)^2 y^*}{b + (1 - m)x^*}\right) - Ad_2(y - y^*)^2 - \frac{rK}{(1 + Ky)(1 + Ky^*)}(x - x^*)(y - y^*),$$

where $\frac{dV}{dt}$ is negative definite if

$$4d_2\left(\mu_2 - \frac{a(1 - m)^2 y^*}{b + (1 - m)x^*}\right) > \frac{r^2 K^2}{(1 + K\frac{r^2}{\eta\mu_2})^2(1 + Ky^*)^2} \quad \text{and}$$

$$(1 - m)^2 a y^* < \mu_2((1 - m)x^* + b).$$

□

6 Periodic orbits

In this section, system (1) is explored for the appearance or disappearance of periodic orbits. For this, we use the Dulac–Bendixon and Poincare–Bendixon criteria.

Theorem 10. *The system* (1) *will not admit a periodic solution in* $\mathbb{R}_+^2$ *if*

$$\frac{\eta\mu_2^2}{r} + \frac{\mu_2}{d_2} > \frac{ar}{b^2}.$$

Proof. Let

$$f(x,y) = \frac{rx}{1+Ky} - \mu_1 x - \mu_2 x^2 - \frac{a(1-m)x(t)y(t)}{b+(1-m)x}$$
$$g(x,y) = \frac{ca(1-m)x(t)y(t)}{b+(1-m)x} - (d_1 x + qE)y - d_2 y^2.$$

Consider a Dulac function as $h(x,y) = \frac{1}{xy}$. Then

$$\begin{aligned}\triangledown &= \frac{\partial(fh)}{\partial(x)} + \frac{\partial(fg)}{\partial(y)}\\ &= -\left(\frac{\mu_2}{y} - \frac{a(1-m)^2}{(x(1-m)+b)^2} + \frac{1}{x}d_2\right).\end{aligned}$$

Note that $\triangledown < 0$ for

$$\frac{\mu_2}{y} + \frac{d_2}{x} > \frac{a(1-m)^2}{((1-m)x)^2 + b}. \tag{7}$$

In particular, $\min.(\frac{\mu_2}{y} + \frac{d_2}{x}) > \max.(\frac{a(1-m)^2}{(b+(1-m)x)^2})$.

Clearly, $\frac{a(1-m)^2}{(b+(1-m)x)^2} < \frac{a}{b^2}$. Also, y and x are substituted by $\frac{r^2}{\eta\mu_2}$ and $\frac{r}{\mu_2}$.

Thus, for $\frac{\eta\mu_2^2}{r} + \frac{\mu_2}{d_2} > \frac{ar}{b^2}$, $\triangledown < 0$. Hence, by the Dulac–Bendixon criterion system (1) will have no periodic orbits for this condition [28]. □

Theorem 11. *Model system* (1) *will possess at least one periodic orbit for* $\det U^* < 0$ *or* $\det U^* > 0$ *and* $\operatorname{trace} U^* > 0$.

Proof. For the stated conditions. we have, $B_2 < 0$ or $B_2 > 0$ and $B_1 < 0$.

By Theorem 8, the equilibrium point E^* is unstable in this case.

System (1) and (2) has a solution set that is positive invariant and bounded. As a result, no equilibrium point can belong to the limit set.

Hence, it is concluded by the Poincare–Bendixon criterion that at least one periodic solution is admitted by the system [14]. □

7 Bifurcations

Bifurcation analysis is a mathematical tool used in dynamical systems theory to study the qualitative changes in the dynamics of the system as one or more parameters are varied. Here, we perform bifurcation analysis of the system to check the possibility of the occurrence of transcritical and Hopf bifurcations.

7.1 Transcritical bifurcation

A transcritical bifurcation occurs when a fixed point remains present across the entire range of a parameter, never disappearing. However, this fixed point undergoes a stability exchange with another fixed point as the parameter is adjusted. In this subsection, we examine system (1) for the existence of a transcritical bifurcation w. r. t. parameter E. Let $X = (x, y)$, so the system (1) can be expressed as $X' = f(X), f = (g, h)^T$. We have

$$f(x,y) = \begin{pmatrix} g(x,y) \\ h(x,y) \end{pmatrix} = \begin{pmatrix} \frac{rx}{1+Ky} - \mu_1 x - \mu_2 x^2 - \frac{axy}{b+x} \\ \frac{caxy}{b+x} - (d_1 x + qE)y - d_2 y^2 \end{pmatrix}$$

$$f_E = \begin{pmatrix} 0 \\ -qy \end{pmatrix}, \quad Df_E = \begin{pmatrix} 0 & 0 \\ 0 & -q \end{pmatrix}.$$

Transcritical bifurcation at E_0 w. r. t. harvesting parameter E

Taking one of the eigenvalues of J_0, $-d_1 - qE = 0$.

Corresponding to zero eigenvalue, the eigenvectors of J_0 and J_0^T are

$$v = \begin{pmatrix} v_1 \\ v_2 \end{pmatrix} = \begin{pmatrix} 1 \\ 0 \end{pmatrix} \quad \text{and} \quad w = \begin{pmatrix} w_1 \\ w_2 \end{pmatrix} = \begin{pmatrix} 1 \\ 0 \end{pmatrix}.$$

It is calculated that 1. $w^T f_E(E_0, E = E_{TC}) = 0$.
2. $w^T Df_E(E_0, E = E_{TC})v = (1\ 0) \begin{pmatrix} 0 & 0 \\ 0 & -q \end{pmatrix} \begin{pmatrix} 1 \\ 0 \end{pmatrix} = 0$.
Therefore, a transcritical bifurcation is not observed around zero steady state.

Transcritical bifurcation at E_1 w. r. t. harvesting parameter E

For a transcritical bifurcation at E_1, taking one of its eigenvalues

$$\frac{ca(1-m)(r-\mu_1)}{b\mu_2 + (1-m)(r-\mu_1)} - d_1 - qE = 0.$$

The eigenvalues of J_1 and J_1^T, associated with zero eigenvalue are

$$v = \begin{pmatrix} v_1 \\ v_2 \end{pmatrix} = \begin{pmatrix} -\frac{rK}{\mu_2} - \frac{a(1-m)}{b\mu_2+(1-m)(r-\mu_1)} \\ 1 \end{pmatrix} \quad \text{and} \quad w = \begin{pmatrix} w_1 \\ w_2 \end{pmatrix} = \begin{pmatrix} 0 \\ 1 \end{pmatrix}.$$

It is observed that:

1. $w^T f_E(E_1, E = E_{TC}) = 0;$
2. $w^T Df_E(E_1, E = E_{TC})v = -q \neq 0;$
3. $w^T D^2 f(E_1, E = E_{TC})(v, v) = \neq 0.$

The transversality conditions for the occurrence of transcritical bifurcations hold good. Hence, by Sotomayor's theorem [20, 22], a transcritical bifurcation exists at E_1 when

$$E = E_{TC} = \frac{1}{q}\left[\frac{ca(1-m)(r-\mu_1)}{(b\mu_2 + (1-m)(r-\mu_1)} - d_1\right].$$

Using the values of parameters from Table 1

$$E_{TC} = \frac{1}{0.04}\left[\frac{(0.4)(10)(1-0.2)(2-0.03)}{(5)(0.1) + (1-0.2)(2-0.03)} - (0.5)\right] = 63.41.$$

This is verified from Figure 8.

7.2 Hopf bifurcation

A Hopf bifurcation is said to occur when a limit cycle or periodic solution, around an equilibrium point, emerges or fades away as some parameter value varies. In this subsection, we explore the possibility of the occurrence of limit cycles around E^*, in the model system (1) via a Hopf bifurcation.

Theorem 12. *For the bifurcation threshold* $K = K^*$ *and* $B_2 > 0$, *the system* (1) *exhibits a Hopf bifurcation about the interior equilibrium point* $E^* = (x^*, y^*)$ *provided the following conditions hold* (1) trace $U^*|_{K=K^*} = 0$. (2) det $.U^*|_{K=K^*} > 0$. (3) $\frac{d(\text{trace } U^*)}{dK}|_{K=K^*} \neq 0$.

Proof. The Jacobian matrix U^* will have purely imaginary eigenvalues if $B_1 = 0, B_2 > 0$, i. e., trace $U^* = 0$, det $U^* > 0$. This gives the threshold value of the fear parameter

$$K = K^* = \frac{1}{y^*}\left(\frac{r}{\mu_1 + 2\mu_2 x + \frac{b(1-m)y^*}{(b+(1-m)x^*)^2} + d_2 y^*} - 1\right).$$

At $K = K^*$, trace $U^* = 0$. The characteristic equation becomes

$$\lambda^2 + B_2 = 0.$$

Hence, the eigenvalues $= \pm\sqrt{B_2}i$, which are purely imaginary. Also, we have

$$\frac{d(\text{trace}\, U^*)}{dK} = \frac{-ry^*}{(1+Ky^*)^2} < 0.$$

It can be observed that

(a) If $K > K^*$, then $B_1 > 0$, $B_2 > 0$. Hence, the interior point E^* is locally asymptotically stable.

(b) If $K < K^*$, then $B_1 < 0$, $B_2 > 0$ and hence, E^* will be unstable. Thus, by the Andronov–Hopf bifurcation theorem, the system undergoes a Hopf bifurcation at the coexisting equilibrium point $E^* = (x^*, y^*)$ as a pair of complex conjugate eigenvalues crosses the imaginary axis, leading to the emergence of a limit cycle, indicating a transition in the dynamics of the system. □

8 Numerical simulations

To verify the analytical findings and assess the effects of fear, harvesting, and refuge by phytoplankton in the system, numerical simulations were performed using MATLAB. The parameters outlined in Table 1 were utilized for these simulations.

8.1 Impact of fear factor on dynamics of system

The fear of being preyed upon can be viewed as an indirect effect of predators on their victims. Researchers have come to the realization that a prey–predator model should incorporate not only direct mortality but also the influence of fear [1, 2, 15, 17]. The mere presence of predators can lead to a change in behavior of prey and the overall system dynamics. For the proposed system, it is seen that the level of fear in predators can modify the stability of the positive equilibrium. In Figures 3 and 4, at $K = 0.2$ the unstable limit cycles shift to periodic oscillations at $K = 1.1$, subsequently leading the system to a subcritical Hopf bifurcation at $K^* = 1.49$ and finally a stable state $K = 3.6 > K^*$. Stability is observed in the system after crossing the threshold value K^* permitting persistence in the system. Figure 5 shows the Hopf point and the bifurcation diagram of both populations with respect to parameter K. From an ecological perspective, it can be inferred that low levels of fear within the prey population lead to oscillations, whereas higher levels induce stability within the system.

8.2 Modulation of system dynamics by parameter *m*

Fear creates a form of nonconsumptive predation pressure, where predators influence prey populations by their presence alone. The availability of refuge influences how

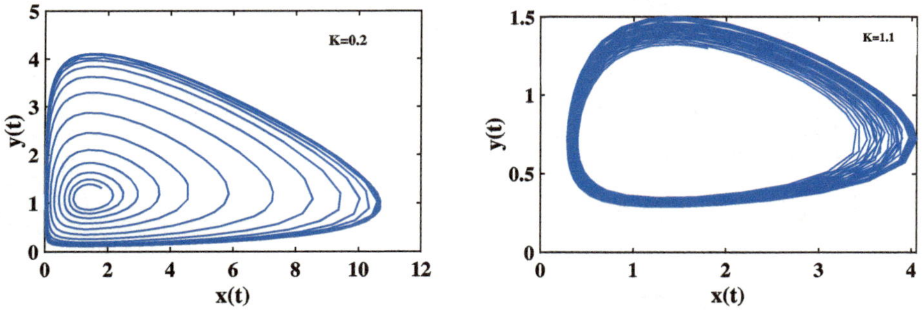

Figure 3: Unstable limit cycles and oscillations in the system.

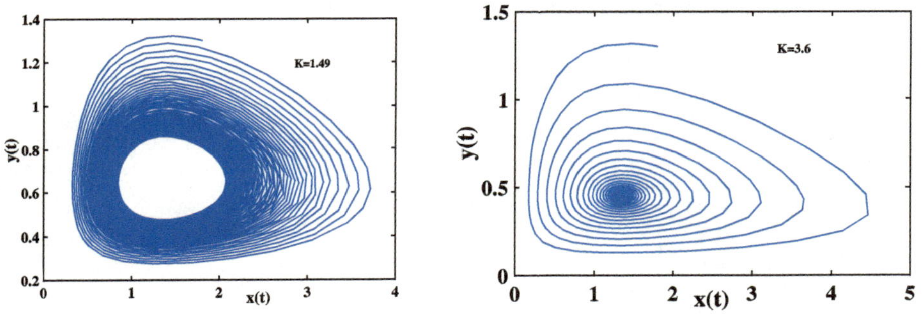

Figure 4: Phase diagrams showing transformation in the system from periodic oscillations to stability with the rise in the level of the fear parameter K.

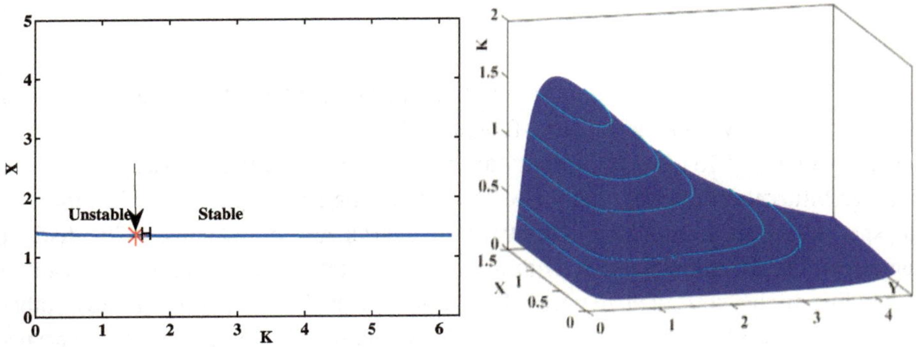

Figure 5: Depiction of system behavior via bifurcation diagrams with respect to the fear parameter K.

prey species balance the costs of fear with the need to forage or reproduce. Prey that can access a refuge are less exposed to predators, which can lower their stress and fear responses. In this way, refuges provide a physical and psychological buffer for prey

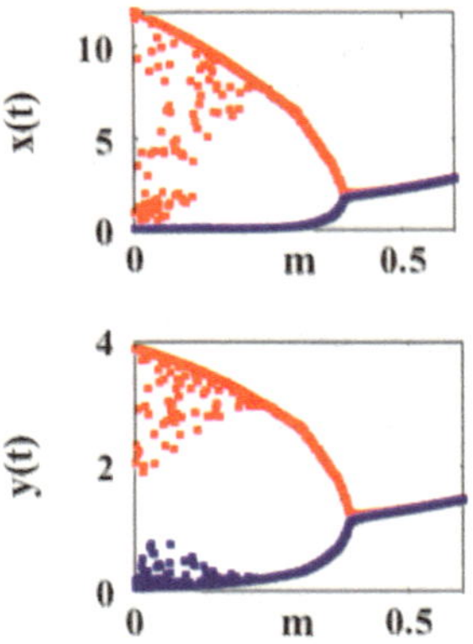

Figure 6: Bifurcation diagram of both populations w. r. t. parameter *m*.

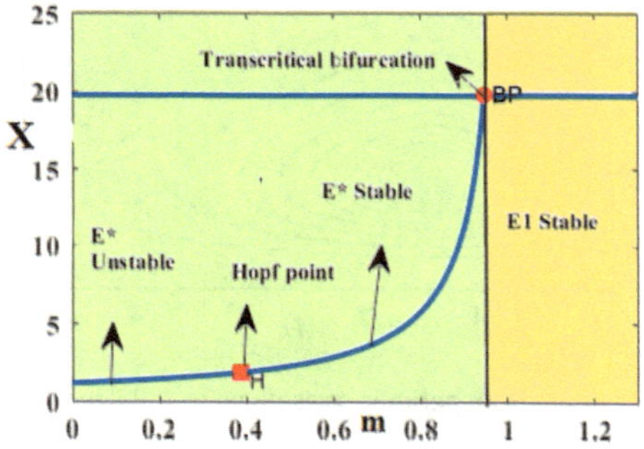

Figure 7: One-parameter bifurcation of the system w. r. t. parameter *m* showing stability of different steady states.

species. Figure 6 illustrates a bifurcation diagram that explores the dynamics of both predator and prey populations as a function of the parameter m, which represents refuge availability. To deepen the exploration of how the system behaves relative to m, another bifurcation diagram highlights the role of the refuge parameter in influencing species coexistence and the stability of the overall system in Figure 7. The distinct regions in the diagram reveal how different steady states become stable or unstable depending on the value of m. In particular, the green zone in the diagram shows a stable coexistence of both predator and prey populations, with a Hopf bifurcation point occurring at $m_h = 0.38$. Below this value of refuge, the system remains oscillatory. However, for values of m between threshold m_h and 0.946, both species are able to coexist. At the transcritical bifurcation point $m_t = 0.946$, equilibrium E_1 becomes stable, marking a shift where predators start to fade from the system. The orange zone in the diagram represents the stability of E_1. Beyond m_t, predators go extinct, leaving only prey to survive. Figure 8 further visualizes the impact of increasing refuge on the prey and predator

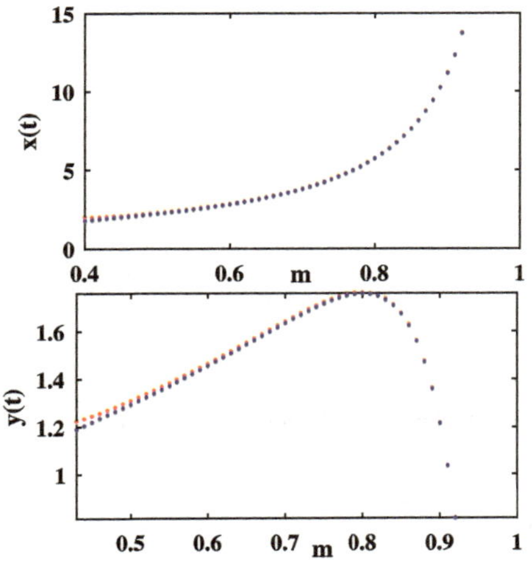

Figure 8: Population densities of prey and predator populations w. r. t. *m*.

populations. As more refuge becomes available, the prey population experiences a clear rise, while the predator population initially increases but then drops off sharply. This reflects how increasing protection (refuge) for prey alters the balance of the ecosystem, eventually tipping it in favor of the prey at the expense of the predators.

8.3 Impact of harvesting effort on the dynamics of system

In prey–predator interactions, the harvesting may help in controlling the system dynamics leading to sustainability in the system. However, for the persistence of species, harvesting needs to be optimal keeping in view the available resources. For the proposed system, linear harvesting of predator population has been considered and it is observed that harvesting should not be too small or too large. Figure 9 shows one-parameter bifurcation of the system w. r. t. the harvesting parameter. The bifurcation diagram is split into different regions. The blue zone shows the positive equilibrium, indicating that the system destabilizes at low harvesting levels. For the harvesting level between 0 to 22.5, the positive equilibrium remains unstable and undergoes a Hopf bifurcation at Hopf point $E^* = 22.5$. After this, the system stabilizes, allowing the coexistence of both species. As harvesting increases further, the populations can no longer coexist. A transcritical bifurcation occurs at $E_t = 63.4$, leading to the stabilization of a predator-free equilibrium, represented by the pink zone. As a result, harvesting above this critical value (E_t) drives the predator species to extinction, while for $0 \leq E \leq 63.4$, coexistence of both species is maintained.

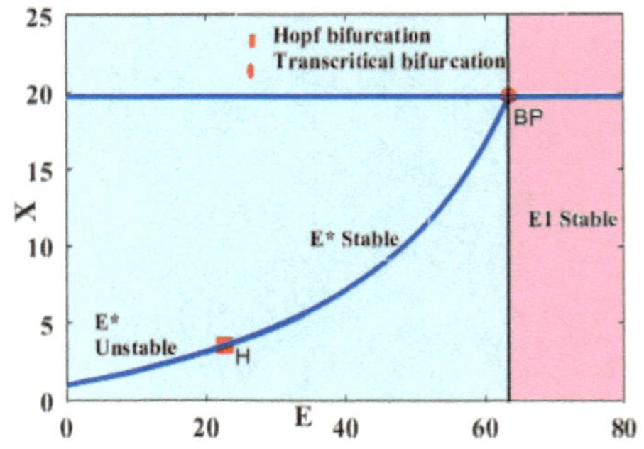

Figure 9: One-parameter bifurcation of the system depicting stability of the steady states w. r. t. E.

9 Conclusion and discussion

In this study, we investigate a prey–predator model that incorporates the effects of fear induced in the prey by the presence of predators, alongside the protective influence of prey refuge. Both species experience intraspecific competition, and a linear harvesting strategy is applied to the predator population. Predation follows a Holling type-II functional response. We establish the positivity and boundedness of the solutions to guarantee the biological feasibility of the model. Furthermore, we derive sufficient conditions for the persistence of both prey and predator species. Our analysis reveals that the prey population remains persistent if its intrinsic growth rate exceeds a critical threshold, while the predator population is sustainable as long as the combined effects of harvesting effort and natural mortality remain below a certain threshold. These findings clearly align with ecological expectations. Further, it is observed that threshold values for population persistence are influenced by the levels of fear in the prey and the extent of the refuge available to them. Hence, it can be concluded that fear and refuge parameters govern the persistence of the system. System dynamics unveiled three distinct equilibrium points: the trivial $E_0(0,0)$, $E(x_1,0)$, and coexisting equilibrium $E^*(x^*,y^*)$. Stability conditions for each were determined through a thorough analysis, revealing $E_0(0,0)$ to be a saddle point when $E(x_1,0)$ does not exist, locally asymptotically stable otherwise; the unique positive equilibrium point attains local asymptotic stability under the condition given in Theorem 7. Performing a Hopf bifurcation analysis with the fear parameter as a bifurcation parameter, we identified a critical value $K^* = 1.49$ for the fear factor, beyond which the system transitions to stability through an Andronov–Hopf bifurcation. The study concludes that a fear factor exceeding K^* effectively excludes population fluctuations and leads the system towards stability, consistent with the previous research by Wang et al. [18]. We also performed global stability analysis of a positive equilibrium by suitably selecting a Lyapunov function. By applying the Poincaré–Bendixon theorem and the Dulac–Bendixon criterion, we determined the existence and the nonexistence of periodic orbits. We discussed the impact of the refuge parameter by examining its influence on the system dynamics. In the range $0 \le m \le 0.38$ the system experiences periodic oscillations, followed by stable coexistence in the range $0.38 \le m \le 0.94$. However, increasing the refuge protection further negatively affects predators, resulting in their extinction and only prey survival. The impact of the harvesting effort, represented by E, on coexistence and system stability was also considered. The persistence of the system was found in the range $0 \le E \le 63.4$, with stability of the coexisting equilibrium evident in $22.5 \le E \le 63.4$, emphasizing the importance of harvesting for ecosystem sustainability. Predator species became extinct when harvesting efforts exceeded $E = 63.4$.

The study gives an understanding of the role of various internal and external factors that may be helpful in sustaining and the persistence of the ecosystem. It is observed that very small prey refuges tend to destabilize predator–prey interactions, whereas moderate refuges support stable coexistence. The system remains unstable at low levels of fear but achieves stability once fear surpasses a certain threshold. Excessive harvesting disrupts stability and raises the risk of ecosystem collapse, while optimal harvesting can maintain sustainable yields and protect both populations. The interaction among harvesting, the availability of refuges, and the fear instilled by predators can result in a variety of ecological outcomes, such as extinction, persistence, chaotic variations, or stable coexistence, depending on their levels. To maintain ecological balance, it is crucial to identify the critical thresholds of harvesting pressure, refuge distribution, and fear effects. These elements underscore the importance of careful regulation to prevent system collapse and ensure long-term sustainability.

In the future, we intend to extend the model by introducing delay in gestation and fear response of prey, which may be more realistic from a biological perspective.

Bibliography

[1] Bollens, S. M., & Stearns, D. E. (1992). Predator-induced changes in the diel feeding cycle of a planktonic copepod. *Journal of Experimental Marine Biology and Ecology*, *156*, 179–186.

[2] Cassini, M. H. (1991). Foraging under predation risk in the wild guinea pig *Cavia aperea*. *Oikos*, *62*, 20–24.

[3] Chinnathambi, R., & Rihan, F. A. (2018). Stability of fractional-order prey–predator system with time-delay and Monod–Haldane functional response. *Nonlinear Dynamics*, *92*(4), 1637–1648.

[4] Creel, S., Christianson, D., Liley, S., & Winnie, J. A. (2007). Predation risk affects reproductive physiology and demography of elk. *Science*, *315*, 960.

[5] Das, A., & Samanta, G. P. (2018). Modelling the fear effect on a stochastic prey–predator system with additional food for the predator. *Journal of Physics A: Mathematical and Theoretical*, *51*(46), 465601.

[6] Dill, L. M., & Houtman, R. (1989). The influence of distance to refuge on flight-initiation distance in the gray squirrel (*Sciurus carolinensis*). *Canadian Journal of Zoology*, *67*, 232–235.

[7] Dong, Y., Wu, D., Shen, C., & Ye, L. (2022). Influence of a fear effect and predator-taxis sensitivity on dynamical behaviour of a predator–prey model. *Zeitschrift für Angewandte Mathematik und Physik*, *73*(1), 1–17.

[8] Eggers, S., Griesser, M., Nystrand, M., & Ekman, J. (2006). Predation risk induces changes in nest-site selection and clutch size in the Siberian jay. *Proceedings of the Royal Society B: Biological Sciences*, *273*, 701–706.

[9] Hassell, M. P. (1978). *The Dynamics of Arthropod Predator–Prey Systems*. Princeton University Press.

[10] Holmes, W. G. (1991). Predator risk affects foraging pikas: observational and experimental evidence. *Animal Behaviour*, *42*, 111–119.

[11] Hua, F., Sieving, K. E., Fletcher, R. J., & Wright, C. A. (2014). Increased perception of predation risk to adults and offspring alters avian reproductive strategy and performance. *Behavioral Ecology*, *25*(3), 509–519.

[12] Kar, T. K. (2005). Stability analysis of a prey–predator model incorporating a prey refuge. *Communications in Nonlinear Science and Numerical Simulation*, *10*(6), 681–691.

[13] Křivan, V. (2013). Behavioural refuges and predator–prey coexistence. *Journal of Theoretical Biology*, *339*, 112–121.

[14] Kumar, A., & Dubey, B. (2019). Modelling the effect of fear in a prey–predator system with prey refuge and gestation delay. *International Journal of Bifurcation and Chaos*, *29*(14), 1950195.

[15] Lima, S. L. (1998). Nonlethal effects in the ecology of predator–prey interactions. *BioScience*, *48*(1), 25–34.

[16] Lima, S. L., & Dill, L. M. (1990). Behavioural decisions made under the risk of predation: a review and prospectus. *Canadian Journal of Zoology*, *68*, 619–640.

[17] Macháček, J. (1991). Indirect effect of planktivorous fish on the growth and reproduction of *Daphnia galeata*. *Hydrobiologia*, *225*, 193–197.

[18] Naji, R. K., & Majeed, J. S. (2016). The dynamical analysis of a prey–predator model with a refuge-stage structure prey population. *International Journal of Differential Equations*, 2010464. https://doi.org/10.1155/2016/2010464.

[19] Panday, P., Pal, N., Samanta, S., & Chattopadhyay, J. (2018). Stability and bifurcation analysis of a three-species food chain model with fear. *International Journal of Bifurcation and Chaos*, *28*(01), 1850009.

[20] Perko, L. (2013). *Differential Equations and Dynamical Systems* (3rd ed.). Springer.

[21] Pijanowska, J. (1993). Diel vertical migration in zooplankton: fixed or inducible behavior? *Archiv für Hydrobiologie, Beihefte Ergebnisse der Limnologie*, *39*, 89–97.

[22] Rana, N., Kumar, R., & Sarkar, A. (2019). Role of fear factor in controlling chaos of a plankton–fish system with prey refuge and an additional food. *The European Physical Journal Plus*, *139*(1), 104.

[23] Raw, S. N., Mishra, P., Kumar, R., & Thakur, S. (2017). Complex behaviour of prey–predator system exhibiting group defence: a mathematical modelling study. *Chaos, Solitons & Fractals*, *100*, 74–90.

[24] Sarkar, K., & Khajanchi, S. (2020). Impact of fear effect on the growth of prey in a predator–prey interaction model. *Ecological Complexity*, *42*, 100826.

[25] Sarwardi, S., Mandal, P. K., & Ray, S. (2012). Analysis of a competitive prey–predator system with a prey refuge. *Biosystems*, *110*, 133–148.

[26] Wang, X., Zanette, L., & Zou, X. (2016). Modelling the fear effect in predator–prey interactions. *Journal of Mathematical Biology*, *75*(5), 1179–1204.

[27] Wang, X., & Zou, X. (2017). Modelling the fear effect in predator–prey interactions with adaptive avoidance of predators. *Bulletin of Mathematical Biology*, *79*, 13–25.

[28] Yadav, S., Tripathi, J. P., Bhuria, S., Tiwari, S. K., Tripathi, D., Tiwari, V., Upadhyay, R. K., & Kang, Y. (2023). Ecological system with fear-induced group defence and prey refuge. arXiv preprint. https://doi.org/10.48550/arXiv.2302.04594.

[29] Zanette, L. Y., White, A. F., Allen, M. C., & Clinchy, M. (2011). Perceived predation risk reduces the number of offspring songbirds produce per year. *Science*, *334*, 1398–1401.

[30] Zhang, H., Cai, Y., Fu, S., & Wang, W. (2019). Impact of fear effect in a prey–predator model incorporating a prey refuge. *Applied Mathematics and Computation*, *356*, 328–337.

Gagandeep Kaur, Rajni, and Jagtar Singh Sivia

Deep learning convolutional neural networks (CNNs) for the detection of leaf blast disease of rice

Abstract: Automatic disease detection using deep learning convolutional neural networks (CNNs) holds immense significance in modern agriculture, transforming the approach to plant diseases that are recognized, examined, and managed. By leveraging the supremacy of CNNs, farmers and agriexperts can precisely and quickly detect diseases, enabling timely intervention to alleviate crop losses and ensure food safety. These advanced artificial intelligence systems provide a scalable and efficient solution to the complex task of disease identification, enabling early detection and targeted treatments, ultimately enhancing crop yields and sustainable agricultural practices. We developed and compared two deep learning CNNs (viz. AlexNet and Xception) models after training selected a dataset (480 images) of rice leaf blast disease images collected from the Kaggle online database. A total dataset was classified in the ratio of 80:20 (training-to-testing). AlexNet was developed at a spatial resolution of 16×16 pixels ($\text{AlexNet}_{16\times16}$), while the Xception model was developed at a higher spatial resolution of 224×224 pixels ($\text{Xception}_{224\times224}$). These results revealed that while both models demonstrate consistent computational performance, the $\text{Xception}_{224\times224}$ model significantly outperformed $\text{AlexNet}_{16\times16}$ in terms of accuracy and efficiency. The $\text{Xception}_{224\times224}$ model showed a steady decrease in epoch time from 441.32 to 429.87 s, indicating improved computational efficiency as training progresses. In contrast, the $\text{AlexNet}_{16\times16}$ model exhibits stable epoch times (29.85 to 32.25 s). The $\text{Xception}_{224\times224}$ model outperformed the $\text{AlexNet}_{16\times16}$ model with accuracy reaching 94.87 %, compared with 68.74 % for $\text{AlexNet}_{16\times16}$. The $\text{Xception}_{224\times224}$ model also surpassed the $\text{AlexNet}_{16\times16}$ model in precision, recall, F1-score, sensitivity, and specificity; thereby, highlighting its enhanced ability to accurately identify and classify rice leaf blast disease. These findings indicate that $\text{Xception}_{224\times224}$ was not only more efficient, but also more reliable and effective in disease detection, making it a superior choice for practical applications in agricultural diagnostics compared to $\text{AlexNet}_{16\times16}$.

Keywords: CNNs, recognition accuracy, deep learning models, Xception model, rice ecosystems

MSC 2020: 68T07, 68T10, 68U10, 62H35, 92B20

Gagandeep Kaur, Rajni, Department of Electronics and Communication Engineering, Shaheed Bhagat Singh State University, Ferozepur, 152001, Punjab, India, e-mail: ergagan84@gmail.com
Jagtar Singh Sivia, Yadavindra Department of Engineering (YDOE), Punjabi University South Campus, Talwandi Sabo, Bathinda, 151501, Punjab, India

https://doi.org/10.1515/9783112234525-011

1 Introduction

Rice leaf blast disease caused by Magnaporthe oryzae (also known as *Pyricularia oryzae*) presents a significant threat to global rice production, impacting food security and agricultural sustainability [14]. Therefore, accurate and timely identification of rice leaf blast disease is crucial for managing outbreaks and mitigating crop losses in different rice ecosystems [3]. Traditional methods for detecting rice leaf blast rely on visual inspection and manual diagnosis, which are often labor-intensive and prone to errors [5]. As the demand for more accurate and scalable solutions grows, there has been increasing attention in leveraging progressive computational methods, particularly deep learning and convolutional neural networks (CNNs), to enhance disease detection capabilities [5, 10].

The CNNs have arisen as influential tools in image recognition and classification tasks [6], offering significant improvements in accuracy and efficiency over the traditional techniques [7, 9]. Among these, the AlexNet and Xception models emerged as robust models for their accuracy in various image classification challenges. AlexNet, introduced by Krizhevsky et al. [8], revolutionized the field with its deep architecture and use of 'Rectified Linear Units' (ReLU) for activation, and thereby, demonstrating exceptional performance on large-scale image datasets. On the other hand, the Xception model, proposed by Chollet [2] built on the concept of depthwise distinguishable convolutions, helped to expressively lessen the number of parameters and computational complication, while maintaining high accuracy.

The present study investigates the application of two CNNs (e. g., AlexNet and Xception) for detecting rice leaf blast disease. By employing these models, we aimed to explore their effectiveness in automating disease detection through image analysis, thereby enhancing both the speed and accuracy of diagnosis. The study focused on comparing their performance based on several metrics, e. g., accuracy, precision, recall, F1-score, sensitivity, and specificity. Additionally, we evaluated the computational efficiency of each model by analyzing training times and their impact on model performance. The present study would help identify the most effective and efficient approach for integrating deep learning technologies into agricultural disease management practices, thereby advancing the capabilities of precision agriculture and contributing to improved crop health and yield.

2 Methodology

The present study involved a machine learning system to categorize and elucidate rice blast disease based on disease infected symptoms on infested leaves (Figure 1). Moreover, a dataset on healthy (disease-free) plant leaves was used for the development of disease detection models. The classification of leaf blast disease symptoms was performed using CNNs, viz. the AlexNet [8] and Xception models [2], using a dataset of selected 480 diseased plant images from an online (https://www.kaggle.com/) database. Additionally, we explored preparing the training and testing datasets by preprocessing the images, while adhering to an 80:20 ratio for training and testing [11]. A generalized overview of the methodology adopted for the detection of leaf blast disease of rice is illustrated in Figure 1. The visual examination of the dataset images, which depict disease symptoms, identified three distinct clusters, prompting the application of k-means clustering ($k = 3$) for segmentation. This approach enabled the segmentation of images into relevant regions, allowing for the precise identification of diseased areas. After segmentation, image features were mined, which were followed by color and texture feature extraction by using a cooccurrence matrix method, as described by Yakkundimath et al. [12].

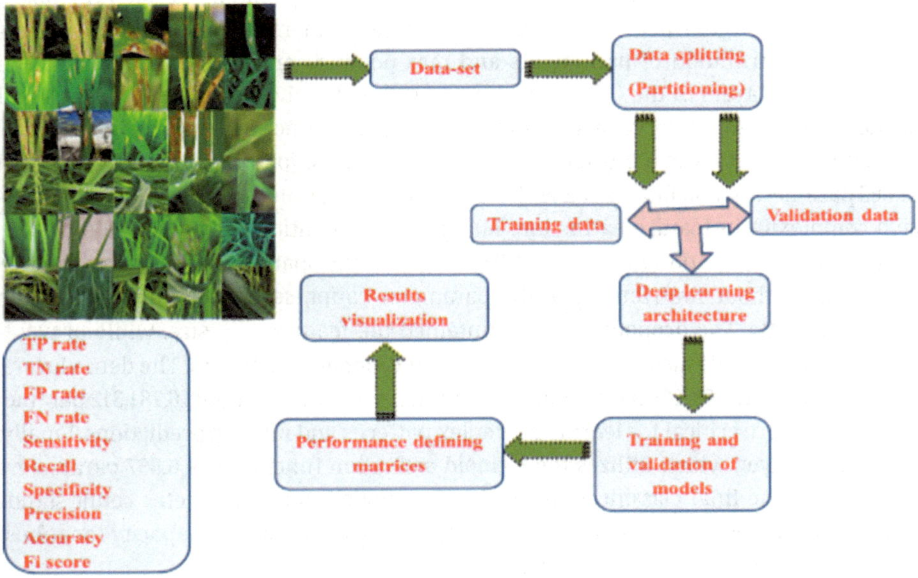

Figure 1: Generalized overview of the methodology adopted used for the detection of leaf blast disease of rice.

The parameters of accuracy, precision, recall, sensitivity, specificity, and F-measure [4] for the proposed model were calculated using the following equations:

$$\text{Accuracy}(\%) = \frac{\text{True positive(TP)} + \text{True negative(TN)}}{\text{Total predictions}} \times 100 \tag{1}$$

$$\text{Precision}(\%) = \frac{\text{True positive(TP)}}{\text{True positive(TP)} + \text{False positive(FP)}} \times 100 \tag{2}$$

$$\text{Recall}(\%) = \frac{\text{True positive(TP)}}{\text{True positive(TP)} + \text{False negative(FN)}} \times 100 \tag{3}$$

$$F - \text{measure} = \frac{2 \times \text{Precision} \times \text{Recall}}{\text{Precision} + \text{Recall}} \times 100 \tag{4}$$

$$\text{Sensitivity} = \frac{\text{True positive(TP)}}{\text{True positive(TP)} + \text{False negative(FN)}} \tag{5}$$

$$\text{Specificity} = \frac{\text{True negative(TN)}}{\text{True negative(TN)} + \text{False positive(FP)}}. \tag{6}$$

3 Results and discussion

3.1 Breakdown of each layer's parameters for the AlexNet model

The AlexNet model, specifically tailored at a spatial resolution of 16 × 16 pixels ($\text{AlexNet}_{16\times16}$) for the detection of leaf blast disease in rice, demonstrated a well-structured distribution of parameters and max-pooling configurations (Table 1). The convolutional layers in the CNNs were configured with ReLU activation functions and 'same' padding to preserve spatial dimensions, while processing the input. The initial convolutional layer (Conv#2D) has 9,936 parameters, followed by Conv2#D-1 with 11,365 parameters, and Conv2#D-2 with 7,256 parameters, reflecting the models' progressive complexity in feature extraction. Max-pooling operations, e. g., Max_Pooling#2D, Max_Pooling#2D_1, and Max_Pooling#2D_2 reduced the spatial dimensions of the feature maps without contributing to the parameter count, serving to downsample the input efficiently. The dropout layers maintained the feature map size, while preventing overfitting, while also not introducing any additional parameters. The dense layers, included one with 104,143 parameters and another with a substantial 16,781,312 parameters, which were critical for learning complex patterns and making predictions. Finally, the dense-2 layer, which utilizes the sigmoid activation function has 6,857 parameters and outputs the final classification results. The model's total parameter count stands at 11,041,512, all of which are trainable, underscoring the network's capacity for robust feature learning and disease detection.

3.2 Performance appraisal of the $\text{AlexNet}_{16\times16}$ model

The dataset provided outlines the time consumed for modeling output as a function of the number of epochs in the context of rice leaf blast disease detection using an AlexNet model with a spatial resolution of 16 × 16 pixels (Figure 2). The recorded times for each epoch, ranging from approximately 29.85 to 32.25 s, exhibited a relatively stable pattern, indicating minimal variation in computational time as a function of number of epochs. Specifically, the time for the 1st epoch was 32.25 s, which slightly decreased (or stabilized) in subsequent epochs, with values such as 30.55, 30.89, and consistently around 29.85 to

Table 1: Number of parameters and max-pooling related details for the AlexNet model (AlexNet$_{16\times16}$) developed for the detection of leaf blast disease of rice.

Layer type	Activation function	Padding	Max-pooling output shape	Number of training parameters
Conv#2D	ReLU	Same padding	(None, 16, 16, 32)	9,936
Conv2#D-1	ReLU	Same padding	(None, 6, 6, 32)	11,365
Conv2#D-2	ReLU	Same padding	(None, 2, 2, 32)	7,256
Max_Pooling#2D	–	–	(None, 6, 6, 32)	0
Max_Pooling#2D_1	–	–	(None, 2, 2, 32)	0
Max_Pooling#2D_2	–	–	(None, 1, 1, 32)	0
Dropout	–	–	(None, 1, 1, 32)	0
Dense	ReLU	–	(None, 1, 1, 4096)	1,04,143
Dense-1	–	–	(None, 1, 1, 4096)	16,78,1312
Flatten layer	–	–	(None, 4096)	0
Dense-2	Sigmoid	–	(None, 2)	6,857
Total parameters	–	–	–	11,04,1512
Trainable parameters	–	–	–	11,04,1512

30.53 s for epochs 4 through 10. Such stability suggests that the computational demand remains relatively constant throughout the training process. The marginal fluctuations observed are within a narrow range, reflecting the efficient performance of the AlexNet model at the specified spatial resolution. Overall, these results indicate that the time complexity for training the model does not significantly increase with the number of epochs, supporting the model's consistent performance in the detection task.

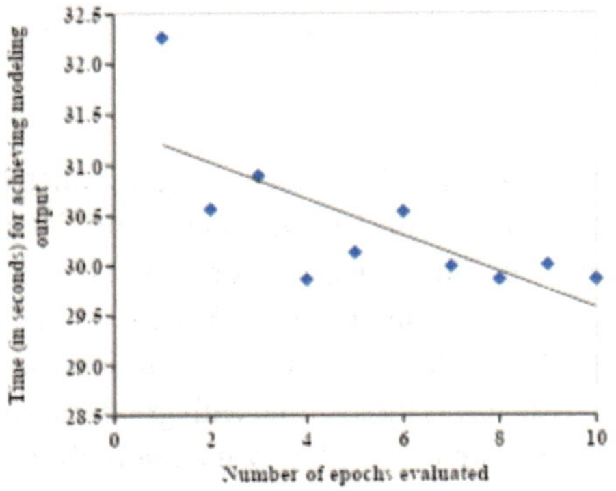

Figure 2: Relationship between the number of epochs evaluated and time (in s) required for the AlexNet$_{16\times16}$ model for executing model output in the detection of leaf blast disease of rice.

3.3 Training and validation loss and accuracy for the AlexNet model across epochs

Figure 3a shows the performance dynamics of the AlexNet$_{16\times16}$ model in detecting rice leaf blast disease over ten epochs. Training loss and accuracy metrics demonstrate a positive trend; the training loss decreases from 0.5874 at epoch 1 to 0.4865 by epoch 10, reflecting improved model performance over the epochs. Validation loss also showed a notable decline from 0.6654 at epoch 1 to 0.5532 by epoch 10, underscoring the model's enhanced generalization capability on an unseen dataset. A steady reduction in validation loss suggests that the AlexNet$_{16\times16}$ model benefits from prolonged training, achieving better fit and robustness in its predictions. Overall, the results highlight the model's effective learning progression and its capacity to deliver increasingly accurate and reliable disease detection.

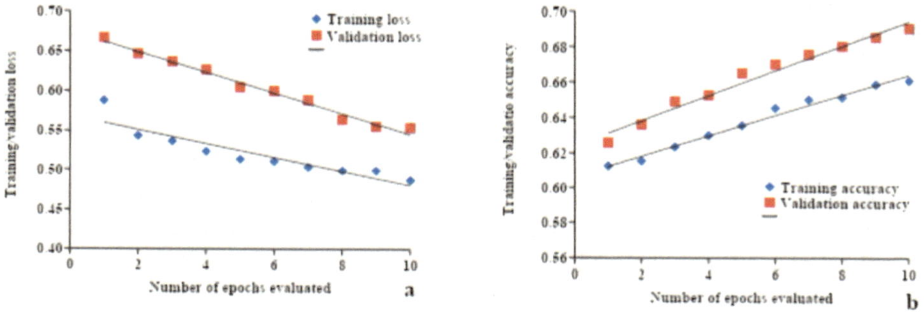

Figure 3: Relationship between the number of epochs evaluated and training/validation loss (a) and training/validation accuracy (b) for the AlexNet$_{16\times16}$ model for executing model output in the detection of leaf blast disease of rice.

In addition, training accuracy increased from 61.24 to 66.07 %, illustrating enhanced model performance. Validation accuracy exhibited a similar trend, increasing from 62.54 % at epoch 1 to 69.05 % at epoch 10 (Figure 3b). This consistent improvement in both validation loss and accuracy signifies that the AlexNet model was effective in learning and generalizing better with additional epochs. These results collectively indicate that the model benefits from extended training, which was discernible with greater reliability achieved by high accuracy in detecting rice leaf blast disease, while maintaining computational efficiency.

3.4 Performance appraisal of the Xception$_{224\times224}$ model

Figure 4 illustrate the time consumption per epoch for the Xception$_{224\times224}$ model designed for detecting rice leaf blast disease. The time required per epoch exhibited a gradual decrease over the course of 10 epochs, starting from 441.32 s at epoch 1 and steadily declining to 429.87 s by epoch 10. Such progressive reduction in epoch time reflects a slight but consistent improvement in computational efficiency as the training progresses. The decrease in time per epoch suggests that the model's learning process becomes increasingly efficient over time, potentially due to optimizations in computational operations or reduced complexity in feature updates as training stabilizes. This trend was indicative of the $Xception_{224\times224}$ model's capability to effectively manage computational resources, while maintaining performance in the disease detection task. Overall, these results underscore a positive trajectory in training efficiency, enhancing the model's operational effectiveness throughout the training process.

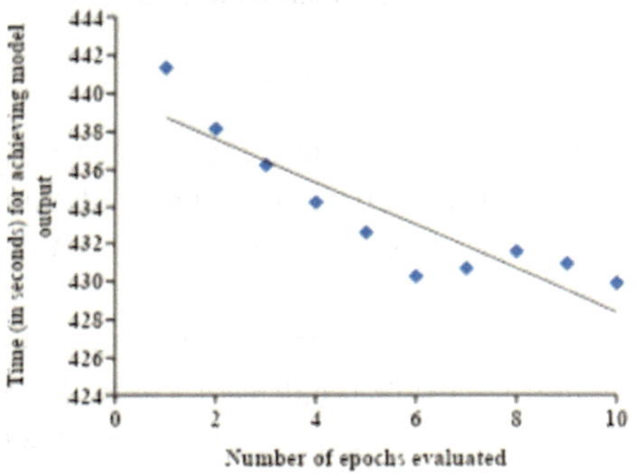

Figure 4: Relationship between the number of epochs evaluated and time (in s) required for the $Xception_{224\times224}$ model for executing model output in the detection of leaf blast disease of rice.

3.5 Training and validation loss and accuracy for the $Xception_{224\times224}$ model across epochs

These results provide a comprehensive view of training and validation loss for the $Xception_{224\times224}$ model's performance in detecting rice leaf blast disease evaluated across ten epochs (Figure 5a). The dataset revealed a consistent decline in both training loss and validation loss as a function of number of epochs. Starting with a training loss of 0.087 and a validation loss of 0.0445 at epoch 1, the $Xception_{224\times224}$ showed a marked improvement over time. By epoch 10, training loss has decreased significantly to 0.032, while validation loss also diminished to 0.02854. This downward trend in both metrics indicates that the $Xception_{224\times224}$ model not only fits the training data increasingly well, but also generalizes effectively to unseen data.

A gradual reduction in training loss from 0.087 to 0.032 demonstrated enhanced model learning and optimization, while the validation loss reduction from 0.045 to 0.029 reflects improved performance on the validation set. The stable and minor fluctuations in epoch time, ranging from 441.32 s at epoch 1 to 429.87 s at epoch 10, further suggest that

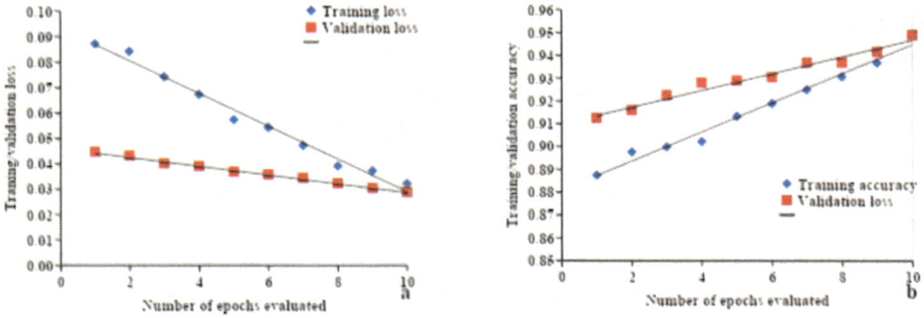

Figure 5: Relationship between the number of epochs evaluated and training/validation loss (a) and training/validation accuracy (b) for the Xception$_{224\times224}$ model for executing model output in the detection of leaf blast disease of rice.

the model's training efficiency improves over epochs. Overall, these results highlight the Xception$_{224\times224}$ model's robust capability in minimizing both training and validation losses, thereby, underscoring its efficacy in detecting rice leaf blast disease with high accuracy and reliability.

Both training and validation accuracy metrics exhibited a progressive improvement, highlighting the model's growing effectiveness over time (Figure 5b). Training accuracy starts at 88.74 % in epoch 1 and steadily rises to 94.87 % by epoch 10, indicating an increasing ability to correctly classify training samples. Simultaneously, validation accuracy showed a similar upward trajectory, beginning at 91.23 % and reaching 94.87 % by epoch 10. Such alignment in the trends of training and validation accuracy reflects the model's consistent improvement not only in learning the training dataset, but also in generalizing to unseen data. The concurrent increase in both accuracy measures suggests that the model was effectively capturing the underlying patterns associated with rice leaf blast disease and was well-calibrated for making accurate predictions on new data. A slight fluctuation in epoch time suggested a small variation in computational efficiency, while maintaining robust performance. Overall, the results underscore the Xception$_{224\times224}$ model's efficacy and reliability, achieving high accuracy in both training and validation phases, thereby, demonstrating a successful approach to disease detection.

3.6 Relative performance appraisal of the AlexNet and Xception$_{224\times224}$ models

Table 2 provides a comparative analysis of the AlexNet$_{16\times16}$ and Xception$_{224\times224}$ models for the detection of rice leaf blast disease. The Xception$_{224\times224}$ model significantly outperformed the AlexNet$_{16\times16}$ model across various performance metrics. Specifically, Xception$_{224\times224}$ achieved an accuracy of 94.87 %, markedly higher (by 38 %) than

AlexNet$_{16\times16}$'s accuracy of 68.74 %. Precision and recall values further illustrated the superiority of the Xception$_{224\times224}$ model, with precision at 95.32 % and recall at 0.879, compared to AlexNet$_{16\times16}$'s 79.62 % precision and 0.551 recall. The F1-score, which balances precision and recall, was notably lower for the Xception$_{224\times224}$ model at 0.945, in contrast to AlexNet$_{16\times16}$'s F1-score of 1.094.

Table 2: Relative performance appraisal of two deep learning convolutional neural networks (CNNs), viz. the AlexNet and Xception (Xception$_{224\times224}$) models that were developed for the detection of leaf blast disease of rice.

Parameter	Value of model parameter	
	AlexNet$_{16\times16}$	Xception$_{224\times224}$
Accuracy (%)	68.74	94.87
Precision (%)	79.62	95.32
Recall	0.551	0.879
F1-score	1.094	0.945
Sensitivity	0.551	0.879
Specificity	0.861	0.969
True positive rate	0.516	0.879
False positive rate	0.132	0.004
True negative rate	0.819	0.987
False negative rate	0.421	0.011

Sensitivity, reflecting the model's ability to correctly identify positive cases, was superior in Xception$_{224\times224}$ (0.879) compared to the AlexNet$_{16\times16}$ (0.551) model. Furthermore, the Xception$_{224\times224}$ model demonstrated higher specificity (0.969) and a lower false positive (FP) rate (0.004) than AlexNet$_{16\times16}$, which has a specificity of 0.861 and a FP rate of 0.132. The true negative (TN) rate was substantially higher for Xception$_{224\times224}$ at 0.987, while AlexNet$_{16\times16}$ stands at 0.819. Conversely, Xception$_{224\times224}$ exhibits a significantly lower false negative (FN) rate (0.011) compared to AlexNet$_{16\times16}$'s FN rate of 0.421. These metrics collectively highlight Xception$_{224\times224}$'s advanced capability in accurately detecting leaf blast disease, exhibiting higher reliability and performance across all evaluated parameters compared to the AlexNet$_{16\times16}$ model. Mohanty et al. [9] developed a smartphone-based disease identification system leveraging deep learning techniques, particularly the AlexNet and GoogLeNet architectures, trained on an extensive dataset comprising both healthy and infected plant leaves. Their [9] methodology enabled them to achieve an impressive accuracy of approximately 99.35 % in diagnosing 26 different diseases across 14 crop species. However, better performance of the Xception model could be ascribed to its design having a deeper architecture and more complex structure compared to AlexNet. It includes a larger number of layers and a more sophisticated network configuration, which enables it to learn more abstract and nuanced features from the input data [13]. Such depth contributes to improved performance in tasks such as image classification and disease detection [1].

4 Conclusions

The comparative analysis of the $\text{AlexNet}_{16\times16}$ and $\text{Xception}_{224\times224}$ models for detecting rice leaf blast disease revealed substantial differences in performance, thereby, highlighting the superior efficacy of the $\text{Xception}_{224\times224}$ model. The $\text{AlexNet}_{16\times16}$ model maintained stable epoch duration, ranging from approximately 29.85 to 32.25 s, reflecting consistent computational efficiency throughout the training process. In contrast, the $\text{Xception}_{224\times224}$ model exhibited a gradual decrease in epoch time, indicating a progressive improvement in computational efficiency as training progresses. The $\text{Xception}_{224\times224}$ model achieved a significantly higher accuracy (of 94.87 %) as compared to $\text{AlexNet}_{16\times16}$ models' accuracy (of 68.74 %), underscoring its superior capability in correctly identifying leaf blast disease cases. The sensitivity and specificity of the $\text{Xception}_{224\times224}$ model outperformed those of $\text{AlexNet}_{16\times16}$, which indicates that $\text{Xception}_{224\times224}$ was more adept at identifying TPs and TNs, resulting in a lower FP rates. The higher TN rate for $\text{Xception}_{224\times224}$ further emphasizes its reliability in correctly identifying nondiseased instances, compared to the $\text{AlexNet}_{16\times16}$ model. These results underscore the $\text{Xception}_{224\times224}$ model's advanced capabilities in both detection accuracy and computational efficiency, offering a more reliable and effective solution for rice leaf blast disease detection. The superior performance across all evaluated metrics highlights the model's robustness and suitability for practical applications in agricultural disease detection, making it a preferred choice over the $\text{AlexNet}_{16\times16}$ model.

Bibliography

[1] Bhagwat, R., & Dandawate, Y. (2021). Comprehensive multilayer convolutional neural network for plant disease detection. *International Journal of Advanced Computer Science and Applications*, *12*(1), 204–211. https://doi.org/10.14569/IJACSA.2021.0120125.

[2] Chollet, F. (2017). Xception: deep learning with depthwise separable convolutions. In *Proceedings of the IEEE Conference on Computer Vision and Pattern Recognition* (pp. 1251–1258). https://arxiv.org/abs/1610.02357.

[3] Deng, Y. W., Zhai, K. R., Xie, Z., Yang, D. Y., Zhu, X. D., & Liu, J. Z. (2017). Epigenetic regulation of antagonistic receptors confers rice blast resistance with yield balance. *Science*, *355*, 962–965. https://doi.org/10.1126/science.aai8898.

[4] Gokula, K. V., Deepa, J., Rao, P. V., Divya, V., & Kaviarasan, S. (2022). An automated segmentation and classification model for banana leaf disease detection. *Journal of Applied Biology and Biotechnology*, *10*(1), 213–220. https://doi.org/10.7324/JABB.2021.100126.

[5] Kaur, G., Rajni, & Sivia, J. S. (2024). Development of deep and machine learning convolutional networks of variable spatial resolution for automatic detection of leaf blast disease of rice. *Computers and Electronics in Agriculture*, *224*, 109210. https://doi.org/10.1016/j.compag.2024.109210.

[6] Kaur, G. (2022). Yellow rust disease of wheat detection using support vector machine and Naïve Bayes algorithm integrated with KNN classifier. Paper presented at the *National Seminar of Electronics and Telecommunication Engineering Division Board*, Institution of Engineers Local Centre Bathinda, Punjab, India.

[7] Kaur, G. (2020). Detection of bacterial leaf blight disease of cotton (*Gossypium hirsutum* L.) using convolution neural network (CNN): Simulations with support vector machine (SVM) and Naïve Bayes algorithms. *Design Engineering*, *5*, 2384–2400.

[8] Krizhevsky, A., Sutskever, I., & Hinton, G. E. (2012). ImageNet classification with deep convolutional neural networks. In *Proceedings of the 25th International Conference on Neural Information Processing Systems* (pp. 1097–1105).

[9] Mohanty, S. P., Hughes, D. P., & Salathé, M. (2016). Using deep learning for image-based plant disease detection. *Frontiers in Plant Science*, *7*, 1419. https://doi.org/10.3389/fpls.2016.01419.

[10] Schmidhuber, J. (2015). Deep learning in neural networks: an overview. *Neural Networks*, *61*, 85–117. https://doi.org/10.1016/j.neunet.2014.09.003.

[11] Shrivastava, P. (2013). A survey of image classification based techniques. *International Journal of Engineering Research and Technology*, *2*, 1987–1992. https://doi.org/10.17577/IJERTV2IS90779.

[12] Yakkundimath, R., Saunshi, G., & Palaiah, S. (2021). Automatic methods for classification of visual-based viral and bacterial disease symptoms in plants. *International Journal of Information Technology*. https://doi.org/10.1007/s41870-021-00701-2.

[13] Sujatha, R., Chatterjee, J. M., Jhanjhi, N. Z., & Brohi, S. N. (2021). Performance of deep learning vs. machine learning in plant leaf disease detection. *Microprocessors and Microsystems*, *80*, 103615. https://doi.org/10.1016/j.micpro.2020.103615.

[14] Zhao, D., Feng, S., Cao, Y., Yu, F., Guan, Q., Li, J., Zhang, G., & Xu, T. (2022). Study on the classification method of rice leaf blast levels based on fusion features and adaptive-weight immune particle swarm optimization extreme learning machine algorithm. *Frontiers in Plant Science*, *13*, 879668. https://doi.org/10.3389/fpls.2022.879668.

Baljinder Kour, Savita Arora, and Ramanpreet Kaur

Soliton solutions of the time-fractional Kadomtsev–Petviashvilli equation using the $(\frac{G'}{G^2})$-expansion method

Abstract: In this chapter, the solutions of the (2+1)-dimensional Kadomtsev–Petviashvili (KP) equation are obtained through the $(\frac{G'}{G^2})$-expansion method. This method reduces the complexity of the transformed ordinary differential equation of the nonlinear fractional partial differential equations by extracting the solution in the form of the expansion series $(\frac{G'}{G^2})$. The obtained solutions are categorized into three types, including trigonometric function solutions, hyperbolic function solutions, and rational function solutions. The physical behavior of the obtained solutions is depicted using different contours. Applying this method, we produce some new soliton solutions, which are useful for depicting wave progression when influenced by both dispersion effects and nonlinear dynamics.

Keywords: (2 + 1)-Dimensional Kadomtsev–Petviashvili (KP) equation, nonlinear fractional partial differential equations, $(\frac{G'}{G^2})$-expansion method, soliton solutions

MSC 2020: 34A08, 35C08, 35C09, 35C10

1 Introduction

Fractional differential equations (FDEs) have gained considerable importance in recent times. Although fractional calculus has a history as long as traditional calculus, it has not gained the same level of popularity in science. FDEs play a vital role in providing the equations for the rate of change of a continuously changing quantity. They are used to elaborate multifaceted systems having a current state depending on the system's entire history. In recent years, the scope of fractional calculus has developed in many fields of engineering and science, including Cosmology, Acoustics, Viscoelasticity, Pharmacokinetics, Neuron Modeling, Fractional-Order PID Controllers, Power-Law Dynamics, etc. The development of solutions for FDEs is an essential research direction within the field of FDEs [1–3]. The generalized version of classical differential equations of integral orders is known as the FDEs. Nonlinear fractional partial differential equations (NLFPDEs) are not only used in mathematics but also in various fields like physics, mechanics, material science, etc. [4–7]. Nowadays, one of the tempting tasks for researchers is to find

Baljinder Kour, Savita Arora, Ramanpreet Kaur, Akal University, Talwandi Sabo, Bathinda, Punjab, India, e-mail: baljindervirk172@gmail.com

https://doi.org/10.1515/9783112234525-012

the solutions of NLFPDEs, which require strong mathematical tools [8, 10–13]. NLFPDEs with time as one of the independent variables have become a useful tool for describing the natural phenomena in science and engineering. Many researchers have proposed various techniques for obtaining solutions of NLFPDES.

The KP equation is one of the significant models used to study ion-acoustic wave phenomena [9, 14]. The (2 + 1)-dimensional KP equation illustrates the progression of wave packets in three-dimensional space. Its rich structure captures the relation between dispersion and nonlinear wave propagations. This equation is expressed in the form [15]

$$u_{tx}^{\alpha} - uu_{xx} - u_x^2 - u_{xxxx} = u_{yy}, \tag{1}$$

where, $0 < \alpha \leq 1$, $D_t^{\alpha}u(x,y,t) = u_t^{\alpha}$ and D_t^{α} represents the fractional derivative with respect to (w. r. t.) time t for order α. $u(x,y,t)$ is a wave function having independent variables x, y, and t (time variable). The term uu_{xx} accounts for nonlinear effects. The KP equation was developed by Boris Kadomtsev and Vladimir Petviashvili.

In recent years, numerous authors have investigated solutions of the KP equation using various methods. In [16], the exact solution of the KP equation is obtained using the $\exp(-\Phi(\xi))$-expansion method. In [17], solitary wave solutions are obtained using the novel generalized $(\frac{G'}{G})$-expansion method. In [18], the Lie symmetry, the extended tanh method, and the homotopy perturbation method are used to find exact and approximate analytic solutions of the KP equation. Now, we will obtain soliton solutions of the KP equation by applying the $(\frac{G'}{G^2})$-expansion method.

2 Preliminaries

In this section, we will discuss basic definitions and the procedure of the $(\frac{G'}{G^2})$-expansion method.

Definition 1 (Conformable fractional derivative). The conformable fractional derivative of a function $g(t)$ of order α, where $\alpha \in (0,1]$ [28], is defined as

$$\mathcal{D}^{\alpha}g(t) = \lim_{\epsilon \to 0} \frac{g(t + \epsilon t^{1-\alpha}) - g(t)}{\epsilon}.$$

Here, for $\alpha \in (0,1]$, the expression generalizes the derivative concept by considering the change in the function $g(t)$ over a small increment ϵ.

Definition 2 (Fractional power series). For $0 \leq k - 1 < \alpha \leq k$ and $z \geq z_0$, the series expansion defined by

$$\sum_{k=0}^{\infty} a_k (z - z_0)^{k^{\alpha}} = a_0 + a_1(z - z_0)^{\alpha} + a_2(z - z_0)^{2^{\alpha}} + \cdots$$

is called the fractional power series at $z = z_0$, where z is a variable and the a_ns are constant coefficients.

The $(\frac{G'}{G^2})$-expansion method

The $(\frac{G'}{G^2})$ is a recently proposed method that can be applied to find soliton solutions of NLFPDEs [19–22]. The soliton solution of the (1 + 1)-dimensional classical Boussinesq equation was given by Aljahdaly by the use of the $(\frac{G'}{G^2})$ method [23]. To obtain the soliton solutions of the Modified Benjamin–Bona–Mahony (MBBM) and Ostrovsky–Benjamin–Bona–Mahony (OBBM) equations Zhouzheny used the $(\frac{G'}{G^2})$ method [24]. The extended $(\frac{G'}{G^2})$-expansion method was used by Gepreed in order to obtain the solutions of the two integral members of Kadomtsev–Petviashvilli (KP) equations. Along with the fractional complex transform, Mohyud-Din and Bibi used the $(\frac{G'}{G^2})$ method to obtain the solution of the Zakharov–Kuznetsov–Benjamin–Bona–Mahony (ZKBBM) equations [25], and for solving the Schrodinger equation, the $(\frac{G'}{G^2})$-expansion method was used [26].

Procedure of the $(\frac{G'}{G^2})$-expansion method

The $(\frac{G'}{G^2})$-expansion method is a mathematical technique used to solve NLFPDEs [27]. The method involves substituting an assumed power-series expansion of the solution into the partial differential equation (PDE). The resulting equations are then solved recursively to obtain a soliton solution. The $(\frac{G'}{G^2})$-expansion method has been widely used to solve various NLFPDEs in physics and engineering. The $(\frac{G'}{G^2})$ is a method [23, 28] that can be applied to find soliton solutions of NLFPDEs.

The detailed procedure of the method is explained as follows:

Consider a NLFPDE, of three independent variables x, y, t and a dependent variable u

$$F(u, , u_t^{\alpha}, u_x, u_y, u_{tx}^{\alpha}, \ldots) = 0,$$

where $U = u(x, y, t)$, is an unknown function of independent variables x, y, t and u_t^{α} is the fractional derivative of u w. r. t. time t. F is a polynomial of U.

Step 1: We convert a nonlinear partial differential equation (NLPDE) into an ordinary differential equation (ODE), such that $u(x, y, t) = U(\xi)$, using the traveling wave transformation given as

$$\xi = x + y - \frac{ct^{\alpha}}{\alpha!},$$

where c is a nonzero arbitrary constant.

Therefore, the resulting ODE for $U = U(\xi)$ will be

$$P(U, U', U'', U''', \ldots) = 0. \tag{2}$$

Here, P is a polynomial function of $U(\xi)$.

Step 2: The general solution of ODE, in (2), can be written as

$$U(\xi) = a_0 + \sum_{i=1}^{n} \left[a_i \left(\frac{G'}{G^2} \right)^i + b_i \left(\frac{G'}{G^2} \right)^{-i} \right], \tag{3}$$

where, $G = G(\xi)$ satisfies the following Riccati equation:

$$\left(\frac{G'}{G^2} \right)' = \mu + \lambda \left(\frac{G'}{G^2} \right)^2, \tag{4}$$

where μ ($\neq 1$) and λ ($\neq 0$) are arbitrary integers. The unknown constants are a_0, a_i, b_i, a_n, b_n. In this, a_n and b_n cannot be zero simultaneously, but either a_n or b_n may be zero. The coefficients a_0, a_i, b_i are to be evaluated in the next step.

Step 3: The value of the integer n can be evaluated using the Homogeneous Balance principle. In other words, to find n balancing between the highest-order derivatives and the nonlinear terms appearing in equation (3) is needed. If the degree of $U(\xi)$ is $[U(\xi)] = n$, the degrees of the terms $\frac{d^q U(\xi)}{d\xi^q}$ and $U(\xi)^p \frac{d^q U(\xi)^s}{d\xi^q}$ are

$$\deg\left[\frac{d^q U(\xi)}{d\xi^q} \right] = n + q,$$
$$\deg\left[U(\xi)^p \frac{d^q U(\xi)^s}{d\xi^q} \right] = np + s(n + q).$$

Step 4: Now, by using equations (3) and (4) in equation (2), we have a polynomial in $(\frac{G'}{G^2})$. Comparing all the coefficients of the same power $(\frac{G'}{G^2})^k$ ($k = -n, \ldots, -3, -2, -1, 0, 1, 2, 3, \ldots, n$; n is an integer) and setting all the obtained coefficients to zero.

We obtain a system of nonlinear algebraic equations for the unknown coefficients a_0, a_i, b_i ($i = 1, 2, \ldots, n$) and c. Then, the resulting algebraic system can be solved for the unknown constants using software such as Mathematica.

The general solution of equation (4) can be categorized into the following three cases, where C and D are arbitrary constants.

Case 1: If $\mu\lambda > 0$, then

$$\left(\frac{G'}{G^2} \right) = \sqrt{\frac{\mu}{\lambda}} \left(\frac{C \operatorname{Cos}(\sqrt{\mu\lambda}\xi) + D \operatorname{Sin}(\sqrt{\mu\lambda}\xi)}{D \operatorname{Cos}(\sqrt{\mu\lambda}\xi) - C \operatorname{Sin}(\sqrt{\mu\lambda}\xi)} \right). \tag{5}$$

Case 2: If $\mu\lambda < 0$, then

$$\left(\frac{G'}{G^2} \right) = -\frac{\sqrt{|\mu\lambda|}}{\lambda} \left(\frac{C \operatorname{Sinh}(2\sqrt{|\mu\lambda|}\xi) + C \operatorname{Cosh}(2\sqrt{|\mu\lambda|}\xi) + D}{C \operatorname{Sinh}(2\sqrt{|\mu\lambda|}\xi) + C \operatorname{Cosh}(2\sqrt{|\mu\lambda|}\xi) - D} \right). \tag{6}$$

Case 3: If $\mu = 0$ and $\lambda \neq 0$, then

$$\left(\frac{G'}{G^2}\right) = -\frac{-C}{\lambda(C\xi + D)}. \tag{7}$$

We obtain the soliton solution by using the values of a_0, a_i, b_i $(i = 1, 2, 3, 4, \ldots, n)$.

3 Application of the $(\frac{G'}{G^2})$-expansion method

Consider the Kadomtsev–Petviashvili equation (KP) [15]

$$u^{\alpha}_{tx} - uu_{xx} - u_x^2 - u_{xxxx} = u_{yy}. \tag{8}$$

Here, x, y, and t $(t > 0)$ are independent variables, while u is a dependent variable and u^{α}_t is the fractional derivative of u w. r. t. time t.

To solve the KP equation through the modified $(\frac{G'}{G^2})$ method, by considering the wave transformation method $u(x, y, t) = U(\xi)$, where the wave variable is $\xi = x + y - \frac{ct^{\alpha}}{\alpha!}$, and we can convert equation (8) into the form of an ODE

$$-cU'' - UU'' - (U')^2 - U'''' = U''$$

and after simplifying, we obtain

$$(1 + c)U'' + UU'' + (U')^2 + U'''' = 0, \tag{9}$$

then integrating (9) twice w. r. t. ξ, we obtain

$$(c + 1)U + U'' + \frac{U^2}{2} = 0. \tag{10}$$

Now, to find N, we use the homogeneous balance method. The maximum degree of it is $N + m$, the maximum degree of $(\frac{G'}{G^2})$ in $\frac{d^m(u)}{d\xi^m}$ is $N + m$, and the maximum degree of $(\frac{G'}{G^2})$ in U^p is Np. On comparing the highest-order term and nonlinear term, we find $2(N + 1) = N + 4$, which gives $N = 2$. The solution can therefore be expressed as

$$U(\xi) = a_0 + a_1\left(\frac{G'}{G^2}\right) + a_2\left(\frac{G'}{G^2}\right)^2 + b_1\left(\frac{G'}{G^2}\right)^{-1} + b_2\left(\frac{G'}{G^2}\right)^{-2}, \tag{11}$$

where, a_0, a_1, a_2, b_1, b_2 are arbitrary parameters and $U = U(\xi)$ satisfies the simple Riccati equation

$$\left(\frac{G'}{G^2}\right)' = \mu + \lambda\left(\frac{G'}{G^2}\right)^2. \tag{12}$$

Using step (3), we obtain a system of equations for a_0, a_1, a_2, b_1, b_2, c, μ, and λ as follows:

$$\left(\frac{G'}{G^2}\right)^0 = \frac{a_0^2}{2} + a_0c + a_0 + a_1b_1 + a_2b_2, \quad \left(\frac{G'}{G^2}\right)^1 = a_0a_1 + a_1c + 2a_1\lambda\mu + a_1 + a_2b_1,$$

$$\left(\frac{G'}{G^2}\right)^2 = a_0a_2 + \frac{a_1^2}{2} + a_2c + 8a_2\lambda\mu + a_2, \quad \left(\frac{G'}{G^2}\right)^3 = a_1a_2 + 2a_1\lambda^2,$$

$$\left(\frac{G'}{G^2}\right)^4 = \frac{a_2^2}{2} + 6a_2\lambda^2, \quad \left(\frac{G'}{G^2}\right)^{-1} = a_0b_1 + a_1b_2 + b_1c + 2b_1\lambda\mu + b_1,$$

$$\left(\frac{G'}{G^2}\right)^{-2} = a_0b_2 + \frac{b_1^2}{2} + b_2c + 8b_2\lambda\mu + b_2, \quad \left(\frac{G'}{G^2}\right)^{-3} = b_1b_2 + 2b_1\mu^2,$$

$$\left(\frac{G'}{G^2}\right)^{-4} = \frac{b_2^2}{2} + 6b_2\mu^2.$$

By solving the above equations using Mathematica, the following solutions are obtained.
Case 1: $a_0 = 4(\sqrt{22}\lambda\mu - 2\lambda\mu)$, $a_1 = 0$, $a_2 = -12\lambda^2$, $b_1 = 0$, $b_2 = -12\mu^2$, $c = -4\sqrt{22}\lambda\mu - 1$.

Case	a_0	a_1	a_2	b_1	b_2	c	μ	λ
1	$4(\sqrt{22}\lambda\mu - 2\lambda\mu)$	0	$-12\lambda^2$	0	$-12\mu^2$	$-4\sqrt{22}\lambda\mu - 1$	μ	λ
2	$-4(\sqrt{22}\lambda\mu + 2\lambda\mu)$	0	$-12\lambda^2$	0	$-12\mu^2$	$4\sqrt{22}\lambda\mu - 1$	μ	λ
3	a_0	0	0	0	$-12\mu^2$	$\frac{1}{2}(-a_0 - 2)$	μ	$-\frac{a_0}{16\mu}$
4	a_0	0	$-12\lambda^2$	0	0	$\frac{1}{2}(-a_0 - 2)$	$-\frac{a_0}{16\lambda}$	λ
5	0	0	$-12\lambda^2$	0	0	-1	0	λ

If $\mu\lambda > 0$ (Trigonometric function solution), we obtained the soliton solution (see Figure 1)

$$\begin{aligned} u_1 = {} & -(1.392(10\sin(0.340588(-1.22312t^{0.9} + x + 2)) \\ & + 20\cos(0.340588(-1.22312t^{0.9} + x + 2)))^2) \\ & /((20\sin(0.340588(-1.22312t^{0.9} + x + 2)) \\ & - 10\cos(0.340588(-1.22312t^{0.9} + x + 2)))^2) \\ & - (1.392(20\sin(0.340588(-1.22312t^{0.9} + x + 2)) \\ & - 10\cos(0.340588(-1.22312t^{0.9} + x + 2)))^2) \\ & /((10\sin(0.340588(-1.22312t^{0.9} + x + 2)) \\ & + 20\cos(0.340588(-1.22312t^{0.9} + x + 2)))^2) \\ & - 3.10435. \end{aligned} \tag{13}$$

If $\mu\lambda < 0$ (Hyperbolic function solution), we obtain the soliton solution (see Figure 2)

$$
\begin{aligned}
u_2 = &-(0.01824(2\sinh(0.0779744(1.1592t^{0.4} + x + 20)) \\
&+ 2\cosh(0.0779744(1.1592t^{0.4} + x + 20)) - 5)^2) \\
&/((2\sinh(0.0779744(1.1592t^{0.4} + x + 20)) \\
&+ 2\cosh(0.0779744(1.1592t^{0.4} + x + 20)) + 5)^2) \\
&- (0.01824(2\sinh(0.0779744(1.1592t^{0.4} + x + 20)) \\
&+ 2\cosh(0.0779744(1.1592t^{0.4} + x + 20)) + 5)^2) \\
&/((2\sinh(0.0779744(1.1592t^{0.4} + x + 20)) \\
&+ 2\cosh(0.0779744(1.1592t^{0.4} + x + 20)) - 5)^2) \\
&+ 0.0406777. \qquad (14)
\end{aligned}
$$

If $\mu = 0$ and $\lambda \neq 0$ (Rational function solution), we obtain the soliton solution (see Figure 3)

$$
u_3 = 0. - \frac{1200.}{(20 - 10(1.03975t^{0.9} + x + 2))^2}. \qquad (15)
$$

Case 2: $a_0 = -4(\sqrt{22}\lambda\mu + 2\lambda\mu)$, $a_1 = 0$, $a_2 = -12\lambda^2$, $b_1 = 0$, $b_2 = -12\mu^2$, $c = 4\sqrt{22}\lambda\mu - 1$.

If $\mu\lambda > 0$ (Trigonometric function solution), we obtain the soliton solution (see Figure 4)

$$
\begin{aligned}
u_4 = &-\frac{0.48(0.1\cos(0.2(0.259453t^{0.9} + x + 2)) - 0.2\sin(0.2(0.259453t^{0.9} + x + 2)))^2}{(0.1\sin(0.2(0.259453t^{0.9} + x + 2)) + 0.2\cos(0.2(0.259453t^{0.9} + x + 2)))^2} \\
&- \frac{0.48(0.1\sin(0.2(0.259453t^{0.9} + x + 2)) + 0.2\cos(0.2(0.259453t^{0.9} + x + 2)))^2}{(0.1\cos(0.2(0.259453t^{0.9} + x + 2)) - 0.2\sin(0.2(0.259453t^{0.9} + x + 2)))^2} \\
&- 1.07047. \qquad (16)
\end{aligned}
$$

If $\mu\lambda < 0$ (Hyperbolic function solution), we obtain the soliton solution (see Figure 5)

$$
\begin{aligned}
u_5 = &-(0.01824(2\sinh(0.0779744(1.02852t + x + 20)) \\
&+ 2\cosh(0.0779744(1.02852t + x + 20)) - 5)^2) \\
&/((2\sinh(0.0779744(1.02852t + x + 20)) \\
&+ 2\cosh(0.0779744(1.02852t + x + 20)) + 5)^2) \\
&- (0.01824(2\sinh(0.0779744(1.02852t + x + 20)) \\
&+ 2\cosh(0.0779744(1.02852t + x + 20)) + 5)^2) \\
&/((2\sinh(0.0779744(1.02852t + x + 20)) \\
&+ 2\cosh(0.0779744(1.02852t + x + 20)) - 5)^2) \\
&+ 0.0406777. \qquad (17)
\end{aligned}
$$

If $\mu = 0$ and $\lambda \neq 0$ (Rational function solution), we obtain the soliton solution (see Figure 6)

$$u_6 = -\frac{4800}{(-20(1.12706t^{0.4} + x + 20) - 50)^2}. \tag{18}$$

Case 3: $a_1 = 0, a_2 = 0, b_1 = 0, b_2 = -12\mu^2, c = \frac{1}{2}(-a_0 - 2), \lambda = -\frac{a_0}{16\mu}$.

If $\mu\lambda > 0$ (Trigonometric function solution), we obtain the soliton solution (see Figure 7)

$$u_7 = 70 + \frac{105(40\cosh(\frac{1}{2}\sqrt{\frac{35}{2}}(36t + x + 2)) + 50i\sinh(\frac{1}{2}\sqrt{\frac{35}{2}}(36t + x + 2)))^2}{2(-50\cosh(\frac{1}{2}\sqrt{\frac{35}{2}}(36t + x + 2)) + 40i\sinh(\frac{1}{2}\sqrt{\frac{35}{2}}(36t + x + 2)))^2}. \tag{19}$$

If $\mu\lambda < 0$ (Hyperbolic function solution), we obtain the soliton solution (see Figure 8)

$$\begin{aligned} u_8 = 2. - (1.5(2\sinh(0.707107(2.07951t^{0.9} + x + 20)) \\ + 2\cosh(0.707107(2.07951t^{0.9} + x + 20)) + 5)^2) \\ /((2\sinh(0.707107(2.07951t^{0.9} + x + 20)) \\ + 2\cosh(0.707107(2.07951t^{0.9} + x + 20)) - 5)^2). \end{aligned} \tag{20}$$

Case 4: $a_1 = 0, a_2 = -12\lambda^2, b_1 = 0, b_2 = 0, c = \frac{1}{2}(-a_0 - 2), \mu = -\frac{a_0}{16\lambda}$.

If $\mu\lambda > 0$ (Trigonometric function solution), we obtain the soliton solution (see Figure 9)

$$\begin{aligned} u_9 = 70 - \Bigg(105\Bigg(-10\sinh\Bigg(\sqrt{\frac{35}{2}}(37.4311t^{0.9} + x + 2)\Bigg) \\ - 10\cosh\Bigg(\sqrt{\frac{35}{2}}(37.4311t^{0.9} + x + 2)\Bigg) + 80\Bigg)^2\Bigg) \\ /\Bigg(2\Bigg(-10\sinh\Bigg(\sqrt{\frac{35}{2}}(37.4311t^{0.9} + x + 2)\Bigg) \\ - 10\cosh\Bigg(\sqrt{\frac{35}{2}}(37.4311t^{0.9} + x + 2)\Bigg) - 80\Bigg)^2\Bigg). \end{aligned} \tag{21}$$

If $\mu\lambda < 0$ (Hyperbolic function solution), we obtain the soliton solution (see Figure 10)

$$u_{10} = 1 - \frac{3(2\sinh(\frac{1}{2}(1.69059t^{0.4} + x + 20)) + 2\cosh(\frac{1}{2}(1.69059t^{0.4} + x + 20)) - 5)^2}{4(2\sinh(\frac{1}{2}(1.69059t^{0.4} + x + 20)) + 2\cosh(\frac{1}{2}(1.69059t^{0.4} + x + 20)) + 5)^2}. \tag{22}$$

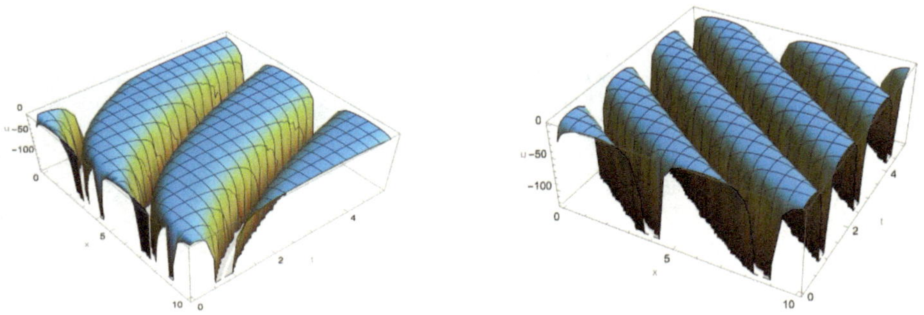

Figure 1: The solution of equation (13) (u_1), where the values of $C = -10, D = 20, \lambda = 0.029, \mu = 4$, and $y = 2$, the graph (a) is for $\alpha = 0.4$ and the graph (b) is for $\alpha = 0.9$.

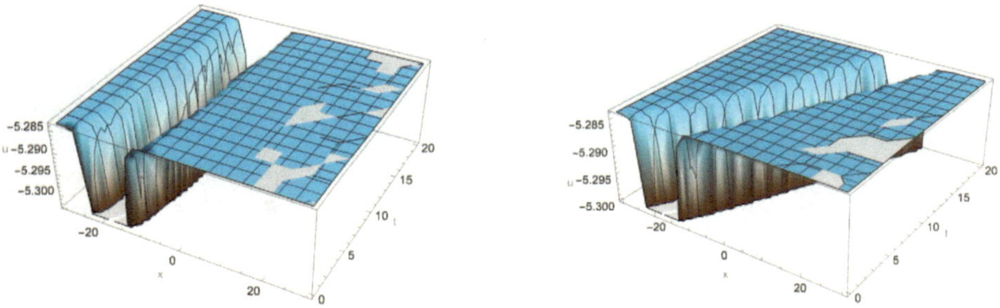

Figure 2: The solution of equation (14) (u_2), where the values of $C = 2, D = -5, \lambda = -0.19, \mu = 0.8$, and $y = 20$, the graph (a) is for $\alpha = 0.4$ and the graph (b) is for $\alpha = 0.9$.

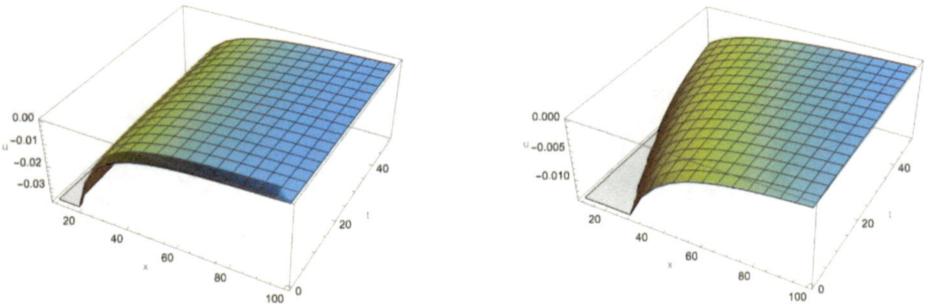

Figure 3: The solution of equation (15) (u_3), where the values of $\lambda = 0.029, \mu = 0, C = -10, D = 20$, and $y = 2$, the graph(a) is for $\alpha = 0.4$ and the graph (b) is for $\alpha = 0.9$.

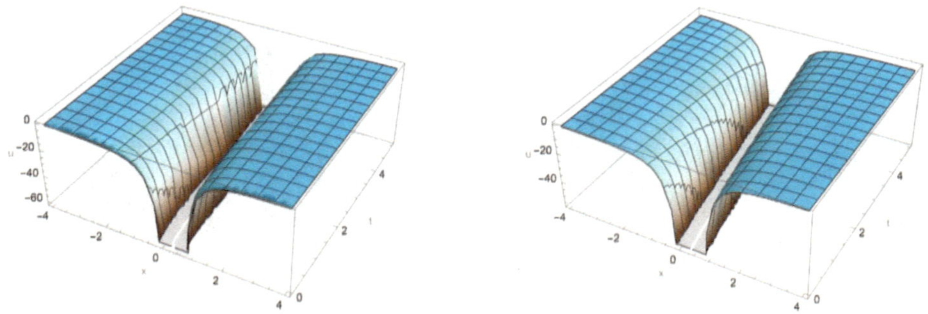

Figure 4: The solution of equation (16) (u_4), where the values of $C = 1, , D = 1, \lambda = -0.1, \mu = -0.4$, and $y = 2$, the graph (a) is for $\alpha = 0.4$ and the (b) is for $\alpha = 0.9$.

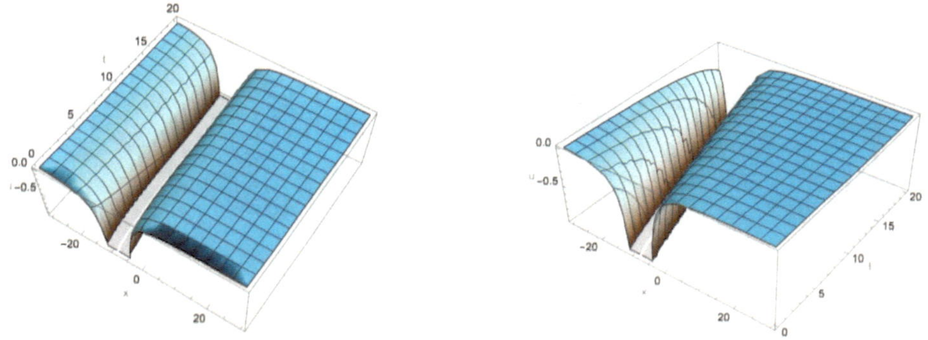

Figure 5: The solution of equation (17) (u_5), where the values of $C = 2, D = -5, \lambda = -0.019, \mu = 0.08$, and $y = 20$, the graph (a) is for $\alpha = 0.4$ and the graph (b) is for $\alpha = 0.9$.

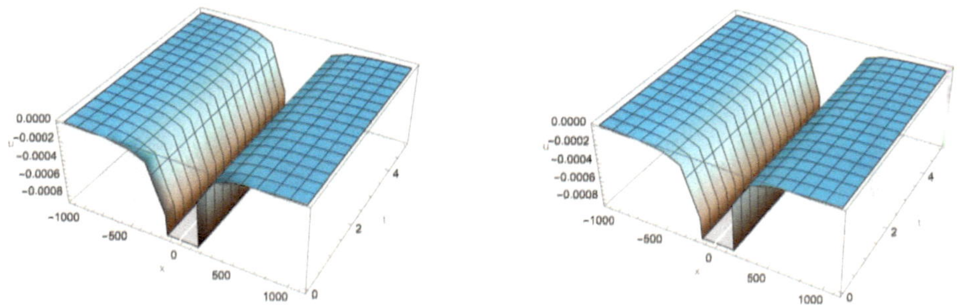

Figure 6: The solution of equation (18) (u_6), where the values of $C = -20, D = -50, \Lambda = 20, \mu = 0$, and $y = 20$, the graph (a) is for $\alpha = 0.4$ and the graph (b) is for $\alpha = 0.9$.

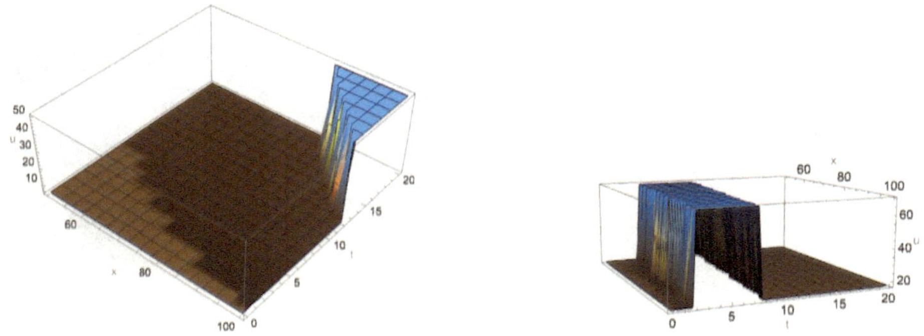

Figure 7: The solution of equation (19) (u_7), where the values of $C = -10$, $D = 20$, $\mu = -4$, $a_0 = 70$, and $y = 2$, the graph (a) is for $\alpha = 0.4$ and the graph (b) is for $\alpha = 0.9$.

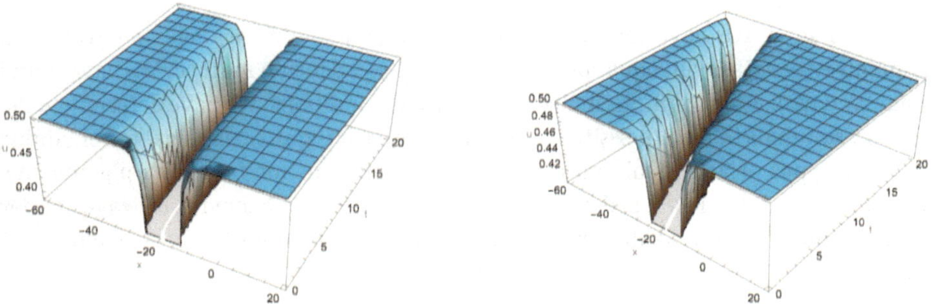

Figure 8: The solution of equation (20) (u_8), where the values of $C = 2$, $D = -5$, $\mu = 0.8$, $a_0 = 2$, and $y = 20$, the graph (a) is for $\alpha = 0.4$ and the graph (b) is for $\alpha = 0.9$.

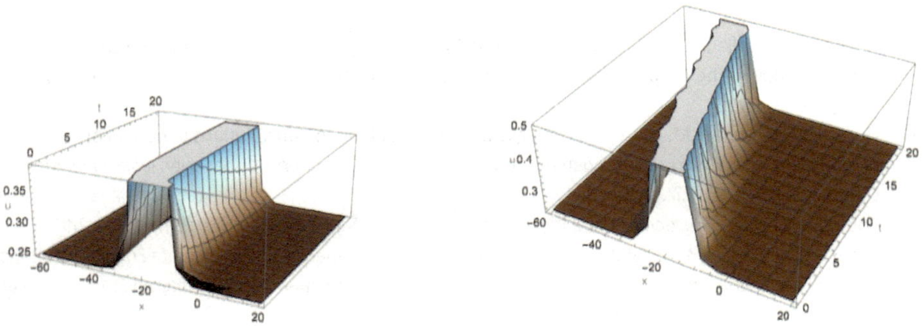

Figure 9: The solution of equation (21) (u_9), where the values of $C = -20$, $D = 30$, $\lambda = -30$, $a_0 = 20$, and $y = 2$, the graph (a) is for $\alpha = 0.4$ and the graph (b) is for $\alpha = 0.9$.

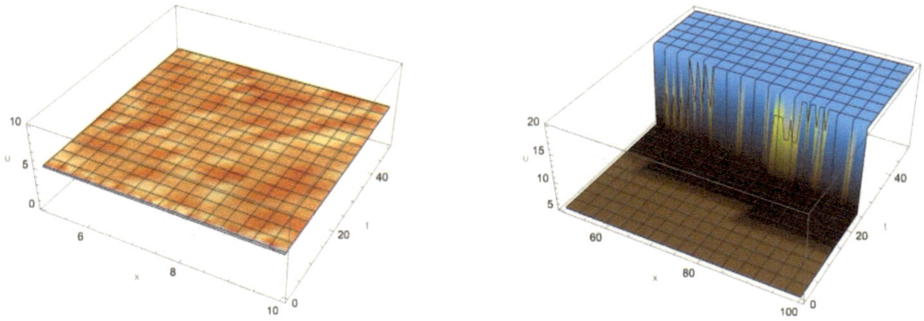

Figure 10: The solution of equation (22) (u_{10}), where the values of $C = 2$, $D = -5$, $\lambda = 9$, $a_0 = 1$, and $y = 20$, the graph (a) is for $\alpha = 0.4$ and the graph (b) is for $\alpha = 0.9$.

4 Conclusion

In this chapter, we have found solutions of the (2 + 1)-dimensional time-fractional Kadomtsev–Petviashvilli equation by using the $(\frac{G'}{G^2})$-expansion method. We have found singular dark soliton solutions in the trigonometric conditions. We have determined the solitary solutions in the other situations. This approach is quite beneficial for obtaining different types of analytical and approximation solutions for fractional phenomena. The capacity of the developed solutions to characterize the behavior of waves under the consequences of weak dispersion and nonlinear interactions has been given significance.

Bibliography

[1] Kiryakova, V. S. (1994). *Generalized fractional calculus and applications* (Vol. 301). New York: John Wiley & Sons, Inc.

[2] Farooq, K., Abdullah, A. R., Hussain, E., et al. (2025). Chaos, Lyapunov exponent, and sensitivity demonstration of the coupled nonlinear integrable model with soliton solutions. *AIMS Mathematics*, *10*(10), 22929–22957.

[3] Omar, F. M., & Murad, M. A. S. (2026). Analytical study of optical soliton solutions in nonlinear Schrödinger model with self-steepening, self-frequency shift, and Kerr nonlinearity. *Modern Physics Letters B*, *40*(01), 2550262.

[4] Podlubny, I. (1999). *Fractional differential equations* (Vol. 198). San Diego, CA: Academic Press, Inc.

[5] Schurz, H., & Talafha, A. M. (2021). Existence, uniqueness, and energy of approximate Fourier solutions of modified stochastic sine-Gordon equation with power law nonlinearity in 1D. *International Journal of Applied and Computational Mathematics*, *7*(4), 126.

[6] Jhangeer, A., Beenish, T. A. M., et al. (2024). Analytical and dynamical analysis of nonlinear Riemann wave equation in plasma system. *Arab Journal of Basic and Applied Sciences*, *31*(1), 536–553.

[7] Murad, M. A. S., Tedjani, A. H., Li, Z., et al. (2025). Optical solutions to time-fractional improved (2 + 1)-dimensional nonlinear Schrödinger equation in optical fibers. *Scientific Reports*, *15*(1), 29208.

[8] Sirisubtawee, S., & Koonprasert, S. (2018). Exact traveling Wave solutions of certain nonlinear partial differential equations using the ($\frac{G'}{G^2}$)-expansion method. *Advances in Mathematical Physics, 2018*(1), 7628651.

[9] Khan, K., & Akbar, M. A. (2015). Exact traveling wave solutions of Kadomtsev–Petviashvili equation. *Journal of the Egyptian Mathematical Society, 23*(2), 278–281.

[10] Malik, S., Boulaaras, S. M., Talafha, A. M., et al. (2024). Soliton dynamics of the fourth-order nonlinear Boussinesq equation arising in shallow-water. *Optical and Quantum Electronics, 56*(3), 379.

[11] Sağlam, F. N. K., Movaheedi, Z., Talafha, A. M., et al. (2025). Different types of soliton solutions for the resonant nonlinear Schrödinger equation with parabolic law nonlinearity via Kumar–Malik approach. *Scientific Reports*.

[12] Murad, M. A. S. (2025). Optical solutions to conformable nonlinear Schrödinger equation with cubic–quintic–septimal in weakly non-local media by new Kudryashov approach. *Modern Physics Letters B, 39*(21), 2550063.

[13] Mahmood, S. S., & Murad, M. A. S. (2025). Dynamics of nonlinear wave of the (3 + 1)-dimensional Date–Jimbo–Kashiwara–Miwa equation with variable coefficients in inhomogeneous media. *The European Physical Journal D, 79*(12), 149.

[14] Talafha, A. M., Jhangeer, A., & Kazmi, S. S. (2023). Dynamical analysis of (4 + 1) dimensional Davey Srewartson Kadomtsev Petviashvili equation by employing lie symmetry approach. *Ain Shams Engineering Journal, 14*(11), 102537.

[15] Kumar, S., Kour, B., Yao, S. W., et al. (2021). Invariance analysis, exact solution and conservation laws of (2 + 1) dim fractional Kadomtsev–Petviashvili (KP) system. *Symmetry, 13*, 477.

[16] Aslan, I. (2012). On the application of the Exp-function method to the KP equation for N-soliton solutions. *Applied Mathematics and Computation, 219*(6), 2825–2828.

[17] Alam, M. N., & Tunc, C. (2020). New solitary wave structures to the (2 + 1)-dimensional KD and KP equations with spatio-temporal dispersion. *Journal of King Saud University Science, 32*(8), 3400–3409.

[18] Gai, L., Bilige, S., et al. (2016). The Lie symmetry, the extended tanh method, and the homotopy perturbation method are used to find exact and approximate analytic solutions of the KP equation. *SpringerPlus, 5*, 1–14.

[19] Arshed, S., & Sadia, M. (2018). ($\frac{G'}{G^2}$)-expansion method: new traveling wave solutions for some nonlinear fractional partial differential equations. *Optical and Quantum Electronics, 50*(3), 123.

[20] Meng, Y. (2018). Expanded $\frac{G'}{G^2}$-expansion method to solve separated variables for the 2 + 1-dimensional NNV equation. *Advances in Mathematical Physics, 2018*(1), 9248174.

[21] Chen, J. P. (2012). The ($\frac{G'}{G^2}$)-expansion method and its applications to coupled nonlinear Klein–Gordon equation. *Journal of South China Normal University, 44*(2).

[22] Ali, M. N., et al. (2019). On the analytical solutions of conformable time-fractional extended Zakharov–Kuznetsov equation through ($\frac{G'}{G^2}$)-expansion method and the modified Kudryashov method. *Boletín de la Sociedad Española de Matemática Aplicada, 76*(1), 15–25.

[23] Aljahdaly, N. H. (2019). Some applications of the modified ($\frac{G'}{G^2}$)-expansion method in mathematical physics. *Results in Physics, 13*, 102272.

[24] Zhouzheng, K. (2015). ($\frac{G'}{G^2}$)-expansion solutions to MBBM and OBBM equations. *Journal of Partial Differential Equations, 28*(2), 158–166.

[25] Mohyud-Din, S. T., & Bibi, S. (2018). Exact solutions for nonlinear fractional differential equations using ($\frac{G'}{G^2}$)-expansion method. *Alexandria Engineering Journal, 57*(2), 1003–1008.

[26] Sahu, I., & Jena, S. R. (2025). On the extended ($\frac{G'}{G^2}$)-expansion method for an optical solution of the Schrodinger equation in quantum mechanics. *Zeitschrift für Naturforschung A, 80*(9), 773–787.

[27] Kadkhoda, N. (2017). Application of G'G2-expansion method for solving fractional differential equations. *International Journal of Applied and Computational Mathematics, 3*, 1415–1424.

[28] Kaewta, S., Sirisubtawee, S., et al. (2021). Applications of the ($\frac{G'}{G^2}$)-expansion method for solving certain nonlinear conformable evolution equations. *Fractal and Fractional, 5*(3), 88.

Inderpreet Kaur and Swati

Numerical investigation of the modified Burgers' equation using the quintic Hermite collocation scheme

Abstract: The numerical approximations of the modified Burgers' equation are provided via the quintic Hermite collocation scheme, a robust numerical technique. The finite difference method is followed to temporally discretize the modified Burgers' equation. The chapter's main objective is to analyze the method's convergence and stability. By applying the Von Neumann method, the process is found to be utterly stable. Errors have been assessed based on dispersion and dissipation. L_2- and L_∞-norms have been evaluated to check the efficiency of the scheme. To preserve the precision, simplicity, and applicability of the method for resolving the modified Burgers' equation, the outcomes are displayed in tabular and graphic formats.

Keywords: Finite difference, modified Burgers' equation, stability analysis, Euclidean and supremum norms, Von Neumann approach

MSC 2020: 35K15, 65M06, 65M12, 65M22

1 Introduction

For academicians and scientists, it might be difficult to model practical issues and use partial differential equations (PDEs). The modeling of nonlinear differential equation systems leads to a large variety of issues. The well-known nonlinear partial differential equation known as Burgers' equation has been used to model a variety of physical processes [1]. It is a condensed version of the fluid dynamics widely used Navier–Stokes equation. With applications to fluid dynamics, elasticity, heat conduction, shock wave theory and traffic flow, this mathematical model plays the significant role. In an effort to achieve better degrees of accuracy, researchers try to tackle the issues analytically or numerically utilizing various techniques and equations. Numerous methods are often employed in numerical analysis, including pseudospectral delta-shaped basis and barycentric rational interpolation [2], iterative differential quadrature [3], operator splitting finite element method [4], the radial basis function-generated finite difference method [5], the Sumudu transform-based technique [6], a finite element method [7], B-splines [8–11], the Petrov–Galerkin method [12], the Haar-wavelet method [13], etc.

Inderpreet Kaur, Swati, CUIET, Department of Applied Sciences, Chitkara University, Punjab, India, e-mail: inder3003@gmail.com

https://doi.org/10.1515/9783112234525-013

This study seeks to determine the approximate solution of the modified Burgers' equation

$$\vartheta_t = \epsilon\vartheta_{xx} - \vartheta^{\delta}\vartheta_x, \tag{1}$$

where $a \le x \le b, t \ge 0, 0 \le \epsilon \le 1; \delta \ge 0$.

Initially

$$\vartheta(x, t_0) = \mu(x). \tag{2}$$

The boundary conditions will be given by

$$\vartheta(a, t) = g_1(t); \vartheta(b, t) = g_2(t); \quad t \in [t_0, T], \tag{3}$$

where $\vartheta(x, t)$ represents the space parameter in x and time in t. It is a nonlinear convection–dispersion equation that was initially originated by [1] and then adapted in the turbulence theory by Burgers [14]. In the present study two cases will be discussed; $\delta = 1, 2$. This model equation may be used to handle a wide range of transport problems, such as reflection of ions at quasiperpendicular shocks, transport of turbulence, nonlinear waves in a medium with absorption at low frequencies, transport and dispersion of contaminants in rivers, etc. The present study contains several sections. The quintic Hermite spline is concisely elucidated in Section 2. Section 3 addresses temporal discretization for the concerned equation. The Von Neumann approach has been assessed in Section 4, to validate the stability of the technique. In Section 5, errors have been evaluated in terms of dispersion and dissipation. L_2- and L_∞-norms are presented in Section 6. The formulation of numerical examples is reported in Section 7. Lastly, the results are concluded.

2 Quintic Hermite splines

A Hermite collocation method known as the Quintic Hermite collocation method (QHCM) [19–22] employs basis functions as Hermite interpolating polynomials [15, 16, 18]. In this process, the finite element method is coupled with orthogonal collocation. By Hermite interpolating polynomials of order $1 + 2e$ ($e > 0$), the trial function in Hermite collocation is approximated. It is a development of Lagrange interpolation, which also interpolates the derivatives of the function at each node when employing polynomials. In general, a unique polynomial of degree $\sum m_i + 1$ exists for real values of x_i's and all positive integers m_i's that at each node solves the problem under consideration. The quintic Hermite interpolating polynomial of order 5 was limited in the current investigation by setting $e = 2$.

Hermite interpolating polynomials of order 5 can be formulated as

$$\vartheta(x,t) = \sum_{s=1}^{6} \mathbb{P}_s(x) r(x_s) + \bar{\mathbb{P}}_s(x) r'(x_s) + \bar{\bar{\mathbb{P}}}_s(x) r''(x_s); \quad x \in [x_{l-1}, x_{l+1}], \tag{4}$$

where, $r(x_s)$, $r'(x_s)$, and $r''(x_s)$ are coefficients pertaining to the quintic Hermite interpolating polynomial and $\mathbb{P}_s(x)$, $\bar{\mathbb{P}}_s(x)$, and $\bar{\bar{\mathbb{P}}}_s(x)$ represent Hermite interpolating polynomials of order 5 that can be written as

$$\mathbb{P}_s(x) = \begin{cases} 6(\frac{x_{l+1}-x}{x_{l+1}-x_l})^5 - 15(\frac{x_{l+1}-x}{x_{l+1}-x_l})^4 + 10(\frac{x_{l+1}-x}{x_{l+1}-x_l})^3, & x_l \le x \le x_{l+1} \\ 6(\frac{x-x_{l-1}}{x_l-x_{l-1}})^5 - 15(\frac{x-x_{l-1}}{x_l-x_{l-1}})^4 + 10(\frac{x-x_{l-1}}{x_l-x_{l-1}})^3, & x_{l-1} \le x \le x_l \\ 0, & \text{elsewhere} \end{cases} \tag{5}$$

$$\bar{\mathbb{P}}_s(x) = \begin{cases} 3\frac{(x_{l+1}-x)^5}{(x_{l+1}-x_l)^4} - 7\frac{(x_{l+1}-x)^4}{(x_{l+1}-x_l)^3} + 4\frac{(x_{l+1}-x)^3}{(x_{l+1}-x_l)^2}, & x_l \le x \le x_{l+1} \\ -3\frac{(x-x_{l-1})^5}{(x_l-x_{l-1})^4} + 7\frac{(x-x_{l-1})^4}{(x_l-x_{l-1})^3} - 4\frac{(x-x_{l-1})^3}{(x_l-x_{l-1})^2}, & x_{l-1} \le x \le x_l \\ 0, & \text{elsewhere} \end{cases} \tag{6}$$

$$\bar{\bar{\mathbb{P}}}_s(x) = \begin{cases} 0.5\frac{(x_{l+1}-x)^5}{(x_{l+1}-x_l)^3} - \frac{(x_{l+1}-x)^4}{(x_{l+1}-x_l)^2} + 0.5\frac{(x_{l+1}-x)^3}{(x_{l+1}-x_l)}, & x_l \le x \le x_{l+1} \\ 0.5\frac{(x-x_{l-1})^5}{(x_l-x_{l-1})^3} - \frac{(x-x_{l-1})^4}{(x_l-x_{l-1})^2} + 0.5\frac{(x-x_{l-1})^3}{(x_l-x_{l-1})}, & x_{l-1} \le x \le x_l \\ 0, & \text{elsewhere,} \end{cases} \tag{7}$$

where,

$$\begin{aligned} &\mathbb{P}_s(x_l) = \delta_{ls}, \quad \mathbb{P}'_s(x_l) = 0, \quad \mathbb{P}''_s(x_l) = 0, \quad l,s = 1,2,\dots,6 \\ &\bar{\mathbb{P}}_s(x_l) = 0, \quad \bar{\mathbb{P}}'_s(x_l) = \delta_{ls}, \quad \bar{\mathbb{P}}''_s(x_l) = 0, \quad l,s = 1,2,\dots,6 \\ &\bar{\bar{\mathbb{P}}}_s(x_l) = 0, \quad \bar{\bar{\mathbb{P}}}'_s(x_l) = 0, \quad \bar{\bar{\mathbb{P}}}''_s(x_l) = \delta_{ls}, \quad l,s = 1,2,\dots,6. \end{aligned}$$

The quintic Hermite collocation method partitioned the domain into finite components known as subdomains. The approximate solution can be initialized within these elements to obtain numerical values up to a specified accuracy. The collocation is implemented within each component by incorporating a novel variable $\zeta = \frac{x-x_y}{h_y}$ and the polynomials [19] can be represented in the simplified form

$$\begin{aligned} H_1 &= (1 - 10\zeta^3 + 15\zeta^4 - 6\zeta^5) \\ H_2 &= h_y(\zeta - 6\zeta^3 + 8\zeta^4 - 3\zeta^5) \\ H_3 &= h_y^2(0.5\zeta^2 - 1.5\zeta^3 + 1.5\zeta^4 - 0.5\zeta^5) \\ H_4 &= h_y^2(0.5\zeta^3 - \zeta^4 + 0.5\zeta^5) \\ H_5 &= (10\zeta^3 - 15\zeta^4 + 6\zeta^5) \\ H_6 &= h_y(-4\zeta^3 + 7\zeta^4 - 3\zeta^5). \end{aligned} \tag{8}$$

The quintic Hermite interpolating polynomials also adhere to the following relation, i. e., $H_1(\zeta) = H_5(1-\zeta)$, $H_2(\zeta) = -H_6(1-\zeta)$, $H_3(\zeta) = H_4(1-\zeta)$.

3 Temporal discretization

Using a weighted finite difference operator in the time direction and the time average for the nonlinear as well as the second-order derivative term in equation (1)

$$\frac{\vartheta^{\ell+1}-\vartheta^{\ell}}{\Delta t} = \epsilon\theta[\vartheta_{xx}^{\ell+1}+\vartheta_{xx}^{\ell}] - (1-\theta)[(\vartheta^{\delta}\vartheta_x)^{\ell+1} + (\vartheta^{\delta}\vartheta_x)^{\ell}] \tag{9}$$

$0 \le \theta \le 1$, $\ell = 1, 2, \dots, n$.

Linearizing the nonlinear term [2]

$$(\vartheta^{\delta}\vartheta_x)^{\ell+1} = (\vartheta^{\delta})^{\ell}(\vartheta_x)^{\ell+1} + \delta(\vartheta^{\delta-1})^{\ell}(\vartheta_x)^{\ell}\vartheta^{\ell+1} - \delta(\vartheta^{\delta})^{\ell}(\vartheta_x)^{\ell}. \tag{10}$$

Substituting equation (10) into equation (9) and simplifying we obtain

$$\begin{aligned}\vartheta^{\ell+1} &- \Delta t\epsilon\theta\vartheta_{xx}^{\ell+1} + \Delta t(1-\theta)(\vartheta^{\ell})^{\delta}\vartheta_x^{\ell+1} + \delta\Delta t(1-\theta)(\vartheta^{\ell})^{\delta-1}\vartheta_x^{\ell}\vartheta^{\ell+1} \\ &= \Delta t(1-\theta)\delta(\vartheta^{\ell})^{\delta}\vartheta_x^{\ell} + \vartheta^{\ell} + \epsilon\Delta t\theta\vartheta_{xx}^{\ell} - \Delta t(1-\theta)(\vartheta^{\ell})^{\delta}\vartheta_x^{\ell}.\end{aligned} \tag{11}$$

Initial and boundary conditions can be discretized as $\vartheta(\xi_j h_\ell + x_\ell) = \mu(\xi_j h_\ell + x_\ell)$ and $\vartheta(a, t_\ell) = g_1(t_\ell)$; $\vartheta(b, t_\ell) = g_2(t_\ell)$.

From equation (11)

$$\Rightarrow -\alpha_1\vartheta_{xx}^{\ell+1} + \alpha_2\vartheta_x^{\ell+1} + (1+\alpha_3)\vartheta^{\ell+1} = (\beta_1-\beta_3)\vartheta_x^{\ell} + \beta_2\vartheta_{xx}^{\ell} + \vartheta^{\ell}, \tag{12}$$

where

$$\begin{aligned}&\alpha_1 = \Delta t\epsilon\theta; \quad \alpha_2 = \Delta t(1-\theta)(\vartheta^{\ell})^{\delta}; \quad \alpha_3 = \delta\Delta t(1-\theta)(\vartheta^{\ell})^{\delta-1}\vartheta_x^{\ell} \quad \text{and} \\ &\beta_1 = \Delta t(1-\theta)\delta(\vartheta^{\ell})^{\delta}; \quad \beta_2 = \epsilon\Delta t\theta; \quad \beta_3 = \Delta t(1-\theta)(\vartheta^{\ell})^{\delta}.\end{aligned}$$

Applying Quintic Hermite collocation in equation (12) at the η element

$$\begin{aligned}&\frac{-\alpha_1}{h^2}\sum_{i=1}^{6} H_{ij}''\vartheta_{i,\ell+1}^{\eta} + \frac{\alpha_2}{h}\sum_{i=1}^{6} H_{ij}'\vartheta_{i,\ell+1}^{\eta} + (1+\alpha_3)\sum_{i=1}^{6} H_{ij}\vartheta_{i,\ell+1}^{\eta} \\ &\quad = \frac{\beta_1-\beta_3}{h}\sum_{i=1}^{6} H_{ij}'\vartheta_{i,\ell}^{\eta} + \frac{\beta_2}{h^2}\sum_{i=1}^{6} H_{ij}''\vartheta_{i,\ell}^{\eta} + \sum_{i=1}^{6} H_{ij}\vartheta_{i,\ell}^{\eta}.\end{aligned}$$

Here, $H_{ij} = H_i(\xi_j)$, where ξ_j is the jth interior collocation at the η subdomain. After applying QHCM the resulting system of equations can be expressed as

$$\mathbb{P}\mathbb{V} = \mathbb{Q}. \tag{13}$$

Here, $\mathbb{V} = \vartheta^{\eta}_{i,\ell+1}$ is the column vector of unknown coefficients to be determined and of order $4ne \times 1$. Moreover, $\mathbb{P}$ and $\mathbb{Q}$ are coefficient matrices of order $4ne \times 4ne$ and $4ne \times 1$, respectively.

$$\mathbb{P} = \begin{cases} -\frac{\alpha_1}{h^2}\sum_{i=2}^{6} H''_{ij} + \frac{\alpha_2}{h}\sum_{i=2}^{6} H'_{ij} + (1+\alpha_3)\sum_{i=2}^{6} H_{ij}; & \eta = 1 \text{ element} \\ -\frac{\alpha_1}{h^2}\sum_{i=1}^{6} H''_{ij} + \frac{\alpha_2}{h}\sum_{i=1}^{6} H'_{ij} + (1+\alpha_3)\sum_{i=1}^{6} H_{ij}; & \eta^{\text{th}} \text{ element} \\ -\frac{\alpha_1}{h^2}\sum_{i=1,\neq 5}^{6} H''_{ij} + \frac{\alpha_2}{h}\sum_{i=1,\neq 5}^{6} H'_{ij} + (1+\alpha_3)\sum_{i=1,\neq 5}^{6} H_{ij}; & \eta = ne \text{ element} \end{cases} \tag{14}$$

$$\mathbb{Q} = \begin{cases} \frac{\beta_1-\beta_3}{h}\sum_{i=2}^{6} H'_{ij} + \frac{\beta_2}{h^2}\sum_{i=2}^{6} H''_{ij} + \sum_{i=2}^{6} H_{ij}; & \eta = 1 \text{ element} \\ \frac{\beta_1-\beta_3}{h}\sum_{i=1}^{6} H'_{ij} + \frac{\beta_2}{h^2}\sum_{i=1}^{6} H''_{ij} + \sum_{i=1}^{6} H_{ij}; & \eta^{\text{th}} \text{ element} \\ \frac{\beta_1-\beta_3}{h}\sum_{i=1,\neq 5}^{6} H'_{ij} + \frac{\beta_2}{h^2}\sum_{i=1,\neq 5}^{6} H''_{ij} + \sum_{i=1,\neq 5}^{6} H_{ij}; & \eta = ne \text{ element.} \end{cases} \tag{15}$$

4 Stability analysis

The stability analysis has been carried out with the Von Neumann approach [5]. From equation (11), let

$$\begin{aligned} &\max \vartheta^{\ell} \to \varsigma \quad \text{and} \quad \max \vartheta^{\ell}_{x} \to \varsigma_x \\ &\Rightarrow \vartheta^{\ell+1} - \Delta t \epsilon \theta \vartheta^{\ell+1}_{xx} + \Delta t(1-\theta)\varsigma^{\delta}\vartheta^{\ell+1}_{x} + \delta \Delta t(1-\theta)\varsigma^{\delta-1}\varsigma_x \vartheta^{\ell+1} \\ &\quad = \Delta t(1-\theta)\delta\varsigma^{\delta}\vartheta^{\ell}_{x} + \vartheta^{\ell} + \epsilon \Delta t \theta \vartheta^{\ell}_{xx} - \Delta t(1-\theta)\varsigma^{\delta}\vartheta^{\ell}_{x}. \end{aligned}$$

Apply Von Neumann's method for each r and using

$$\begin{aligned} &\vartheta^{\ell}_{r} = \vartheta^{\ell}(\xi_r) = \alpha^{\ell} e^{i\beta\xi_r} \\ &\Rightarrow \alpha^{\ell+1} e^{i\beta\xi_r}\left[1 + \epsilon\Delta t\theta\beta^2 + \delta\Delta t(1-\theta)\varsigma^{\delta-1}\varsigma_x + i\beta\Delta t(1-\theta)\varsigma^{\delta}\right] \\ &\quad = \alpha^{\ell} e^{i\beta\xi_r}\left[1 - \epsilon\Delta t\theta\beta^2 + i\Delta t(\delta-1)(1-\theta)\varsigma^{\delta}\beta\right]. \end{aligned} \tag{16}$$

Simplifying the above equation

$$\Rightarrow \alpha = \frac{p+iq}{r+is},$$

where $p = 1 - \epsilon\Delta t\theta\beta^2$, $q = \Delta t(\delta-1)(1-\theta)\varsigma^{\delta}\beta$, $r = 1 + \epsilon\Delta t\theta\beta^2 + \delta\Delta t(1-\theta)\varsigma^{\delta-1}\varsigma_x$, and $s = \beta\Delta t(1-\theta)\varsigma^{\delta}$ Now, take the absolute value of α

$$\Rightarrow |\alpha|^2 = \frac{p^2+q^2}{r^2+s^2}.$$

For unconditional stability, $|\alpha| \le 1$, which means $r^2 + s^2 - p^2 - q^2 \ge 0$

$$r^2 + s^2 - p^2 - q^2 = \delta^2(\Delta t)^2(1-\theta)^2\varsigma^{2\delta-2}\varsigma_x^2 + 4\epsilon\Delta t\theta\beta^2 + 2\epsilon(\Delta t)^2\beta^2\theta\delta(1-\theta)\varsigma^{\delta-1}\varsigma_x + 2\delta\Delta t(1-\theta)\varsigma^{\delta-1}\varsigma_x + \beta^2(\Delta t)^2(1-\theta)^2\varsigma^{2\delta}(2\delta-\delta^2).$$

Hence, $r^2 + s^2 - p^2 - q^2 \ge 0$ for $\Delta t \ge 0$; $0 \le \epsilon \le 1$; $\varsigma \ge 0$; $\varsigma_x \ge 0$, $0 \le \theta \le 1$; $\delta \ge 0$.

Hence, the stability condition is substantiated.

5 Error evaluation

Using the scrutiny of fallacy carried out by [23], the errors have to be evaluated for the followed quintic Hermite splines. Three different types of error measurements are taken into account, generated by dispersion, dissipation, and total mean square error. Let $\vartheta(x_i, t)$ be the exact solution and ϑ_i^ℓ be the numerical solution computed by the quintic Hermite splines. Generally, the approximate error can be evaluated as

$$E_i = |\vartheta(x_i, t) - \vartheta_i^\ell|.$$

To demonstrate the total mean square error, the following formula can be used:

$$E_T = \frac{1}{n}\sum_i [\vartheta(x_i, t) - \vartheta_i^\ell]^2, \tag{17}$$

where n represents the number of node points. Introducing the difference mean squared error and ρ, the correlation coefficient, equation (17) can be expressed as

$$E_T = [\sigma(\vartheta(x_i, t)) - \sigma(\vartheta_i^\ell)]^2 + [(\bar{\vartheta}(x_i, t)) - (\bar{\vartheta}_i^\ell)]^2 + 2(1-\rho)\sigma(\vartheta(x_i, t))\sigma(\vartheta_i^\ell). \tag{18}$$

If $\vartheta(x_i, t)$ and ϑ_i^ℓ are perfectly positive correlated, then the error due to dissipation is given by

$$E_1 = [\sigma(\vartheta(x_i, t)) - \sigma(\vartheta_i^\ell)]^2 + [(\bar{\vartheta}(x_i, t)) - (\bar{\vartheta}_i^\ell)]^2.$$

For a correlation coefficient not equal to 1, the additional error is due to the dispersion which is given by

$$E_2 = 2(1-\rho)\sigma(\vartheta(x_i, t))\sigma(\vartheta_i^\ell).$$

6 L_2- and L_∞-norms

The L_2-norm and L_∞-norm have been investigated via the following formulation:

$$L_2\text{-norm},\quad \|\vartheta\|_2 = \sqrt{\sum_{\eta} h_\ell \sum_{i} w_i(\vartheta(\xi_i,t) - \vartheta^{\eta}(\xi_i,t))^2} \tag{19}$$

$$L_\infty\text{-norm},\quad \|\vartheta\|_\infty = \max\left|\vartheta(\xi_i,t) - \vartheta^{\eta}(\xi_i,t)\right|. \tag{20}$$

Here, $w_i^{\prime}$ is the weight function and $\vartheta(\xi_i, t)$ is the exact solution, whereas $\vartheta^{\eta}(\xi_i, t)$ is the approximate solution using QHCM.

Lemma 1 ([17]). *Let $\vartheta(x,t)$ be the solution of the considered generalized modified Burgers' equation, such that $\vartheta(a,t) \geq 0$ and $\vartheta(b,t) \geq 0$, then $\wp\vartheta(x,t) \geq 0$, $\wp$ being the differential operator, for all $x \in (a,b)$ indicates that $\vartheta(x,t) \geq 0$ for $x \in [a,b]$.*

Theorem 1 (Weierstrass approximation theorem). *To any degree of precision on a closed, bounded interval, interpolating polynomials can uniformly approximate any continuous function specified on that interval.*

Lemma 2 ([19]). *Let $\ddot{H}$ be the space of $[0,1]$-defined Hermite interpolating polynomials, then matrix $\mathbf{H}$ is diagonally dominant*

$$\sum_{j=1,i\neq j}^{n} |H_{ij}| < |H_{ii}|, \quad \forall\, i. \tag{21}$$

Lemma 3 ([19]). *Assume that $\ddot{H}$ is the space of $[0,1]$-defined Hermite interpolating polynomials, then*

$$\sum_{i=1}^{n} \left|H_i(x)\right| \leq 6; \quad \forall\, x \in (a,b). \tag{22}$$

Lemma 4 ([19]). *If $\vartheta(x,t)$ is the solution of equation (1) defined on (a,b), then for a constant C independent of v, $|\vartheta(x,t)| \leq C$ and $|\vartheta_t| \leq C$, $\forall (x,t) \in [a,b]$.*

Lemma 5 ([19], Local error estimate). *If $|\vartheta_t^i| \leq C$, $\forall (x,t) \in [a,b] \times (0,T]$, $i = 0,1$, then the local error estimate in the temporal direction is represented as $\|e_{\ell+1}\| \leq C\Delta t^2$, where $e_{\ell+1} = \vartheta(x,t_{\ell+}) - \vartheta^{\ell+1}(x,t)$ is the local truncated error.*

Lemma 6 ([19], Global error estimate). *Assuming E_ℓ to be the estimation of global error, then $\|E_\ell\| \leq C\Delta t$.*

Theorem 2 ([19]). *Assuming $\ddot{H}$ to be the space of $[0,1]$-defined Hermite interpolating polynomials of order 5, let $\bar{\vartheta}(x,t_\ell)$ be the quintic Hermite interpolate of $\vartheta(x,t)$, for $\vartheta(x,t) \in C^6[0,1]$, then*

$$\left\|\vartheta(x,t_\ell) - \bar{\vartheta}(x,t_\ell)\right\|_\infty \leq C\Delta t + n^{-4}\ln^6(n). \tag{23}$$

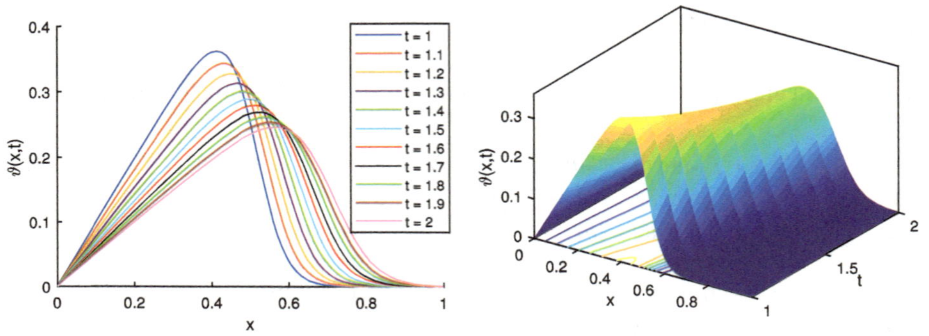

Figure 1: Illustration of numerical solution at $\epsilon = 0.01$ for Problem 1.

7 Numerical examples

Example 1. Consider equation (1) with $\delta = 1$,

$$\vartheta(\xi_j h_\ell + x_\ell) = \mu(\xi_j h_\ell + x_\ell) = \frac{\xi_j h_\ell + x_\ell}{1 + \sqrt{1/\tau}\exp(\frac{(\xi_i h_\ell + x_\ell)^2}{4\epsilon})}.$$

Here, $\tau = \exp(\frac{1}{8\times\epsilon})$. The boundaries can be taken as: $\vartheta(0,t) = g_1(t) = 0, \vartheta(1,t) = g_2(t) = 0$. The numerical example 1 has been solved using quintic Hermite splines. The numerical findings of the L_2-norm and L_∞-norm have been contrasted with the earlier findings [8, 9] for $\epsilon = 0.005$ and are presented in Table 1. The norms in comparison shows the accuracy and the efficiency of the technique. In Table 2, the L_2- and L_∞-norms are shown for $\epsilon = 0.01, 0.001$. The errors in terms of dispersion and dissipation are shown tabularly in Table 3. In Table 4, The precise results and the numerically generated results have been compared for varying ϵ and at $t = 2$. The pictorial representation of the numerical solution is visualized in Figure 1 to Figure 3 for varying ϵ. All the quantitative results shows the efficiency and the accuracy of the method.

Example 2. Consider equation (1) with $\delta = 2$,

$$\vartheta(\xi_j h_\ell + x_\ell) = \mu(\xi_j h_\ell + x_\ell) = \frac{\xi_j h_\ell + x_\ell}{1 + \frac{1}{C_0}\exp(\frac{(\xi_i h_\ell + x_\ell)^2}{4\epsilon})}.$$

The boundaries can be taken as: $\vartheta(0,t) = g_1(t) = 0, \vartheta(1,t) = g_2(t) = 0$. The numerical Example 2 has been solved using quintic Hermite splines for $C_0 = 0.5$ and varying values of ϵ for different time steps. The comparison for these values in terms of L_2- and L_∞-norms has been carried out with various techniques [2–4]. This similitude is represented in Table 5. It has been shown that quintic Hermite splines result in better accuracy as compared to DBF-PSM, BRIM, DQTFDS and OS-FEM. In Table 6, the errors have been represented that shows the efficiency of the quintic Hermite splines. In Table 7, the

Table 1: Comparison of the L_2-norm and L_∞-norm for $\epsilon = 0.005$.

t	QHCM		[8]		CN [8]		[9]	
	L_2	L_∞	L_2	L_∞	L_2	L_∞	L_2	L_∞
1.2	3.714×10^{-5}	1.146×10^{-3}	3.594×10^{-3}	2.389×10^{-3}	3.630×10^{-3}	2.455×10^{-3}	1.461×10^{-2}	9.445×10^{-3}
1.4	2.470×10^{-5}	7.689×10^{-4}	3.779×10^{-3}	2.140×10^{-3}	3.762×10^{-3}	2.201×10^{-3}	1.986×10^{-2}	9.192×10^{-3}
1.6	1.336×10^{-5}	3.191×10^{-4}	3.589×10^{-3}	1.868×10^{-3}	3.546×10^{-3}	1.905×10^{-3}	2.421×10^{-2}	8.531×10^{-3}
1.8	1.368×10^{-5}	4.056×10^{-4}	3.310×10^{-3}	1.653×10^{-3}	3.258×10^{-3}	1.667×10^{-3}	2.781×10^{-2}	1.048×10^{-2}
2	1.415×10^{-5}	4.291×10^{-4}	3.025×10^{-3}	1.431×10^{-3}	2.972×10^{-3}	1.426×10^{-3}	3.069×10^{-2}	1.206×10^{-2}

Table 2: L_2- and L_∞-norms for varying ϵ.

t	$\epsilon = 0.01$		$\epsilon = 0.001$	
	L_2	L_∞	L_2	L_∞
1.2	1.556×10^{-5}	3.929×10^{-4}	1.516×10^{-5}	3.463×10^{-4}
1.4	1.303×10^{-5}	3.132×10^{-4}	6.295×10^{-5}	2.549×10^{-4}
1.6	1.050×10^{-5}	2.216×10^{-4}	1.041×10^{-5}	2.017×10^{-4}
1.8	9.130×10^{-6}	2.210×10^{-4}	8.765×10^{-6}	1.850×10^{-4}
2	7.845×10^{-6}	1.650×10^{-4}	2.452×10^{-5}	7.965×10^{-4}

Table 3: Error measures for $t = 2$ and varying values of ϵ.

η	E_1	E_2
	$\epsilon = 0.01$	
50	1.0182×10^{-9}	7.8092×10^{-9}
100	1.0182×10^{-9}	7.8092×10^{-9}
150	1.0182×10^{-9}	7.8092×10^{-9}
200	1.0182×10^{-9}	7.8092×10^{-9}
	$\epsilon = 0.005$	
50	1.0147×10^{-9}	2.7757×10^{-8}
100	1.0147×10^{-9}	2.7757×10^{-8}
150	1.0147×10^{-9}	2.7757×10^{-8}
200	1.0147×10^{-9}	2.7757×10^{-8}
	$\epsilon = 0.001$	
50	6.3351×10^{-8}	4.3854×10^{-7}
100	6.7353×10^{-8}	4.5876×10^{-7}
150	6.7509×10^{-8}	4.5955×10^{-7}
200	3.4814×10^{-9}	8.2802×10^{-8}

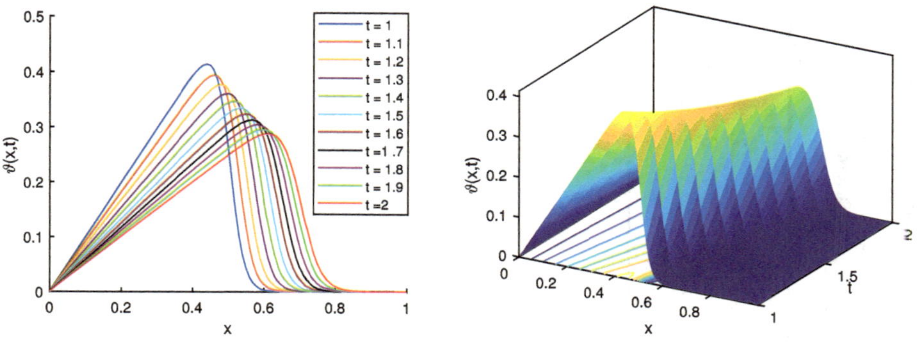

Figure 2: Illustration of numerical solution at $\epsilon = 0.005$ for Problem 1.

Table 4: Comparison of exact and numerical solutions for varying ϵ at $t = 2$ for Problem 1.

x	$\epsilon = 0.01$		$\epsilon = 0.005$		$\epsilon = 0.001$	
	Exact	**QHCM**	**Exact**	**QHCM**	**Exact**	**QHCM**
0.1	4.9846×10^{-2}	4.9821×10^{-2}	5.0000×10^{-2}	4.9975×10^{-2}	5.0000×10^{-2}	4.9975×10^{-2}
0.2	9.9552×10^{-2}	9.9502×10^{-2}	9.9999×10^{-2}	9.9949×10^{-2}	1.0000×10^{-1}	9.9950×10^{-2}
0.3	1.4875×10^{-1}	1.4868×10^{-1}	1.4999×10^{-1}	1.4992×10^{-1}	1.5000×10^{-1}	1.4993×10^{-1}
0.4	1.9605×10^{-1}	1.9595×10^{-1}	1.9994×10^{-1}	1.9984×10^{-1}	2.0000×10^{-1}	1.9990×10^{-1}
0.5	2.3537×10^{-1}	2.3528×10^{-1}	2.4932×10^{-1}	2.4920×10^{-1}	2.5000×10^{-1}	2.4988×10^{-1}
0.6	2.4082×10^{-1}	2.4079×10^{-1}	2.8771×10^{-1}	2.8762×10^{-1}	3.0000×10^{-1}	2.9985×10^{-1}
0.7	1.5569×10^{-1}	1.5586×10^{-1}	1.6656×10^{-1}	1.6699×10^{-1}	2.4908×10^{-1}	2.4988×10^{-1}
0.8	4.3772×10^{-2}	4.3893×10^{-2}	8.3625×10^{-3}	8.4219×10^{-3}	7.1022×10^{-9}	7.3978×10^{-9}
0.9	6.5085×10^{-3}	6.4815×10^{-3}	1.3702×10^{-4}	1.3830×10^{-4}	4.7183×10^{-18}	4.9934×10^{-18}

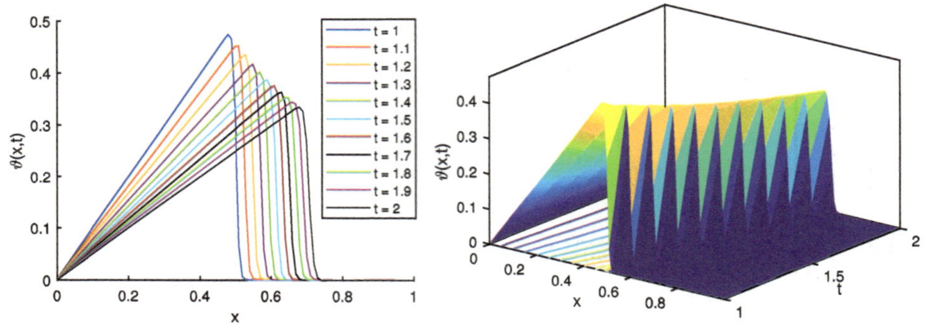

Figure 3: Illustration of numerical solution at $\epsilon = 0.001$ for Problem 1.

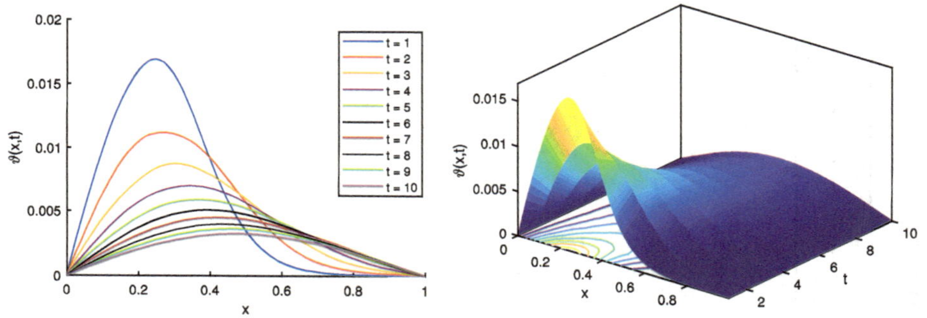

Figure 4: Illustration of numerical solution at $\epsilon = 0.01$ for Problem 2.

obtained results have been compared with the precise ones for varying ϵ and at $t = 2$. The graphic representation of the numerical findings is shown in Figure 4 to Figure 6 that show the smoothness of the solution.

8 Conclusion

The modified Burgers' equation has undergone numerical approximation in this work using quintic Hermite splines. To evaluate the effectiveness and applicability of the approach, the L_2- and L_∞-norms have been evaluated and juxtapositioned with the earlier findings, which shows an improvement in comparison. The method is infallibly stable using the Von Neumann approach. The efficiency as well as accuracy of the method is clearly seen from the dispersion and dissipation errors. Two numerical illustrations have been investigated and the results have been presented in tabular and pictorial form. This method has less computational complication and is easily programmable, which gives accurate results for nonlinear problems.

Table 5: Comparison of the L_2- and L_∞-norms for $\delta = 2$ and varying values of Δt and ϵ.

	$t = 2$		$t = 6$		$t = 10$	
	L_2	L_∞	L_2	L_∞	L_2	L_∞
$\Delta t = 0.01, \epsilon = 0.01$						
QHCM	3.086×10^{-6}	8.507×10^{-5}	1.436×10^{-6}	2.690×10^{-5}	1.997×10^{-6}	5.076×10^{-5}
DBF-PSM [2]	3.811×10^{-4}	7.933×10^{-4}	2.671×10^{-4}	4.459×10^{-4}	1.616×10^{-4}	2.629×10^{-4}
BRIM [2]	4.393×10^{-4}	7.692×10^{-4}	2.794×10^{-4}	4.589×10^{-4}	2.231×10^{-4}	3.247×10^{-4}
DQTFDS [3]	3.834×10^{-4}	8.235×10^{-4}	2.706×10^{-4}	4.662×10^{-4}	1.909×10^{-4}	3.015×10^{-4}
OS-FEM [4]	3.789×10^{-4}	8.163×10^{-4}	3.239×10^{-4}	5.113×10^{-4}	5.395×10^{-4}	1.262×10^{-3}
$\Delta t = 0.001, \epsilon = 0.005$						
QHCM	4.736×10^{-7}	1.319×10^{-5}	6.457×10^{-7}	1.431×10^{-5}	3.860×10^{-7}	6.339×10^{-6}
DBF-PSM [2]	2.712×10^{-4}	7.203×10^{-4}	1.426×10^{-4}	2.321×10^{-4}	1.087×10^{-4}	1.735×10^{-4}
BRIM [2]	2.680×10^{-4}	5.710×10^{-4}	1.682×10^{-4}	3.211×10^{-4}	1.383×10^{-4}	2.222×10^{-4}
DQTFDS [3]	2.337×10^{-4}	5.928×10^{-4}	1.633×10^{-4}	3.313×10^{-4}	1.337×10^{-4}	2.289×10^{-4}
OS-FEM [4]	2.260×10^{-4}	5.792×10^{-4}	1.646×10^{-4}	3.298×10^{-4}	1.394×10^{-4}	2.288×10^{-4}
$\Delta t = 0.01, \epsilon = 0.001$						
QHCM	4.324×10^{-7}	1.406×10^{-5}	2.550×10^{-7}	8.354×10^{-6}	1.533×10^{-7}	4.091×10^{-6}
DBF-PSM [2]	1.217×10^{-4}	4.448×10^{-4}	3.186×10^{-5}	8.839×10^{-5}	2.447×10^{-5}	6.068×10^{-5}
BRIM [2]	8.018×10^{-5}	1.644×10^{-4}	5.361×10^{-5}	1.313×10^{-4}	4.316×10^{-5}	9.577×10^{-5}
DQTFDS [3]	8.316×10^{-5}	2.824×10^{-4}	4.970×10^{-5}	1.513×10^{-4}	4.006×10^{-5}	1.034×10^{-4}
OS-FEM [4]	6.729×10^{-5}	2.591×10^{-4}	4.939×10^{-5}	1.478×10^{-4}	4.069×10^{-5}	1.026×10^{-4}

Table 6: Error measures for $t = 2$, $\delta = 2$ and varying values of Δt and ϵ.

η	E_1	E_2
	$\Delta t = 0.01$, $\epsilon = 0.01$	
50	3.8610×10^{-10}	1.5484×10^{-9}
100	4.0713×10^{-10}	1.0971×10^{-9}
150	6.0509×10^{-10}	8.2530×10^{-10}
200	2.8757×10^{-10}	1.0771×10^{-9}
	$\Delta t = 0.001$, $\epsilon = 0.005$	
50	1.2258×10^{-9}	3.7194×10^{-10}
100	8.3006×10^{-10}	3.3585×10^{-10}
150	7.2319×10^{-10}	3.1906×10^{-10}
200	2.5247×10^{-11}	6.4397×10^{-12}
	$\Delta t = 0.01$, $\epsilon = 0.001$	
50	1.5106×10^{-10}	4.1988×10^{-12}
100	4.7335×10^{-11}	3.0467×10^{-12}
150	3.0342×10^{-11}	2.7668×10^{-12}
200	2.4077×10^{-11}	2.6441×10^{-12}

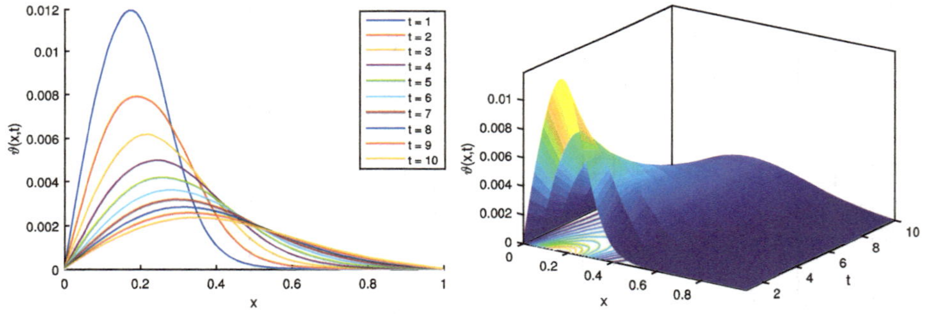

Figure 5: Illustration of numerical solution at $\epsilon = 0.005$ for Problem 2.

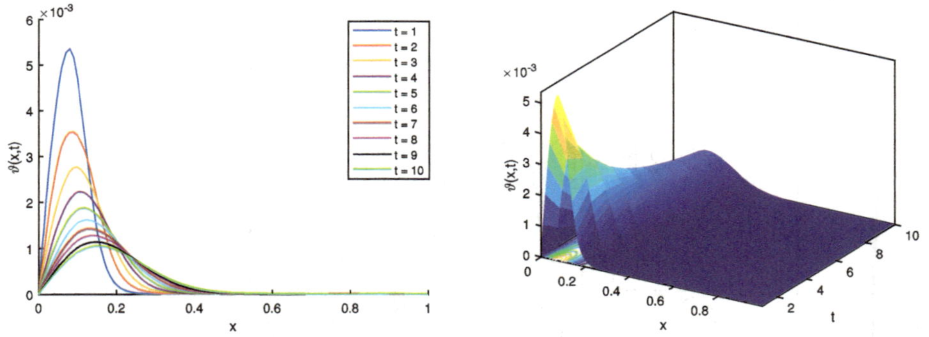

Figure 6: Illustration of numerical solution at $\epsilon = 0.001$ for Problem 2.

Table 7: Comparison of exact and numerical solutions for varying ϵ at $t = 2$ for Problem 2.

x	$\epsilon = 0.01$		$\epsilon = 0.005$		$\epsilon = 0.001$	
	Exact	**QHCM**	**Exact**	**QHCM**	**Exact**	**QHCM**
0.1	1.1891×10^{-2}	1.1805×10^{-2}	1.0795×10^{-2}	1.0782×10^{-2}	4.5989×10^{-3}	4.5848×10^{-3}
0.2	1.7658×10^{-2}	1.7671×10^{-2}	1.1510×10^{-2}	1.1501×10^{-2}	2.3766×10^{-4}	2.4182×10^{-4}
0.3	1.5445×10^{-2}	1.5398×10^{-2}	5.3888×10^{-3}	5.3899×10^{-3}	6.8981×10^{-7}	7.2441×10^{-7}
0.4	9.1327×10^{-3}	9.1463×10^{-3}	1.2868×10^{-3}	1.2885×10^{-3}	1.4575×10^{-10}	1.6050×10^{-10}
0.5	3.8241×10^{-3}	3.8528×10^{-3}	1.7051×10^{-4}	1.7094×10^{-4}	2.3697×10^{-15}	2.8070×10^{-15}
0.6	1.1737×10^{-3}	1.1909×10^{-3}	1.3089×10^{-5}	1.3139×10^{-5}	3.0362×10^{-21}	4.0500×10^{-21}
0.7	2.7048×10^{-4}	2.7673×10^{-4}	5.9213×10^{-7}	5.9534×10^{-7}	3.1045×10^{-28}	2.1480×10^{-28}
0.8	4.7436×10^{-5}	4.8990×10^{-5}	1.5915×10^{-8}	1.6031×10^{-8}	2.5524×10^{-36}	2.2135×10^{-36}
0.9	6.3743×10^{-6}	6.5987×10^{-6}	2.5539×10^{-10}	2.5779×10^{-10}	1.6957×10^{-45}	2.3227×10^{-45}

Bibliography

[1] Bateman, H. (1915). Some recent researches on the motion of fluids. *Monthly Weather Review, 43*, 163–170.

[2] Oruç, Ö. (2020). Two meshless methods based on pseudo spectral delta-shaped basis functions and barycentric rational interpolation for numerical solution of modified Burgers equation. *International Journal of Computer Mathematics, 98*, 461–479.

[3] Aswin, V. S., & Awasthi, A. (2018). Iterative differential quadrature algorithms for modified Burgers equation. *Engineering Computations, 35*(1), 235–250.

[4] Uçar, Y., Yağmurlu, N. M., & Çelikkaya, İ. (2018). Operator splitting for numerical solution of the modified Burgers equation using finite element method. *Numerical Methods for Partial Differential Equations*, 1–15.

[5] Rashidinia, J., & Rasoulizadeh, M. N. (2019). Numerical methods based on radial basis function generated finite difference (RBF-FD) for solution of GkdvB equation. *Wave Motion, 90*, 152–167.

[6] Singh, I., Ansari, N. H., & Singh, G. (2023). Solving PDEs arising in the formation of liquid drop pattern using Sumudu transform based technique. *Partial Differential Equations in Applied Mathematics, 8*, 100578.

[7] Aneja, M., & Sharma, S. (2022). Analysis of heat transfer characteristics of an Al_2O_3–SiO_2/water hybrid nanofluid in a localized heated porous cavity. *Arabian Journal of Science and Engineering, 48*, 1–17.

[8] Parumasur, N., Adetona, R. A., & Singh, P. (2023). Efficient solution of Burgers, modified Burgers and KdV–Burgers equations using B-spline approximation functions. *Mathematics, 11*, 1847.

[9] Raslan, K. R. (2003). A collocation solution for Burgers equation using quadratic B-spline finite elements. *International Journal of Computer Mathematics, 80*(7), 931–938.

[10] Irk, D. (2009). Sextic B-spline collocation method for the modified Burgers equation. *Kybernetes, 38*, 1599–1620.

[11] Lakshmi, C., & Awasthi, A. (2018). Robust numerical scheme for nonlinear modified Burgers equation. *International Journal of Computer Mathematics, 95*, 1910–1926.

[12] Roshan, T., & Bhamra, K. S. (2011). Numerical solutions of the modified Burgers equation by Petrov–Galerkin method. *Applied Mathematics and Computation, 218*, 3673–3679.

[13] Oruç, Ö., Bulut, F., & Esen, A. (2015). A Haar wavelet–finite difference hybrid method for the numerical solution of the modified Burgers equation. *Journal of Mathematical Chemistry, 53*(7), 1592–1607.

[14] Burgers, J. M. (1948). A mathematical model illustrating the theory of turbulence. *Advances in Applied Mechanics, 1*, 171–199.

[15] Dyksen, W. R., & Lynch, R. E. (2000). A new decoupling technique for the Hermite cubic collocation equations arising from boundary value problems. *Mathematics and Computers in Simulation, 54*, 359–372.

[16] Lang, A. W., & Sloan, D. M. (2002). Hermite collocation solution of near-singular problems using numerical coordinate transformations based on adaptivity. *Journal of Computational and Applied Mathematics, 140*, 499–520.

[17] Miller, J. J. H., O'Riordan, E., & Shishkin, G. I. (1996). *Fitted Numerical Methods for Singular Perturbation Problems*. Singapore: World Scientific.

[18] Ricciardi, K. L., & Brill, S. H. (2009). Optimal Hermite collocation applied to a one-dimensional convection–diffusion equation using an adaptive hybrid optimization algorithm. *International Journal of Numerical Methods for Heat and Fluid Flow, 19*, 874–893.

[19] Arora, S., & Kaur, I. (2018). Applications of quintic Hermite collocation with time discretization to singularly perturbed problems. *Applied Mathematics and Computation, 316*, 409–421.

[20] Arora, S., Kaur, I., & Tilahun, W. (2020). An exploration of quintic Hermite splines to solve Burgers' equation. *Arabian Journal of Mathematics, 9*, 19–36.

[21] Arora, S., Kaur, I., & Potuček, F. (2015). Modelling of displacement washing of pulp fibers using the Hermite collocation method. *Brazilian Journal of Chemical Engineering, 32*, 563–575.

[22] Kaur, I., Arora, S., & Bala, I. (2023). An improvised technique of quintic Hermite splines to discretize generalized Burgers–Huxley type equations. *Iranian Journal of Numerical Analysis and Optimization*, *13*, 59–79.

[23] Takacs, L. L. (1985). A two-step scheme for the advection equation with minimized dissipation and dispersion errors. *Monthly Weather Review*, *113*, 1050–1065.

Harpreet Kaur and Atendra Singh Yadav

Mathematical model to analyze the viral infection model in populations using the Monod–Haldane incidence rate

Abstract: This study introduces the Monod–Haldane incidence rate into viral infection models, capturing nonlinear infection dynamics with greater biological realism. The stability analysis of the disease-free equilibrium provides critical insights into control thresholds. The improved mathematical model is recommended to apply this modeling approach to assess intervention strategies, enabling more accurate predictions and effective public health decision-making in managing viral outbreaks.

Keywords: Monod–Haldane incidence rate, stability, epidemic disease, mathematical model

MSC 2020: 92D30, 34D20, 34C23

1 Introduction

Dengue is a mosquito-borne viral infection caused by the Dengue virus, a member of the Flavivirus genus [7]. It is transmitted primarily through the bite of infected Aedes mosquitoes, particularly Aedes aegypti and Aedes albopictus [19, 20]. Clinically, Dengue manifests as a flu-like illness characterized by high fever, severe headache, pain behind the eyes, joint and muscle pain, a rash, and mild bleeding manifestations. Dengue-like illnesses were described as early as 1780, with physician Benjamin Rush describing it as "breakbone fever" due to the intense pain experienced by patients. Dengue's spread intensified due to rapid urbanization, poor mosquito control, and increased global travel [8]. Dengue became endemic in over 100 countries by the 1990s. Dengue remains a global health threat, with around 390 million infections annually, especially in tropical and subtropical regions [2, 3]. Ongoing research focuses on developing better vaccines and mosquito control techniques. Pandey and Phaijoo worked on Dengue infection transmission dynamics using fractional-order mathematical modeling in Nepal [4, 5]. Lipschitz and Banach contraction theorems were used to analyze for a unique solution. The Ulam–Hyers stability is developed in this model. Sensitivity analysis proved that the biting rate and death rate of mosquito are extremely sensitive to the disease transmission [14]. Liu et al. worked on the dynamical behavior of epidemiological models with nonlinear inci-

Harpreet Kaur, Atendra Singh Yadav, Guru Kashi University, Talwandi Sabo, Punjab, India, e-mail: toorharpreet1286@gmail.com

https://doi.org/10.1515/9783112234525-014

dence rate [12, 16, 18]. Their research proposed a generalized incidence function of the form

$$g(I)S = \frac{KI^p S}{1 + \alpha I^q}.$$

Building on prior work in which the term kI reflects the force of infection of the infectious disease and the factor $\frac{1}{1+\alpha I^q}$ captures the psychological effect arising from behavioral changes among susceptibles when the number of infectives is large, this study employs a *Monod–Haldane*-type incidence rate [9]. This formulation yields a more refined and biologically realistic infection term by incorporating both saturation and inhibition mechanisms

$$\mathcal{I}(S,I) = \frac{\beta S I}{1 + \alpha I + \gamma I^2}, \quad \beta > 0,\ \alpha > 0,\ \gamma > 0,$$

where β denotes the effective contact rate, α quantifies saturation effects (e. g., limited effective contacts at higher I), and γ represents inhibition due to crowding/avoidance and other behavioral feedbacks at high prevalence.

2 Mathematical model

According to Wagh et al. [1, 6, 15, 21], we have constructed an improved mathematical model, which is defined as

$$\begin{aligned}
\frac{dS}{dt} &= \Delta - \frac{\beta SE}{1 + \alpha E^2} - \mu S \\
\frac{dE}{dt} &= \frac{\beta SE}{1 + \alpha E^2} - \rho E - \mu E \\
\frac{dI}{dt} &= \rho E - \frac{aI}{1 + bI^2} - \gamma I - \mu I \\
\frac{dR}{dt} &= \frac{aI}{1 + bI^2} + \gamma I - \mu R.
\end{aligned}$$

In this compartmental model, the terms

$$\frac{\beta SE}{1 + \alpha E^2} \quad \text{and} \quad \frac{aI}{1 + bI^2}$$

are the Monod–Haldane (M–H) function. Also, the important parameters are given as follows:

- S: Susceptible Class
- E: Exposed Class
- I: Infected Class

- R: Recovered Class
- β: transmission rate
- α: inhibitory rate
- a: cure rate of the infected people
- b: limited rate in the availability of treatment.

The recovered class is not involved in the first two equations of the improved model. Hence, without loss of generality, the first two equations are analyzed Let, the total population N be given by

$$N = S + R,$$

or

$$\begin{aligned} \frac{dN}{dt} &= \frac{dS}{dt} + \frac{dE}{dt} \\ &= \Delta - \mu S - \rho E - \mu E \\ &= \Delta - \mu S - (\rho + \mu)E \\ &= \Delta - \mu N - \rho E \\ &\leq \Delta - \mu N, \end{aligned}$$

or

$$\frac{dN}{dt} + \mu N \leq \Delta.$$

After solving

$$N \leq N(0)e^{-\mu t} + \frac{\Delta}{\mu}(1 - e^{-\mu t}),$$

hence

$$\lim_{t\to\infty} \sup(N) \leq \frac{\Delta}{\mu}$$
$$\therefore \frac{dN}{dt} < 0 \quad \text{if } N > 0.$$

Therefore, all the outcomes of the model tend towards the region D obtained in Lemma I. Therefore, the region D is positively invariant and the solution of the model is bounded.

Lemma I. *The region*

$$D = \left\{ (S, E) \;\middle|\; 0 < S + E \leq \frac{\Delta}{\mu} \right\}$$

is a positively invariant region of the model.

Proof. This lemma implies that all solutions of the model remain bounded and nonnegative for all $t \geq 0$. Therefore, the improved model is *well-posed* epidemiologically. □

2.1 Disease free-equilibrium point (DFE)

From the first two equations the model has nonnegative equilibria and is defined as

$$\text{DFE} = P\left(\frac{\Delta}{\mu}, 0\right).$$

Using the governing equations, we will obtain the stability of the model. For this, we first find the *basic reproduction number* (R_0)

$$R_0 = \frac{\Delta\beta}{\mu(\rho + \mu)} = 0.000343.$$

If $R_0 < 1$, the improved epidemic model will be *stable*; otherwise, if $R_0 > 1$, the given model will be *unstable* [17].

2.2 Numerical analysis at DFE

Now, we check the behavior of the model at the DFE point

$$\left(\frac{\Delta}{\mu}, 0\right).$$

The Jacobian is given by

$$J = \begin{bmatrix} \Delta - \beta & 0 \\ 0 & 0 \end{bmatrix}.$$

Also

$$|J - I| = \begin{vmatrix} \Delta - \beta - \lambda & 0 \\ 0 & -\lambda \end{vmatrix} = 0.$$

Hence, the eigenvalues are

$$\lambda_1 = 0, \quad \lambda_2 = \Delta - \beta,$$

which have a nonnegative real part. Therefore, the nature of the model is unstable.

Theorem. *If $a \geq 1$, then the system for the first two governing equations does not allow any periodic solution in the interior of the positive region of the S–E plane.*

Proof. Suppose a real-valued function $H(S,E)$ in the interior of the S–E plane is given as

$$H(S,E) = \frac{1 + aE^2}{SE}.$$

$$\text{Also, consider,} \quad h_1 = \Delta - \frac{\beta SE}{1 + aE^2} - \mu S$$

$$h_2 = \frac{\beta SE}{1 + aE^2} - (\rho + \mu)E$$

$$\therefore \ \operatorname{div}(H_1, H_2) = \frac{\partial}{\partial S}(H_1) + \frac{\partial}{\partial E}(H_2)$$

$$= -\frac{\Delta(1 + aE^2)}{S^2 E} - \frac{2a(\rho + \mu)E}{S}.$$

This provides that the above expression can never be equal to zero when $a > 1$, and also the sign of this expression will not change in the positive quadrant of the S–E plane if the restriction $a \geq 1$ holds true.

Therefore, if the epidemic model satisfies this restriction, then the epidemic disease will not spread in the population. □

3 Numerical simulation

Here, the model is analyzed by simulation Using a variety of datasets. The purpose of the simulations is to examine how the system behaves dynamically under various initial conditions and parameter settings. This method sheds light on the existence of threshold factors like the fundamental reproduction number, the stability of equilibrium points, and the potential long-term effects of the disease dynamics.

4 Graphical analysis

To show the graphical analysis of the obtained results, we first selected specific parameter values, as shown in Table 1. Further, the Susceptible (S), Exposed (E), Infected (I), and Recovered (R) classes for the years 2023 and 2024 are shown in Figures 1, 2, 3, 4, 5, 6, 7, 8, 9, and 10, for better understand the disease spread. These graphs show the movement of people between different compartments of the SEIR model over time. The graphical representations show how the model's predictions change by looking at the differences in virus transmission, exposure rates, and recovery patterns between 2023 and 2024.

Table 1: Model parameters and their values with references.

S. No	Name of variable/parameter	Value	Reference [10, 11, 13]
1	S	9651	NICD
2	E	9473	NICD
3	I	9473	NICD
4	R	0.0	NICD
5	β	0.005	Dubey et al. 2013 & 2016
6	α	0.05	Dubey et al. 2013 & 2016
7	A	0.02	Dubey et al. 2013 & 2016
8	B	0.05	Dubey et al. 2013 & 2016
9	μ	7.473	Dubey et al. 2013 & 2016
10	Δ	16.949	Dubey et al. 2013 & 2016
11	ρ	25.58	Dubey et al. 2013 & 2016
12	Γ	1	Dubey et al. 2013 & 2016

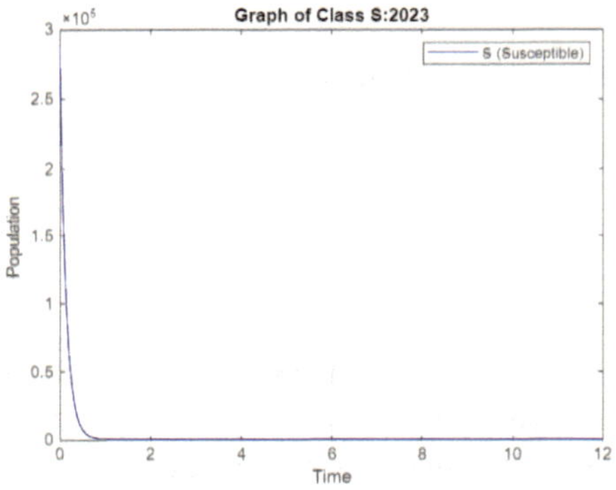

Figure 1: Graph of susceptible class (2023).

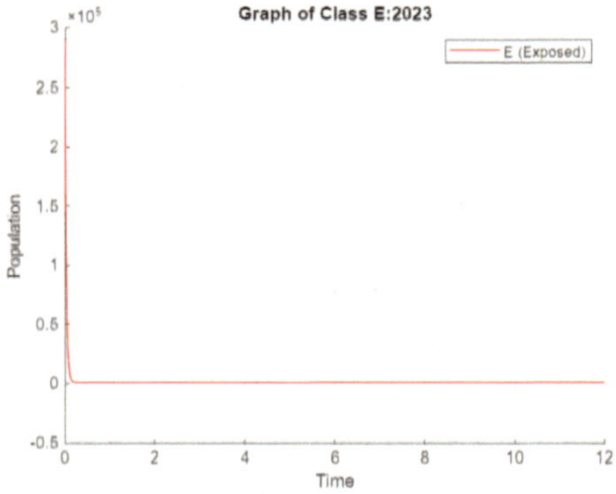

Figure 2: Graph of exposed class (2023).

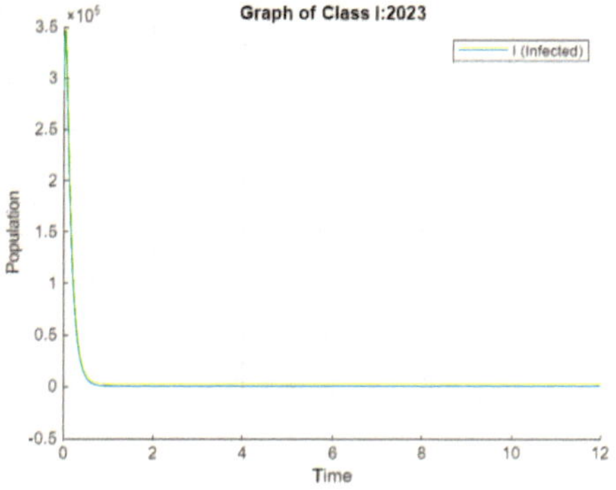

Figure 3: Graph of infected class (2023).

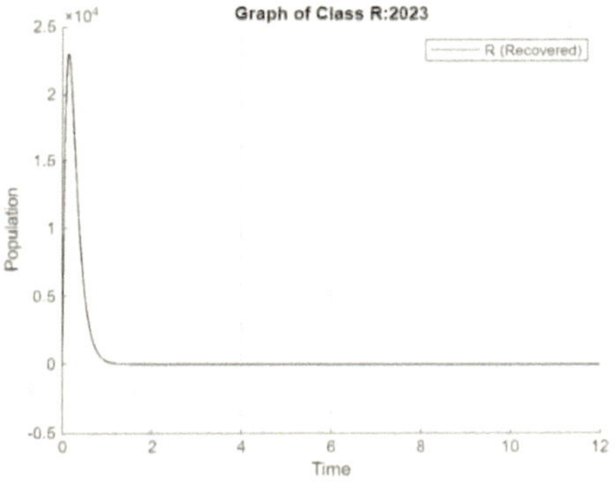

Figure 4: Graph of recovered class (2023).

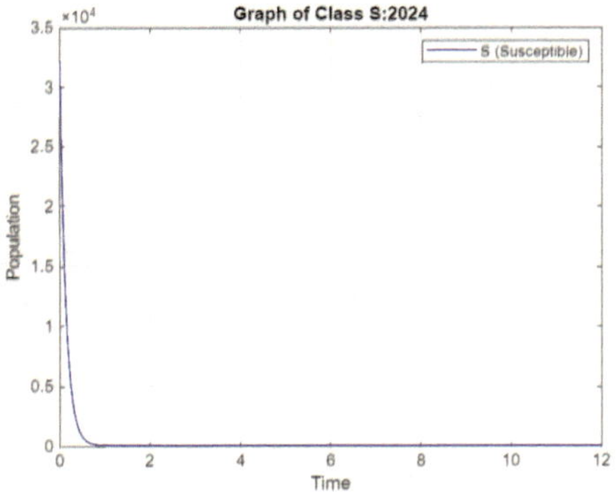

Figure 5: Graph of susceptible class (2024).

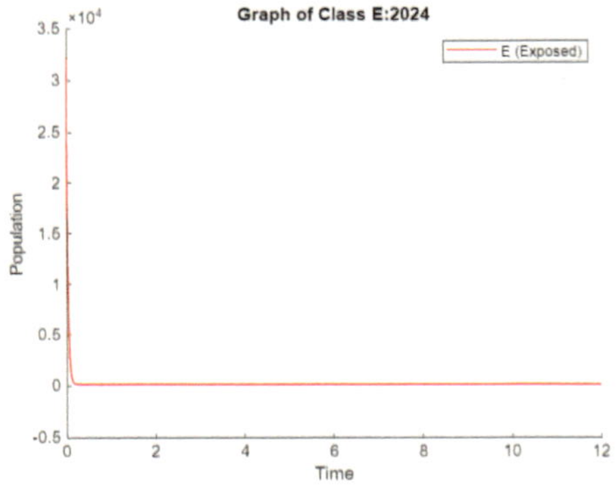

Figure 6: Graph of exposed class (2024).

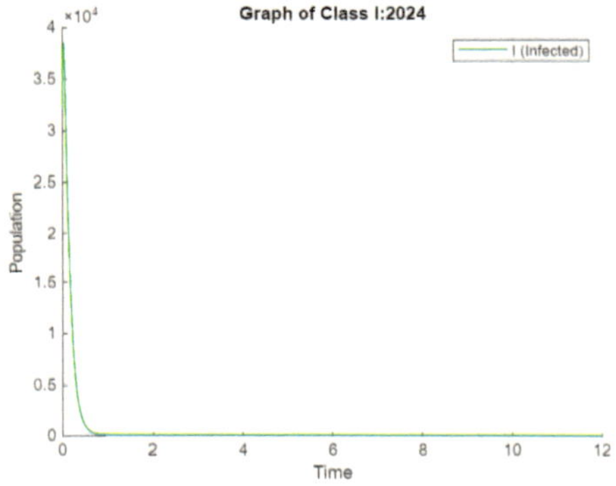

Figure 7: Graph of infected class (2024).

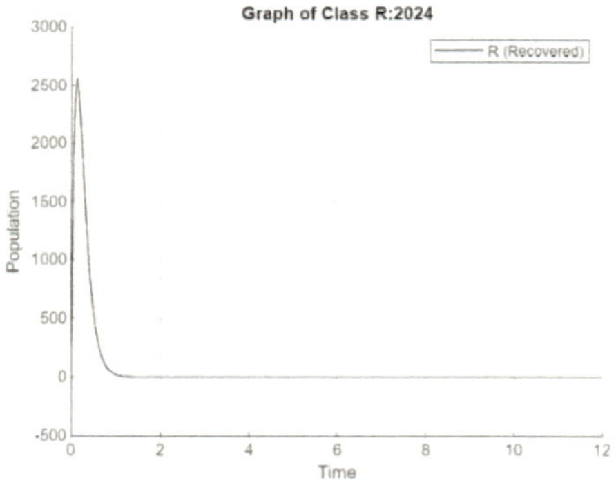

Figure 8: Graph of recovered class (2024).

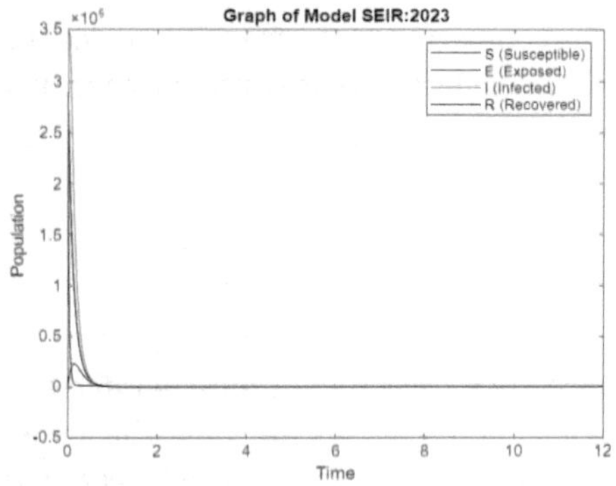

Figure 9: Graph of SEIR (2023).

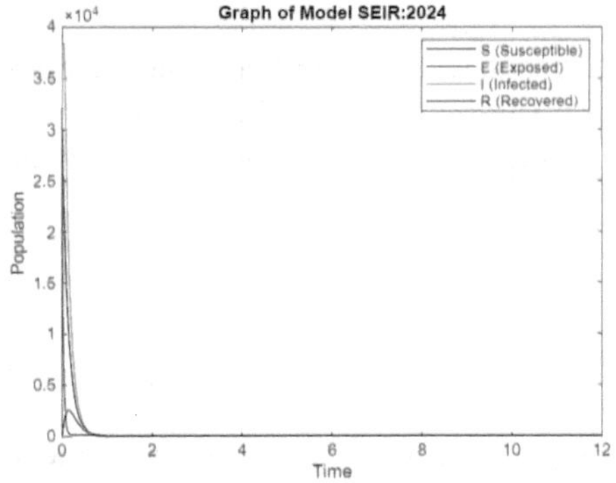

Figure 10: Graph of SEIR (2024).

5 Results and discussion

Using MATLAB, the graphical analysis is presented above for the years 2023 and 2024. The observations are as follows:

1. *Graphs of S for the years 2023 and 2024:* As individuals come into contact with the virus, they transition from being susceptible to exposed.
2. *The number of exposed cases (E) for the years 2023 and 2024:* These individuals remain in the incubation phase, possibly due to early detection and reduced transmission.
3. *Infection class I for the years 2023 and 2024:* The reduction in active infections over time reflects effective treatment and isolation efforts, helping to control the disease spread.
4. *Graphs of R for the years 2023 and 2024:* The constant line at the end implies no further infections, indicating that the disease is progressing toward elimination.

Similarly, the combined representation of all the compartments in the SEIR model are shown for the years 2023 and 2024, respectively. This collective representation confirms a controlled outbreak.

6 Conclusion

In this research, we analyzed a fractional-order SEIR (Susceptible–Exposed–Infected–Recovered) mathematical model incorporating a Monod–Haldane function. The model integrates biologically realistic mechanisms such as infection inhibition and treatment factors. Based on the basic reproduction number, given as

$$R_0 = 0.000343 < 1,$$

the analysis indicates that the disease-free equilibrium is stable, implying that the infection will not persist within the population. This result confirms the effectiveness of the control measures represented in the model and provides valuable insights for public health policymakers.

Bibliography

[1] Abta, A., Kaddar, A., & Alaoui, H. T. (2012). Global stability for delay SIR and SEIR epidemic models with saturated incidence rates. *Electronic Journal of Differential Equations*, *2012*, 1–13.

[2] Adebimpe, O., Bashiru, K. A., & Ojurongbe, T. A. (2015). Stability analysis of an SIR epidemic model with non-linear incidence rate and treatment. *Open Journal of Modelling and Simulation*, *3*, 104–110.

[3] Allen, L. J. S. (1994). Some discrete-time SI, SIR, and SIS epidemic models. *Mathematical Biosciences*, *124*, 83–105.

[4] Anderson, R. M., & May, R. M. (1991). *Infectious Disease of Humans*. New York: Oxford University Press.

[5] Bocharov, G., Volpert, V., Ludewig, B., & Meyerhans, A. (2018). *Mathematical Immunology of Virus Infections* (Vol. 245). Cham, Switzerland: Springer International Publishing.

[6] Capasso, V., & Serio, G. (1978). A generalization of the Kermack–McKendrick deterministic epidemic model. *Mathematical Biosciences*, *42*, 41–61.

[7] Carvalho, S. A., da Silva, S. O., & Charret, I. D. C. (2019). Mathematical modeling of dengue epidemic: control methods and vaccination strategies. *Theory in Biosciences*, *138*(2), 223–239.

[8] Chen, Y., Yang, J., & Zhang, F. (2014). The global stability of an SIRS model with infection age. *Mathematical Biosciences and Engineering*, *11*, 449–469.

[9] Chen, Y., & Liu, Z. (2009). Modelling and analysis of an impulsive SI model with Monod–Haldane functional response. *Chaos, Solitons & Fractals*, *39*(4), 1698–1714.

[10] Dubey, B., Patra, A., Srivastava, P. K., & Dubey, U. S. (2013). Modeling and analysis of a SEIR model with different types of nonlinear treatment rates. *Journal of Biological Systems*, *21*(3), 1350023.

[11] Dubey, P., Dubey, B., & Dubey, U. S. (2016). An SIR model with nonlinear incidence rate and Holling type III treatment rate. In *Applied Analysis in Biological and Physical Sciences: ICMBAA, Aligarh, India, June 2015* (pp. 63–81). Springer India.

[12] Hethcote, H. W., & Driessche, P. V. D. (1991). Some epidemiological models with nonlinear incidence. *Journal of Mathematical Biology*, *29*(3), 271–287.

[13] National Centre for Disease Control (NCDC). Available at: https://ncdc.mohfw.gov.in/.

[14] Idemudia, E. G., & Ojo, O. O. (2023). A comparative analysis of growth models on Nigeria population. *FUDMA Journal of Sciences*, *7*(6), 373–381.

[15] Kermack, W. O., & McKendrick, A. G. (1927). A contribution to the mathematical theory of epidemics. *Proceedings of the Royal Society of London A*, *115*(772), 700–721.

[16] Kumar, A., & Nilam (2019). Mathematical analysis of a delayed epidemic model with nonlinear incidence and treatment rates. *Journal of Engineering Mathematics*, *115*(1), 1–20.

[17] Kumar, V., & Kumar, D. (2018, April). Analysis of epidemic model using basic reproduction number. In *Proceedings of 3rd International Conference on Internet of Things and Connected Technologies (ICIoTCT)* (pp. 26–27).

[18] Liu, W. M., Hethcote, H. W., & Levin, S. A. (1987). Dynamical behavior of epidemiological models with nonlinear incidence rates. *Journal of Mathematical Biology*, *25*, 359–380.

[19] Pandey, H. R., & Phaijoo, G. R. (2023). Analysis of dengue infection transmission dynamics in Nepal using fractional order mathematical modeling. *Chaos, Solitons & Fractals: X*, *11*, 100098.

[20] Srivastav, A. K., & Ghosh, M. (2019). Assessing the impact of treatment on the dynamics of dengue fever: a case study of India. *Applied Mathematics and Computation*, *362*, 124533.

[21] Wagh, C. S., Mahalle, P. N., & Wagh, S. J. (2020). Epidemic peak for COVID-19 in India, 2020. [Conference/Journal details if available].

Salil Kumar, Kamaljit Kaur, Navpreet Sharma, and Sanjeev Gautam

Customer satisfaction towards transportation services in the Tri-city

Abstract: In recent times, rapid growth of the passenger vehicle market has been significantly backed by the rented taxi or cab service companies. The Indian passenger vehicle industry is anticipated to have even more growth potential in the near future. This research concentrates on the customer satisfaction towards the various brands of rented taxi or cab service providers as well as public transport in the region of the Tri-city. There are several factors being studied under this study such as level of comfort, ease of access, security, reliability, safety and convenience, and overall satisfaction towards the service quality of the service providers. Data is analyzed by descriptive research and testing procedures using SPSS software.

Keywords: Transportation, chi-square test, taxi, descriptive measures

1 Introduction

Urban transportation plays a vital role in facilitating daily mobility, especially in metropolitan regions [1]. In India's Tri-city area, encompassing Chandigarh, Mohali, and Panchkula, the demand for efficient and accessible transportation services has grown significantly due to rapid urbanization, increased population density, and evolving commuter preferences. Traditionally, public transport systems such as buses and autorickshaws were the primary modes of travel [2]. However, recent years have witnessed a paradigm shift, with consumers increasingly gravitating toward app-based taxi services and other modern transportation solutions [19].

This shift has been driven by a combination of factors, including the convenience of digital platforms for booking rides, improved service quality, and competitive pricing. Popular app-based taxi services such as Ola, Uber, and local operators now dominate the urban transportation landscape, offering users not just transportation but also a personalized, secure, and a comfortable commuting experience [19].

Customer satisfaction is a critical parameter in the transportation industry, as it directly impacts service retention, brand loyalty, and market growth [18]. Factors such as safety, affordability, driver behavior, ease of booking, vehicle condition, and timeliness significantly influence customer perceptions and preferences. In the context of the

Salil Kumar, Kamaljit Kaur, Department of Mathematics, Sri Guru Gobind Singh College, Sector 26, Chandigarh, 160019, India, e-mail: salil.bansal90@gmail.com
Navpreet Sharma, Sanjeev Gautam, Centre for Distance and Online Education, Panjab University, Sector 14, Chandigarh 160014, India

https://doi.org/10.1515/9783112234525-015

Tri-city region, understanding these factors is especially important, given its unique socioeconomic dynamics and the diverse expectations of its residents.

Despite the growing popularity of app-based taxi services in this region, there is limited research analyzing customer satisfaction specifically in Chandigarh, Mohali, and Panchkula [6]. Existing studies largely focus on urban centers such as Delhi and Mumbai, leaving a research gap that needs to be addressed. This study aims to fill this gap by investigating the key determinants of customer satisfaction in the Tri-city transportation sector [13].

By examining the experiences and preferences of passengers, this research seeks to provide valuable insights into how service providers can enhance the quality of their offerings and better meet the needs of their customers [11]. Additionally, the findings may help policymakers and stakeholders design more efficient urban mobility solutions that align with the expectations of the region's growing population.

2 Review of the literature

Customer satisfaction is a cornerstone of success in the transportation industry, influencing consumer loyalty, service adoption, and overall market growth. Extensive research has explored the factors affecting satisfaction in transportation services, including service quality, reliability, affordability, convenience, and safety [15]. This review synthesizes key studies relevant to the understanding of customer satisfaction, with a focus on their implications for urban transportation systems in regions like the Tri-city area.

Service quality is consistently recognized as a critical determinant of customer satisfaction in transportation services. Parasuraman et al. [9] proposed the SERVQUAL model, identifying reliability, responsiveness, assurance, empathy, and tangibles as core dimensions of service quality. These factors are particularly relevant to modern transportation services, where customer expectations have risen significantly due to technological advancements and competitive market dynamics. In the context of app-based taxi services, Horsu and Yeboah [4] emphasized that high-quality service delivery, including timely arrivals, vehicle cleanliness, and professional driver behavior, play a vital role in enhancing customer satisfaction. Ross [10], in a study conducted in Washington, highlighted the importance of vehicle condition, minimal waiting times, and driver professionalism in shaping positive customer experiences. These findings underscore the importance of operational excellence and consistent service delivery.

Safety is a fundamental aspect influencing customer satisfaction in transportation. Research indicates that passengers prioritize secure travel, particularly in app-based taxi services, which often cater to urban commuters traveling at odd hours. Comfort and reliability are equally significant; customers value smooth and timely rides, as observed in studies by Imran [5] and others focusing on shared taxi services. Reliability, in particular, is tied to the predictability of service quality, including adherence to promised pickup and drop-off times. This factor becomes even more critical in high-density urban areas like the Tri-city region, where traffic congestion and infrastructural challenges can disrupt transportation services [16].

The advent of app-based transportation services has revolutionized urban mobility, offering unparalleled convenience through features such as real-time tracking, cashless payments, and digital receipts. According to Chauhan and Singh [3], these technological innovations have significantly influenced customer satisfaction by enhancing accessibility and transparency. Additionally, ease of online booking and personalized ride options contribute to the appeal of these services, especially in cities with tech-savvy populations like Chandigarh, Mohali, and Panchkula. Price sensitivity is another critical factor in customer satisfaction. Affordability and transparent pricing mechanisms, as discussed by Imran [5], influence the perceived value of services. In a competitive market, where multiple operators vie for consumer attention, offering cost-effective yet high-quality services is essential for maintaining customer loyalty. Sharma and Choudhury [12] examined public transport services in urban India, concluding that digital integration, such as real-time bus tracking, significantly enhances commuter satisfaction. Recent research also emphasizes sustainability in urban transport. Studies suggest that integrating electric buses and nonmotorized transport options could further enhance commuter satisfaction, while reducing environmental impact (Mishra et al. [7]).

Recent studies have explored the relationship between logistics service quality and customer satisfaction using the SERVQUAL model. Yang et al. [17] examined fresh e-commerce logistics, highlighting the significant impact of service quality on customer satisfaction, with customer perceived value, expectations, and emotions acting as mediators. Similarly, Ong et al. [8] analyzed motorcycle taxi services in the Philippines using an integrated Social Exchange Theory (SET) and SERVQUAL approach, identifying trust in service quality as the most influential factor affecting customer satisfaction. Both studies emphasize the crucial role of logistics and transportation service quality in enhancing customer experiences and sustaining business growth.

Despite the growing body of research on customer satisfaction in urban transportation, there is limited focus on specific regions like the Tri-city area. Existing studies predominantly analyze larger metropolitan cities such as Delhi, Mumbai, and Bangalore, leaving a gap in understanding how regional factors, including socioeconomic demographics and local infrastructure, impact satisfaction in smaller urban clusters. Moreover, there is a need for more granular analyses that integrate the unique preferences of Tri-city commuters, including their attitudes toward emerging mobility trends such as shared rides and electric vehicles [14].

Given the Tri-city area's rapid urbanization and evolving transportation landscape, there is a pressing need to assess customer satisfaction in this region. The findings from this review highlight the importance of focusing on service quality, safety, affordability, and technological innovations to meet the needs of urban commuters. Future studies can build on this foundation to provide actionable insights for service providers and policymakers aiming to optimize transportation services in the Tri-city area.

3 Objective of the study

The transportation preferences and satisfaction levels of commuters are influenced by a variety of socioeconomic and demographic factors. In the context of the Tri-city area, understanding these dynamics is crucial for improving service quality and tailoring transportation solutions to meet diverse needs. This study focuses on exploring the interrelationships between key demographic variables, such as qualifications, income, and gender and the mode of transport used, as well as evaluating the levels of convenience and comfort experienced by customers. By analyzing these factors, the study aims to provide insights into behavior and satisfaction, which can inform service providers and policymakers in enhancing urban mobility and customer satisfaction in the Tri-city region. The specific objectives of the study are as follows:

1. To find the relationship between level of qualification and mode of transport used by the respondents.
2. To find the level of convenience and comfort of Tri-city customers with mode of transportation.
3. To find the relationship between annual income and mode of transport used by the respondents.
4. To find the relationship between gender and customer satisfaction.

4 Research methodology

A random sample of 2,658 respondents from the Tri-city region was selected to address the study's objectives. Data were collected using a convenient sampling method and a structured questionnaire designed to gather comprehensive information from participants. The questionnaire comprised 25 questions, organized into distinct categories: demographic characteristics, commute-related attributes, ratings and reviews for public transport and rented cab services, passenger priorities and awareness, and pricing considerations. This structured approach ensures that the study captures diverse perspectives and provides meaningful insights into commuter behavior and satisfaction in the Tri-city area. Descriptive statistics is used to describe the characteristics and picture of the random sample. A Chi-square test evaluates the association and relationship between categorical variables. The p-value is a statistical measure used to determine the significance of results in hypothesis testing. It indicates the probability of obtaining the observed results (or more extreme ones) if the null hypothesis is true. Figure 1 shows the graphical representation of demographic characteristics.

Graphical representation of demographic characteristics

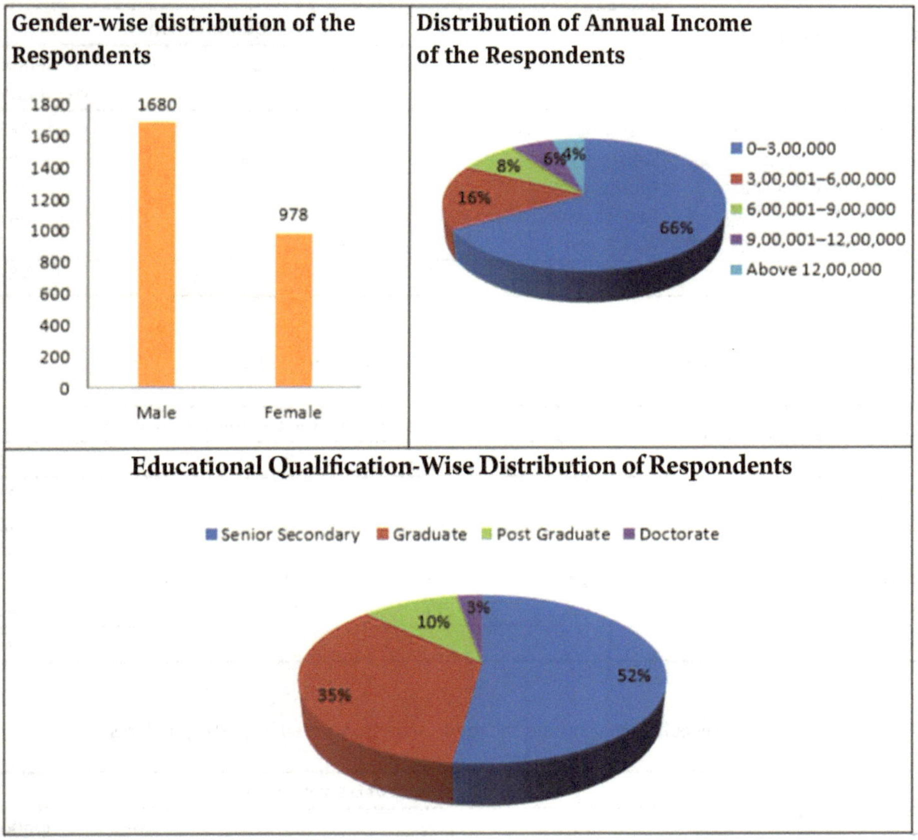

Figure 1: Educational qualification-wise distribution of respondents.

There is no significant relationship between the respondents' level of qualification and their mode of transportation (see Table 1). However, the level of convenience and comfort experienced by Tri-city customers varies significantly with the mode of transport as the p-value is 0.02 (see Table 2). This indicates that cabs provide greater convenience and comfort compared to other forms of public transportation. Additionally, annual income significantly influences the choice of mode of transport among respondents (see Table 3). Lastly, there is no significant relationship between gender and satisfaction levels, as both male and female respondents report similar levels of satisfaction with their chosen mode of transportation as p-value is 0.524) (see Tables 4 and 5).

Table 1: Chi-square test for association between mode of transport and qualification.

Research hypothesis	Mode of transport	Contingency tables			
		Qualification			
		Doctorate	UG	PG	Senior secondary
Relationship between level of qualification and mode of transport used by the respondents	Buses/ Autorickshaws	12	294	102	582
	Cabs	06	78	30	90
	Four-wheeler	42	204	66	198
	Two-wheeler	12	228	66	342
	Walk/Cycle	00	126	6	174
p value					0.82

Table 2: Level of convenience and comfort of Tri-city customers with mode of transportation.

Mode of transportation	Level of convenience and comfort				
	1	2	3	4	5
Buses/Autorickshaws	78	156	372	264	120
Cabs	06	30	84	66	18
Four-wheeler	30	84	252	78	66
Two-wheeler	72	114	282	126	54
Walk/Cycle	00	30	120	96	60
p-value					0.02

Table 3: Relationship between annual income and mode of transport used by the respondents.

Annual income	Mode of transportation				
	Buses/ Autorickshaws	Cabs	Four-wheeler	Two-wheeler	Walk/ Cycle
0–300,000	702	126	258	450	234
300,001–600,000	132	36	78	126	48
600,001–900,000	90	24	48	30	12
900,001–1,200,000	54	6	54	30	6
More than 1,200,000	12	12	72	12	6
p-value					0.03

5 Results of multiple regression analysis

Multiple regression is a statistical technique used to examine the relationship between one dependent variable and two or more independent variables. It extends simple linear regression by incorporating multiple predictors to improve explanatory power. Tables 2

Table 4: Relationship between gender and satisfaction level (1 = lowest, 5 = highest).

Gender	Satisfaction level				
	1	2	3	4	5
Female	96	180	390	192	120
Male	90	234	720	438	198
p-value					0.524

Table 5: Model summary.

Mode of transport	R-squared	Adjusted *R*	Standard error	*F*	Significance *F*
Public transport	0.6614	0.6519	0.851535	70.20001	0.000
Rented Cabs	0.565	0.5512	0.7759	69.62166	0.000

Table 6: Results of multiple regression model (public transport).

Public transport	Coefficients	Standard error	t Stat	p-value	Lower 95 %	Upper 95 %
Intercept	0.5241	0.1682	3.1148	0.00198	0.19322	0.8550
Level of comfort of public transport	0.1073	0.0584	1.8365	0.06709	−0.0076	0.2222
Service provided by public transport	0.2262	0.0596	3.7902	0.00017	0.10884	0.3435
Ease of access provided by public transport	0.2491	0.0565	4.4047	0.000	0.13793	0.3604
Security of self, peers, and luggage in public transport	0.2422	0.0525	4.6071	0.000	0.13885	0.3456
Expenses incurred in a given period	0.0493	0.0435	1.1329	0.25799	−0.0363	0.1350

and 3 represent the association between customer satisfaction and taxi service quality parameters. The predictors were the variables, i. e., mobile app convenience, driver behavior, time reliability, price affordability, and safety, while the dependent variable was general satisfaction of customers.

The goodness-of-fit indicator, R-squared, is 0.6614 for public transport, suggesting that 66.14 % of the variation in customer experiences is explained by service quality parameters, including the level of comfort, services provided, ease of access, and security for passengers and their belongings. For rented cabs, the R-squared value is 0.565, indicating that 56.5 % of the variation in customer experiences is explained by the same set of service quality parameters. The F-statistic has a significance value of 0.000 for both modes of transportation, which is well below the 0.05 threshold, confirming that the models fit the data well and the regression results are statistically significant for both public transport and rented cabs (see Tables 6 and 7).

Table 7: Results of multiple regression model (rental cabs).

Rented cabs	Coefficients	Standard error	t Stat	p-value	Lower 95 %	Upper 95 %
Intercept	0.9758	0.1764	5.5294	0.000	0.62878	1.3229
Level of comfort of public transport	0.1913	0.0583	3.2780	0.00114	0.07653	0.3060
Service provided by public transport	0.2140	0.0627	3.4131	0.00071	0.09073	0.3374
Security of self, peers, and luggage in public transport	0.3226	0.0498	6.4694	0.000	0.22459	0.4207
Ease of access provided by public transport	0.1407	0.0430	3.2734	0.00116	0.05621	0.2253
Expenses incurred in a given period	0.0290	0.0381	0.7596	0.44798	−0.04609	0.1041

For public transport, three hypotheses were accepted, confirming that comfort, service quality, and security positively influence customer satisfaction, highlighting their importance in improving the overall user experience. Additionally, two hypotheses were rejected. The null hypothesis regarding the relationship between comfort levels and customer satisfaction was rejected, indicating that higher comfort significantly enhances satisfaction. Similarly, the null hypothesis concerning expenses incurred for public transport services was also rejected, showing that affordability is crucial, with customers expressing greater satisfaction when services are reasonably priced. In the case of rented cab services, one hypothesis was rejected: the null hypothesis regarding the relationship between expenses incurred and customer satisfaction. This demonstrates that higher costs negatively impact satisfaction, emphasizing the importance of competitive pricing in ensuring a positive customer experience.

The study's results show that cabs offer greater convenience and comfort than public transport, emphasizing the need for modernization through real-time tracking, digital payments, and electric buses. With comfort, service quality, and security significantly influencing satisfaction ($R^2 = 0.6614$), improving traffic flow with dedicated lanes and better signaling can enhance reliability. As affordability impacts satisfaction, policies should enhance accessibility and safety to make public transit more competitive. Promoting sustainable initiatives like electric cabs and cycling/walking infrastructure can further improve urban mobility while meeting customer expectations.

6 Conclusion

The study reveals 57.5 % of the respondents are satisfied with the public transportation in the Tri-city. The results of the study clearly demonstrate that the highest numbers of respondents come from maximum income group. Expenses are a key factor influencing customer satisfaction across public transport and rented cab services. For public transport, comfort, service quality, security, and affordability are critical drivers of satisfaction, while for rented cabs, competitive pricing is essential. To enhance satisfaction, public transport should prioritize comfort, affordability, and service quality, while rented cabs should focus on pricing and reliability. These findings underscore the importance of affordability and service excellence in fostering customer loyalty and positive experiences. spintronic technologies.

Bibliography

[1] Bharadwaj, R., & Mehta, P. (2021). Examining the factors influencing customer satisfaction in app-based taxi services in India. *Journal of Service Research*, *18*(2), 95–112.

[2] Bhat, C. R., & Koppelman, F. S. (1999). A retrospective and prospective survey of time-use research. *Transportation*, *26*(2), 119–139.

[3] Chauhan, A., & Singh, S. (2020). Impact of app-based transportation services on urban mobility: a case study of India. *Journal of Transportation Research*, *8*(3), 205–219.

[4] Horsu, E. N., & Yeboah, S. T. (2015). Influence of service quality on customer satisfaction in the transportation sector: a case of Ghana. *European Journal of Business and Management*, *7*(9), 1–7.

[5] Imran, R. (2014). Shared taxis: understanding customer satisfaction with cost and convenience. *International Journal of Transport Economics*, *41*(2), 179–198.

[6] Jain, A., & Goel, R. (2020). Service quality and consumer satisfaction in the app-based cab industry in Indian metros. *Journal of Marketing Trends*, *12*(4), 45–60.

[7] Mishra, P., Singh, R., & Jha, S. (2023). Sustainable urban transport systems: a pathway to greener cities. *International Journal of Sustainable Transportation*, *17*(2), 120–138.

[8] Ong, A. K. S., German, J. D., Dangaran, P. C., Paz, J. J. B., & Macatangay, R. R. G. (2024). Service quality and customer satisfaction analysis among motorcycle taxi transportation in the Philippines through SERVQUAL dimensions and social exchange theory. *Case Studies on Transport Policy*, *15*, 101139.

[9] Parasuraman, A., Zeithaml, V. A., & Berry, L. L. (1988). SERVQUAL: a multiple-item scale for measuring consumer perceptions of service quality. *Journal of Retailing*, *64*(1), 12–40.

[10] Ross, D. (2015). Exploring factors affecting customer satisfaction in app-based taxi services: a study in Washington. *Urban Mobility Journal*, *12*(4), 45–60.

[11] Seth, N., Deshmukh, S. G., & Vrat, P. (2005). Service quality models: a review. *International Journal of Quality & Reliability Management*, *22*(9), 913–949.

[12] Sharma, K., & Choudhury, S. (2023). Smart mobility solutions and customer satisfaction: evaluating the impact of intelligent transport systems in India. *Asian Journal of Urban Studies*, *12*(1), 58–72.

[13] Smith, J. A., & Walker, P. (2013). Technology and convenience: customer satisfaction in modern transportation systems. *Journal of Urban Transport Systems*, *29*(1), 50–65.

[14] Train, K. E. (2009). *Discrete Choice Methods With Simulation* (2nd ed.). Cambridge University Press.

[15] Zeithaml, V. A., Berry, L. L., & Parasuraman, A. (1996). The behavioral consequences of service quality. *Journal of Marketing*, *60*(2), 31–46.

[16] Wilson, A., Zeithaml, V., Bitner, M. J., & Gremler, D. (2016). *Services Marketing: Integrating Customer Focus Across the Firm* (3rd ed.). McGraw Hill.

[17] Yang, Q., Wang, Z. S., Feng, K., & Tang, Q. Y. (2024). Investigating the crucial role of logistics service quality in customer satisfaction for fresh e-commerce: a mutually validating method based on SERVQUAL and service encounter theory. *Journal of Retailing and Consumer Services*, *81*, 103940.

[18] Yeo, G. T., Thai, V. V., & Roh, S. Y. (2015). An analysis of the relationship between service quality and customer satisfaction in the transportation sector. *Transport Policy*, *45*(2), 123–133.

[19] Zafar, A., & Khan, S. (2019). App-based taxi services and their impact on urban transportation: a case study from India. *Indian Journal of Transport Systems*, *34*(1), 75–88.

Abdallah M. Talafha, Pankaj Kumar, Jaskirat P. Singh, and Dipesh

Investigating allelopathic effects and population density variations in plant competition through mathematical modeling

Abstract: Allelopathy interference between plants has traditionally been debated with respect to the development of allelochemicals. A mathematical competition model is employed to examine the existence of duet competing plant species, focusing specifically on time-dependent variations in their population density and allelopathic effects. The study determines the equilibrium points and utilizes the Routh–Hurwitz theorem to derive certain formations that describe the stability and trajectory of periodic solutions derived from the Hopf bifurcation occurring at the positive equilibrium. Numerical simulations and graphical representations are generated using the “dde23” command in MATLAB software to support the analytical findings.

Keywords: Plant population, allelopathy, delay, stability, Hopf bifurcation

MSC 2020: 92D25, 34K20, 34D20, 34C23

1 Introduction

Allelopathy refers to phenomena in which one plant negatively impacts another plant due to chemical liberation into the aerosphere. Unfortunately, however, there has been considerable variation and confusion in the definition and use of allelechemistry. On the surface of the planet, living creatures choose only those behaviors that increase their chances of surviving and passing their genetics onto the next generation. Fluctuations in plant population densities under reciprocal rivalry are vital elements of the plant physiology analysis. Modifications in the physical environment of the necessary nutrients and the coalescence of these nutrients are crucial parameters to affect the size and density of the plant population [1]. Zhou and Cui describe the stability and Hopf bifurcation study of a delayed ecoepidemiological model [2]. In a delay differential system with delay-dependent parameters, the geometric stability changes requirements [3]. The authors of [4] constructed and modeled single-species growth in which single parts of the

Abdallah M. Talafha, Department of Mathematics and Natural Sciences, Prince Mohammad Bin Fahd University (PMU), AL Khobar, Saudi Arabia
Pankaj Kumar, Department of Mathematics, Lovely Professional University, Phagwara-144411, Punjab, India
Jaskirat P. Singh, Department of Mathematics, Akal University, Talwandi Sabo, Bathinda-151302, Punjab, India
Dipesh, Department of Mathematics, SR University, Warangal-506371, Telangana, India, e-mail: dipeshdalal2411@gmail.com

https://doi.org/10.1515/9783112234525-016

species go through two stages of development: adolescent and robust. The processes of shoot–root interactions and resource partitioning in plant development are mathematically modeled [5]. Igwe described modeling of the increase of differential effects of the first intercompetition on the ecodiversity value; rivalry between two phytoplankton species [6]. Thornely introduced work on mathematical modeling of specific plant development processes, and his framework has been utilized for a wide range of plant biology problems [7]. An essential part of an accord that acts at the single, community, and environs levels is the soil–nutrient–plant interaction. Various plant developments are influenced by nutrients, that have an impact on population growth dynamics, which influences standing crop production. Models were created to explain specific plant development in a crop classified by Benamin and Hardwick based on the premise that supplies are allocated in the right way [8]. Chattopadhyay also suggested a model for the study of the influence of toxic agents on a competitive two-species environment withdrawing a clear account of the amount of the toxicant in the model [9]. Many plants' physiochemical processes, including photosynthetic production, respiratory phenomena, weakening of roots, and others, are impaired by allelochemicals [10]. Huang et al. analyze the global equilibrium of a nonlinear delay differential equation method in relation to population growth [11]. A detailed investigation was carried out to examine the presence of zeros of the empirically given polynomial [12]. Kalra and Kumar conducted a study to investigate the impact of time lag on plant growth under the influence of toxic metals [13]. Ruan incorporated various types of time delays in biological models, which have been extensively explored by several researchers studying the dynamics of populations consisting of a single species [14].

Plant population dynamics are given a temporal delay because, according to the phenomena of population mutation, variation in biomass is impacted by both the previous and current populations. Kalra and Kumar investigate the effect of the components of delay on toxicant-induced dynamics of plant growth [15]. The mechanism of the exponential characteristic equation's zeros was studied in depth [16]. In this environment, it's a valuable note by May that the time delay investigations, in which he found that if a physiological or other process has a potentially stable negativity process, but it is applied with a time lag that is longer than the system's organic frequency magnitude the outcome is instability, not stability [17]. Arino et al. worked on Delay Differential Equations and Applications in mathematics. However, the research on allelopathy, which refers to the phenomenon where a plant species releases chemicals that affect other plant species, has generally received limited attention in the scientific community [18]. Phytoplankton research is a useful concern from an environment view. All of these are creatures that drift easily and move slowly in an aquatic system. The organisms, which are single cells and tiny in size, are extremely useful in sea species. Abbas et al. developed a fractional framework for a distinct phytoplankton category where a species generated allelochemicals that stimulated the growth of another [19]. According to Roy et al. allelopathy is expected to decrease corival displacement and promote phytoplankton diversity in marine ecosystems in which many species fight for a limited number of resources, due to which competition occurs in between species [20]. Grzyb et al. conducted research on the environmental variables that influence the decomposition of crop residues and elaborated the potential uses of crop residues and show how the various parameters, like temperature, soil qualities, the chemical makeup of crop wastes, rain fall, the physical state of microorganism populations, and the carbon–nitrogen balance in land, influence the biological phenomena [21]. We can control the growth rate and path of processing of agricultural wastes in soil by adjusting these parameters. Tanveer et al. conducted an experiment on the allelopathic effects of wasteland weeds on crops through their leaf leachates and rhizosphere soils in agricultural fields [22]. Furthermore, plants' root development was suppressed more by plant phenolic content than their shoot production was suppressed by their mycorrhizae environments. We claim that all methods used in the modeling process are moderate and low such that responses to the IVBP are identical and constant at all significant periods. To date, there has been rare use of delayed differential equations in the domain of allelopathy competitive plant populations. Based on the above, we examine the significance of time delay in the competition among plant communities under the influence of allelopathy, where one population releases allelochemicals that harm other organisms, while the unaffected population does not produce any allelochemicals and realizes that the affected population may face extinction.

2 Mathematical model

The dynamical model is constrained by three state variables: P_1 and P_2 are duet corival plant populations, respectively, and T represents the amount (concentration) of allelochemicals generated via the species P_2. The competition dynamics is described by the following system of equations:

$$\begin{aligned}\frac{dP_1}{dt} &= S - \beta_1 P_1 P_2 - a_1 P_1,\\ \frac{dP_2}{dt} &= \beta_1 P_1(t-\tau) P_2 - b_1 P_2 - a_2 P_2,\\ \frac{dT}{dt} &= b_2 P_2 - c_1 T,\end{aligned} \tag{1}$$

where $P_1(0), P_2(0) > 0$, $T(0) > 0$, $\forall\ t$ & $P_1(t-\tau) = \text{constant}\ \forall t \in [0, \tau]$.

In system (1), S is the availability of the environment factor that has plant population P_1 (constant), β_1 is the rate of mutual competition between plant populations P_1 and P_2, respectively. Moreover, a_1 and a_2 are the natural decay rate of plant populations P_1 and P_2, respectively, b_1 is the environment factor that is stimulating for P_2 (water, nutrient, CO_2 gas, etc.). Further, b_2 and c_1 are the growth rate and natural decay of coefficient of T, respectively. The coefficients a_1, a_2, b_1, b_2, and c_1 have units of cell per hour and the time unit is hours.

These properties would make evaluating the model's suitability for physiological processes much easier. Although it appears to be self-evident that delays exist in nature, they may have a significant influence in system dynamics.

2.1 Boundedness

The solutions boundedness of the model is given as

$$A = P_1 + P_2,$$
$$\frac{dA}{dt} = \frac{d(P_1 + P_2)}{dt} = \frac{dP_2}{dt} + \frac{dP_2}{dt}.$$

Let $\Omega = \min(a_1, a_2, \beta_1, b_1, b_2)$ and $P_1(t - \tau) \sim P_1$

$$\frac{dA(t)}{dt} \leq S - \Omega.$$

As $t \to \infty$ and applying the comparison theorem, we have

$$0 \leq (P_1 + P_2) \leq \frac{S}{\Omega}.$$

Hence, $0 \leq (P_1 + P_2) \leq \frac{S}{\Omega}$ and

$$\frac{dT}{dt} \geq -c_1 T.$$

Using the comparison theorem, when $t \to \infty$, $\frac{dT(t)}{dt} \to 0$

$$T \leq \frac{1}{c_1},$$

hence

$$0 \leq T \leq \frac{1}{c_1}.$$

As a consequence, all the equations of the system of equation (1) lie in the three-dimensional region $D = [(P_1, P_2, T) \in \mathbb{R}^3_+ : 0 \leq P_1 + P_2 \leq \frac{S}{\Omega}, 0 \leq T \leq \frac{1}{c_1}]$ as $t \to \infty$, for all positive initial values.

Here, $P_1(0), P_2(0) > 0$, $T(0) > 0$, $\forall\, t$ & $P_1(t - \tau) = \text{constant}\ \forall t \in [-\tau, 0] \in D \subset \mathbb{R}^3_+$ and $\Omega = \min(a_1, a_2, \beta_1, b_1, b_2)$.

2.2 Positivity of the solutions

Positivity implies that the system is viable or sustainable. One needs to demonstrate that all the solutions are provided by equation (1), where the starting condition $P_1(0)$, $P_2(0) > 0$, $T(0) > 0$, $\forall\ t$ & $P_1(t-\tau)T = \text{constant}\ \forall t \in [0,\tau]$ and $(P_1(t), P_2(t), P_3(t))$ of the model remains positive $\forall\, t > 0$. From system (1)

$$\frac{dT}{dt} \geq -c_1 T,$$
$$T \geq e^{-c_1 t}.$$

Similarly, we can calculate P_1, P_2.

2.3 Equilibrium points

Assume

$$\frac{dP_1^*}{dt} = 0$$
$$\frac{dP_1^*}{dt} = S - \beta_1 P_1 P_2 - a_1 P_1$$
$$S - \beta_1 P_1 P_2 - a_1 P_1 = 0$$
$$P_1^* = \frac{S}{a_1 + \beta_1 P_2}.$$

Similarly, we can calculate P_2^*, T^*

$$P_2^* = \frac{\beta_1 S - a_1 b_1 - a_2 a_3}{\beta_1 b_1 + \beta_1 a_2},$$
$$T^* = \frac{b_2 P_2}{c_1}.$$

2.4 Stability of the equilibrium point $E^*(P_1^*, P_2^*, T^*)$ and a Hopf bifurcation

The balancing competition process of $E^*(P_1^*, P_2^*, T^*)$ of the plant is specified by the system of equation (1) scheme

$$\frac{dP_1^*}{dt} = S - \beta_1 P_1^* P_2^* - a_1 P_1^*,$$
$$\frac{dP_2^*}{dt} = \beta_1 P_1^*(t-\tau) P_2^* - b_1 P_2^* - a_2 P_2^*,$$
$$\frac{dT^*}{dt} = b_2 P_2^* - c_1 T^*.$$

The system's characteristic equation is

$$\lambda^3 + x_1\lambda^2 + x_2\lambda + x_3 + e^{-\lambda\tau}(y_1\lambda^2 + y_2\lambda + y_3) = 0, \tag{2}$$

where

$$\begin{aligned}
x_1 &= b_1 + a_2 + \beta_1 P_2 + a_1 + c_1,\\
x_2 &= \beta_1 b_1 P_2 + \beta_1 a_2 P_2 + a_1 b_1 + a_1 a_2 + c_1 b_1 + c_1 a_2 + a_1 c_1 + \beta_1 c_1 P_2,\\
x_3 &= \beta_1 b_1 c_1 P_2 + \beta_1 a_2 c_1 P_2 + a_1 b_1 c_1 + a_1 a_2 c_1,\\
y_1 &= \beta_1,\\
y_2 &= \beta_1 c_1 - \beta_1 a_1 - \beta_1^2 P_1 P_2 - \beta_1^2 P_2,\\
y_3 &= \beta_1^2 (P_1 - 1) P_2 c_1 - \beta_1 a_1 c_1.
\end{aligned}$$

Clearly, $x_1, x_2, x_3, y_1, y_2, y_3$ are all positive.
With clarity $\lambda = i\phi$ is a solution of equation (2) iff

$$(i\phi)^3 + x_1(i\phi)^2 + x_2(i\phi) + x_3 + e^{-i\phi\tau}(y_1(i\phi)^2 + y_2(i\phi) + y_3) = 0. \tag{3}$$

Separating real and imaginary parts, we obtain

$$x_3 - x_1\phi^2 + (y_3 - y_1\phi^2)\cos(\phi\tau) + y_2\phi\sin(\phi\tau) = 0, \tag{4}$$

$$x_2\phi - \phi^3 + y_2\phi\cos(\phi\tau) - (y_3 - y_1\phi^2)\sin(\phi\tau) = 0. \tag{5}$$

Squaring and adding equation (6) and equation (7) gives the polynomial

$$\phi^6 + (x_1^2 - y_1^2 - 2x_2)\phi^4 + (x_2^2 - y_2^2 + 2y_1y_3 - 2x_1x_3)\phi^2 + (x_3^2 - y_3^2) = 0. \tag{6}$$

Putting

$$j = x_1^2 - y_1^2 - 2x_2, \quad k = x_2^2 - y_2^2 + 2y_1y_3 - 2x_1x_3, \quad l = x_3^2 - y_3^2.$$

Letting $\phi^2 = z$, equation (6) becomes

$$z^3 + jz^2 + kz + l = 0. \tag{7}$$

There exists $z_0 \in (0, \infty)$, $h(z_0) = 0$. If $l < 0$, equation (7) will give at least one positive real root. Define $h(z) = z^3 + jz^2 + kz + l$. Then

$$h(0) = l < 0, \quad \lim_{z\to\infty} h(z) = \infty.$$

If $l \geq 0$, the essential parameter for equation (7) to have positive real roots is

$$D = j^2 - 3k \geq 0, \quad h(z) = z^3 + jz^2 + kz + l.$$

Differentiating

$$h'(z) = 3z^2 + 2jz + k. \tag{8}$$

The roots of $h'(z) = 0$ are

$$z_{1,2} = \frac{-2j \mp \sqrt{4j^2 - 12k}}{6} = \frac{-j \mp \sqrt{D}}{3}. \tag{9}$$

If $D < 0$ then equation (8) has no real roots, hence $h(z)$ is a monotone increasing function in z,

If $h(0) = l \geq 0$, then equation (6) will not contain any actual real roots.

If $D \geq 0$, then $z_1 = \frac{-j+\sqrt{D}}{3}$ is the local minimum of $h(z)$.

If $l \geq 0$, then equation (7) will contain an actual real root if $z_1 > 0$ and $h(z_1) \leq 0$.

Assume $z_1 \leq 0$ or $z_1 > 0$ and $h(z_1) > 0$.

If $z_1 \leq 0$, since $h(z)$ is increasing for $z \geq z_1$ and $h(0) = l \geq 0$, it follows that $h(z)$ does not contain any actual real zero.

If $z_1 > 0$ and $h(z_1) > 0$, since $z_2 = \frac{-j-\sqrt{D}}{3}$ is the local maximum value, which shows $h(z) \leq h(z_2)$ and $h(0) = l \geq 0$, hence $h(z)$ does not give any actual real root.

Lemma 1. *Let z_1 be defined by equation* (9)*:*
(1.1) *If $l < 0$, equation* (7) *has at least one real positive root.*
(1.2) *If $l \geq 0$ and $D < 0$, equation* (7) *has no real root.*
(1.3) *If $l \geq 0$, then equation* (7) *has a positive root if $z_1 > 0$ and $h(z_1) \leq 0$.*

Assume equation (7) has real roots z_1, z_2, z_3. Then, equation (6) yields roots

$$\phi_i = \sqrt{z_i}, \quad i = 1, 2, 3.$$

From equation (7)

$$\sin(\phi\tau) = \frac{x_2\phi - \phi^3}{d},$$

thus

$$\tau = \frac{1}{\phi}\left[\sin^{-1}\left(\frac{x_2\phi - \phi^3}{d}\right) + 2(i-1)\pi\right], \quad i = 1, 2, 3.$$

Let

$$\tau_p^{(i)} = \frac{1}{\phi_p}\left[\sin^{-1}\left(\frac{x_2\phi_p - \phi_p^3}{d}\right) + 2(i-1)\pi\right], \quad p = 1, 2, 3, \quad i = 0, 1, 2, \ldots.$$

The above equation gives the set of purely imaginary roots of equation (6). If

$$\tau = \tau_p^{(i)}, \quad p = 1, 2, 3, \quad i = 0, 1, 2, \ldots,$$

then

$$\lim_{i\to\infty} \tau_p^{(i)} = \infty, \quad p = 1, 2.$$

Define

$$\tau_0 = \tau_{p_0}^{(i_0)} = \min_{1\le p\le 3,\, i\ge 1} \{\tau_p^{(i)}\}, \quad \phi_0 = \phi_{p_0}, \quad z_0 = z_{p_0}. \tag{10}$$

Lemma 2. *Suppose* $x_1 \ge 0$, $(x_3 + d) > 0$, $x_1 x_2 (x_3 - d) > 0$.
(2.1) *If* $l \ge 0$ *and* $D < 0$, *then the roots of equation* (2) *are negative and real for all* $\tau \ge 0$.
(2.2) *If* $l < 0$ *or* $l \ge 0$, $z_1 > 0$, *and* $h(z_1) \le 0$, *then the roots of equation* (2) *are negative and real for all* $\tau \in (0, \tau_0)$.

Proof. When $\tau = 0$, equation (2) becomes

$$\lambda^3 + (x_1 + y_1)\lambda^2 + (x_2 + y_2)\lambda + (x_3 + y_3) = 0. \tag{11}$$

By using the Routh–Hurwitz criteria, the roots of equation (3) will have negative real parts if and only if

$$(x_3 + y_3) \ge 0, \quad (x_1 + y_1)(x_2 + y_2) - (x_3 + y_3) > 0.$$

If $l \ge 0$ and $D = j^2 - 3k < 0$, Lemma 1 (1.2) shows that equation (2) has no roots with a zero real part for all $\tau \ge 0$.

If $l < 0$ or $l \ge 0$, $z_1 > 0$, and $h(z_1) \le 0$, Lemma 1 (1.1) and (1.2) imply that when

$$\tau \ne \tau_p^{(i)}, \quad p = 1, 2, 3, \quad i \ge 1,$$

equation (2) does not have a root with a zero real component, and even the value of τ is minimal, thus it is completely imaginary. □

Lemma 3. *Suppose* $z_0 = \phi_0^2$ *and* $\tau = \tau_0$. *Then*

$$\operatorname{sgn}[\phi'(\tau_0)] = \operatorname{sgn}[f'(z_0)].$$

Proof. Let us assume that

$$\lambda(\tau) = \phi(\tau) + i\varphi(\tau)$$

is a root of equation (2). Let $\phi(\tau_0) = 0$ and $\varphi(\tau_0) = \varphi_0$, hence $\pm i\varphi_0$ are simple purely imaginary roots of equation (2) at $\tau = \tau_0$, and $\lambda(\tau)$ crosses the imaginary axis transversely. Assume that $f'(z_0) \ne 0$.

Substituting $\lambda(\tau)$ into equation (2) and differentiating with respect to λ, we obtain

$$\frac{d\lambda}{d\tau}[3\lambda^2 + 2x_1\lambda + x_2 + (y_1\lambda^2 + y_2\lambda + y_3)(-\tau) + (2y_1\lambda + y_2)e^{-\lambda\tau}] = \lambda(y_1\lambda^2 + y_2\lambda + y_3)e^{-\lambda\tau}.$$

Thus

$$\left(\frac{d\lambda}{d\tau}\right)^{-1} = \frac{(3\lambda^2 + 2x_1\lambda + x_2)e^{-\lambda\tau}}{\lambda(y_1\lambda^2 + y_2\lambda + y_3)} + \frac{2y_1\lambda + y_2}{\lambda(y_1\lambda^2 + y_2\lambda + y_3)} - \frac{\tau}{\lambda}.$$

From equations (5)–(7), we obtain

$$\phi'(\tau_0) = \mathrm{Re}\left[\frac{(3\lambda^2 + 2x_1\lambda + x_2)e^{-\lambda\tau}}{\lambda(y_1\lambda^2 + y_2\lambda + y_3)}\right] + \mathrm{Re}\left[\frac{2y_1\lambda + y_2}{\lambda(y_1\lambda^2 + y_2\lambda + y_3)}\right].$$

Evaluating at $\lambda = i\varphi_0$, this becomes

$$\phi'(\tau_0) = \frac{3\varphi_0^6 + 2j\varphi_0^4 + k\varphi_0^2}{(y_3 - y_1\varphi_0^2)^2 + (y_2\varphi_0)^2}.$$

Since

$$(y_3 - y_1\varphi_0^2)^2 + (y_2\varphi_0)^2 > 0 \quad \text{and} \quad \varphi_0 > 0,$$

it follows that

$$\mathrm{sgn}[\phi'(\tau_0)] = \mathrm{sgn}[f'(z_0)].$$

□

Theorem 1. *Let τ_0, ϕ_0 be defined by equation* (10) *and $z_0 = \phi_0^2$. Let $(x_1+y_1) > 0$, $(x_3+y_3) > 0$, $(x_1 + y_1)(x_2 + y_2) > 0$.*
(i) *If $l \geq 0$ and $D < 0$, then $E^*(P_1^*, P_2^*, T^*)$ is absolutely stable.*
(ii) *If $l < 0$ or ($l \geq 0$, $z_1 > 0$, $f(z_1) \leq 0$), then $E^*(P_1^*, P_2^*, T^*)$ is asymptotically stable for all $\tau \in [0, \tau_0)$.*
(iii) *If the criteria* (ii) *holds at $\tau = \tau_0$ and $f'(z_0) \neq 0$, then a Hopf bifurcation occurs at $E^*(P_1^*, P_2^*, T^*)$.*

Numerical example. The dynamics described by the numerical solution are graphically produced using the `dde23` command in MATLAB software and described by the following set of parameters:

$$S = 0.4, \quad \beta_1 = 1.98, \quad a_1 = 0.59, \quad b_1 = 0.5, \quad a_2 = 0.6, \quad b_2 = 0.9, \quad c_1 = 0.5.$$

The equilibrium point is

$$E^*(P_1^* = 0.2156,\ P_2^* = 0.4511,\ T^* = 1.7289).$$

The system behavior changes with varying delay τ.

3 Result and discussion

Figure 1(a) shows the stability of plant populations under optimal conditions, where no allelopathic effects are present, and the populations remain stable over time. Figure 1(b)

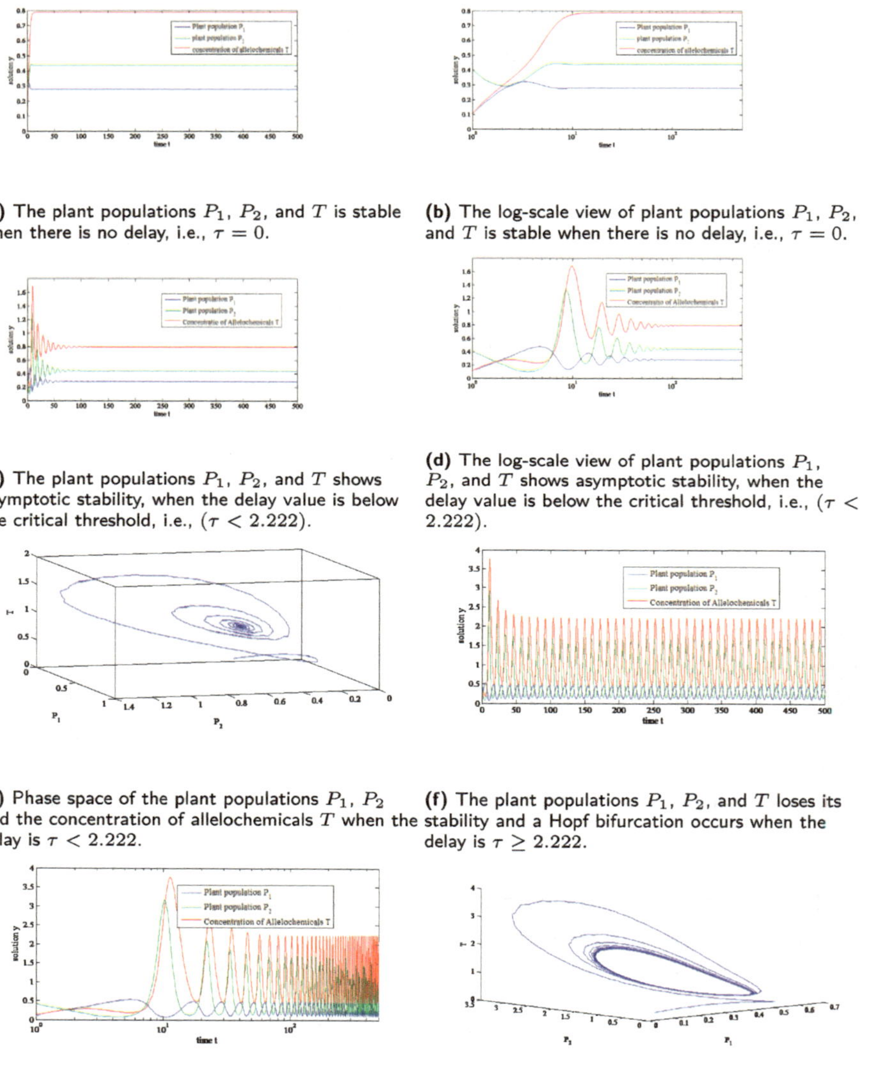

(a) The plant populations P_1, P_2, and T is stable when there is no delay, i.e., $\tau = 0$.

(b) The log-scale view of plant populations P_1, P_2, and T is stable when there is no delay, i.e., $\tau = 0$.

(c) The plant populations P_1, P_2, and T shows asymptotic stability, when the delay value is below the critical threshold, i.e., $(\tau < 2.222)$.

(d) The log-scale view of plant populations P_1, P_2, and T shows asymptotic stability, when the delay value is below the critical threshold, i.e., $(\tau < 2.222)$.

(e) Phase space of the plant populations P_1, P_2 and the concentration of allelochemicals T when the delay is $\tau < 2.222$.

(f) The plant populations P_1, P_2, and T loses its stability and a Hopf bifurcation occurs when the delay is $\tau \geq 2.222$.

(g) The log-scale view of the plant populations P_1, P_2, and T loses its stability and a Hopf bifuraction occurs when the delay is $\tau \geq 2.222$.

(h) Phase space of the plant populations P_1, P_2 and the concentration of allelochemicals T when the delay is $\tau \geq 2.222$.

Figure 1: Numerical solution of system (1) using dde23, with different delay values.

shows the dynamics of the system in a log-scale view in the absence of an allelopathic effect. Figure 1(c) portrays a pivotal shift in the system's behavior. Here, we observe the onset of instability as one plant population begins to release allelochemicals, initiating a cascade effect that impacts both populations, and triggering a decline in overall system stability and manifesting as asymptotic behavior $\tau = 0$ to $\tau < 2.222$. Figure 1(d) provides a logarithmic-scale depiction of the dynamical system's response when the delay in allelochemicals release is below a critical value. This visualization highlights the nuanced changes in stability as the system navigates critical conditions, offering valuable insights into the early warning signs of instability. Figure 1(e) shows the phase view of the dynamical system when $\tau < 2.222$. Figure 1(f) shows a critical juncture where the delay in allelochemical release surpasses the threshold value. At this point, the system undergoes a significant transformation, dynamics show the loss of asymptotic stability and the emergence of a Hopf bifurcation and periodic solutions $\tau \geq 2.222$ to $\tau < 2.222$. This illustrates the intricate interplay between delays and system dynamics. Figure 1(g) shows a logarithmic-scale view of the dynamical system as it crosses the threshold value. Here, we demonstrate how the system's behavior changes when it enters a phase of increased instability. Figure 1(h) depicts the phase view of the dynamical system when the delay is $\tau \geq 2.222$. These graphics serve as an accurate depiction of the complex equilibrium that natural systems maintain and the consequences of the above critical thresholds.

4 Conclusion

This chapter presents an analytical model to examine the impact of delay on the dynamics of two competing plant populations in the presence of allelochemicals. We assume from the literature review that the plant population is influenced by allelochemicals. Delay interrupts the mechanism and initiates dynamic behavior by a Hopf bifurcation with minimal intervals. The stability analysis of equilibrium E^* ($P1^*$, $P2^*$, T^*) is carried out. By applying the Routh–Hurwitz method, we determine that in the absence of delay, the plant population exhibits stability, but upon the occurrence of a delay below the critical value, the plant population loses its stability and moves asymptotically towards stability of the plant population. Once the delay surpasses the critical value, the system experiences a loss of asymptotic stability and it starts oscillating and periodic solutions occur leading to a Hopf bifurcation. The future scope of this model includes the practical use of delay differential equations for verification of the statistical data obtained from the agriculture field. It will help farmers to predict in advance the rate of growth and the factors that adversely affect crops, under intervals of time.

Bibliography

[1] Willis, R. J. (2007). *The History of Allelopathy*. Dordrecht: Springer Netherlands.

[2] Shi, X., Cui, J., & Zhou, X. (2011). Stability and Hopf bifurcation analysis of an eco-epidemic model with a stage structure. *Nonlinear Analysis: Theory, Methods & Applications*, *74*(4), 1088–1106.

[3] Beretta, E., & Kuang, Y. (2002). Geometric stability switch criteria in delay differential systems with delay dependent parameters. *SIAM Journal on Mathematical Analysis*, *33*(5), 1144–1165.

[4] Aiello, W. G., & Freedman, H. I. (1990). A time-delay model of single-species growth with stage structure. *Mathematical Biosciences*, *101*(2), 139–153.

[5] Feller, C., Favre, P., Janka, A., Zeeman, S. C., Gabriel, J. P., & Reinhardt, D. (2015). Mathematical modeling of the dynamics of shoot-root interactions and resource partitioning in plant growth. *PLoS ONE*, *10*(7), e0127905.

[6] Igwe, P. Y., Atsu, J. U., Ekaka-a, E. N., & Nwaoburu, A. O. Modelling the increasing differential effects of the first inter-competition coefficient on the biodiversity value; competition between two phytoplankton species. *International Journal of Advanced Engineering, Management and Science*, *4*(10), 266189.

[7] Thornley, J. H. (1976). Mathematical models in plant physiology (pp. 318).

[8] Benjamin, L. R., & Hardwick, R. C. (1986). Sources of variation and measures of variability in even-aged stands of plants. *Annals of Botany*, 757–778.

[9] Chattopadhyay, J. (1996). Effect of toxic substances on a two-species competitive system. *Ecological Modelling*, *84*(1–3), 287–289.

[10] Cheng, F., & Cheng, Z. (2015). Research progress on the use of plant allelopathy in agriculture and the physiological and ecological mechanisms of allelopathy. *Frontiers of Plant Science*, *6*, 1020.

[11] Huang, G., Liu, A., & Foryś, U. (2016). Global stability analysis of some nonlinear delay differential equations in population dynamics. *Journal of Nonlinear Science*, *26*(1), 27–41.

[12] Ruan, S., & Wei, J. (2003). On the zeros of transcendental functions with applications to stability of delay differential equations with two delays. *Dynamics of Continuous Discrete and Impulsive Systems Series A*, *10*, 863–874.

[13] Kalra, P., & Kumar, P. (2018). The study of time lag on plant growth under the effect of toxic metal: a mathematical model. *Pertanika Journal of Science & Technology*, *26*(3).

[14] Ruan, S. (2006). Delay differential equations in single species dynamics. In *Delay Differential Equations and Applications* (pp. 477–517). Dordrecht: Springer Netherlands.

[15] Dipesh, & Kumar, P. (2022). Modelling the stimulatory and inhibitory allelopathic effects on competing plant populations. *AIP Conference Proceedings*, *2435*, 020044.

[16] Ruan, S. (2001). Absolute stability, conditional stability and bifurcation in Kolmogorov-type predator–prey systems with discrete delays. *Quarterly of Applied Mathematics*, *59*(1), 159–173.

[17] May, R. M. (1973). Time-delay versus stability in population models with two and three trophic levels. *Ecology*, *54*(2), 315–325.

[18] Arino, O., Hbid, M. L., & Ait Dads, E. (Eds.) (2007). *Delay Differential Equations and Applications: Proceedings of the NATO Advanced Study Institute held in Marrakech, Morocco, 9–21 September 2002* (Vol. 205). Springer Science & Business Media.

[19] Abbas, S., Mahto, L., Favini, A., & Hafayed, M. (2016). Dynamical study of fractional model of allelopathic stimulatory phytoplankton species. *Differential Equations and Dynamical Systems*, *24*(3), 267–280.

[20] Felpeto, A. B., Roy, S., & Vasconcelos, V. M. (2018). Allelopathy prevents competitive exclusion and promotes phytoplankton biodiversity. *Oikos*, *127*(1), 85–98.

[21] Grzyb, A., Wolna-Maruwka, A., & Niewiadomska, A. (2020). Environmental factors affecting the mineralization of crop residues. *Agronomy*, *10*(12), 1951.

[22] Hayyat, M. S., Safdar, M. E., Asif, M., Tanveer, A., Ali, L., Qamar, R., et al. H Tarar, Z. (2020). Allelopathic effect of waste-land weeds on germination and growth of winter crops. *Planta daninha*, *38*, e020173626.

Anita Devi, Archna Kumari, and Vijay K. Kukreja

Solution of the fractional Burger–Fisher model using an improvised cubic splines approximation

Abstract: This work aims to develop a robust approach based on an improvised cubic B-splines collocation approximation for the numerical simulation of the one-dimensional Burger–Fisher model, involving the Caputo interpretation of the fractional operator with respect to the temporal domain. In order to carry out the discretization of derivatives in the spatial domain, the proposed approach is considered and a finite difference scheme is utilized to approximate the fractional derivative with respect to the time variable τ. By including the effects of the fractional parameter α, the stability has been studied using the Von Neumann approach. Further, the method has been found to be fourth-order accurate in the spatial dimension, while $(2-\alpha)$th-order accuracy is obtained along the temporal direction. The suggested approach is applied to solve some test problems based on the fractional model. Finally, the computed results presented in tables and graphical illustrations validate the accuracy and effectiveness of the developed approach in dealing with higher-order fractional PDEs.

Keywords: Burger–Fisher model, fractional calculus, Caputo's formulation, improvised spline basis, stability, convergence criteria

MSC 2020: 35G31, 65M70, 76M22, 41A15

1 Introduction

The study of nonlinear convection–diffusion equations plays a crucial role in understanding many physical phenomena in scientific and engineering domains. Classical derivatives are limited to describing the behavior of the function at integer positions only and do not provide insights into its behavior at noninteger positions. The introduction of fractional derivatives generalizes the idea of a traditional derivative to noninteger orders and allows for the simulation of complex dynamics in real-world systems, where future behavior is influenced by the history of past events [1, 2]. The general expression for the time-fractional Burger–Fisher (FBF) equation can be written as

$$ {}^{C}\mathfrak{D}_{\tau}^{\alpha}(y) = \eta\nabla^{2}y - \beta y^{p}\nabla y + \epsilon y(1-y^{p}) + S(\xi,\tau) \tag{1} $$

Anita Devi, Archna Kumari, Vijay K. Kukreja, Department of Mathematics, SLIET Longowal, Punjab, 148106, India, e-mail: vkkukreja@gmail.com

https://doi.org/10.1515/9783112234525-017

subject to the following conditions:

$$y(\xi,0) = y_0(\xi), \tag{2}$$

$$y(\xi,\tau) = \mathfrak{k}_1(\tau), \quad y(\xi,\tau) = \mathfrak{k}_2(\tau), \tag{3}$$

where, η symbolizes the perturbation parameter associated with y, α indicates the order of a fractional derivative, ∇ denotes the gradient operator, p is a positive integer, and $S(\xi,\tau)$ is the nonlinear source term here. The above model is a crucial nonlinear fractional PDE that includes a reaction term, convection phenomena, as well as diffusion mechanisms and is utilized to depict several phenomena arising in the fields of ocean engineering, gas dynamics, fluid mechanics, and many more fields of science and engineering.

At $\epsilon = 0$, equation (1) is reduced to the general form of a fractional Burgers equation that can be represented as

$$ {}^{C}\mathfrak{D}_{\tau}^{\alpha}(y) = \eta\nabla^2 y - \beta y^p \nabla y + S(\xi,\tau). \tag{4}$$

This model is a well-known nonlinear equation that captures complex phenomena arising in applied sciences, such as wave propagation in dissipative phenomena, ionic behavior in plasma, aspects of the physics of solid materials, etc. [3–6]. By considering $\beta = 0$ and $\eta = 1$, equation (1) is converted into the generalized fractional Fisher equation (5), which is a significant nonlinear PDE that is utilized to depict spreading of bacterial colonies, perturbation spreads in excitable media, and so on [7, 8]

$$ {}^{C}\mathfrak{D}_{\tau}^{\alpha}(y) = \nabla^2 y + \epsilon y(1-y^p) + S(\xi,\tau). \tag{5}$$

Within the literature, different researchers have proposed various definitions of fractional-order operators based on generalized functions like the Riemann–Liouville derivative, Caputo's definition, the Atangana–Baleanu definition, the Caputo–Fabrizio fractional operator, etc. For more details, one may refer to [9–12]. According to Caputo's formulation, the temporal derivative of $y(\xi,\tau)$ of order α is given by [1]

$$ {}^{C}\mathfrak{D}_{\tau}^{\alpha}(y(\xi,\tau)) = \begin{cases} \frac{1}{\Gamma(1-\alpha)} \int_0^{\tau} (\tau-u)^{q-\alpha-1} \frac{\partial^q y(\xi,u)}{\partial u^q}\, du, & q-1 < \alpha < q \\ \frac{\partial^q y(\xi,\tau)}{\partial \tau^q}, & \alpha = q \in \mathbb{N}. \end{cases} \tag{6}$$

In the literature, various analytical and numerical schemes have been developed to solve models based on the fractional-order PDEs. The solitary wave profile for the generalized form of the fractional-order Burger–Fisher model is derived in [13] by utilizing a high-order ODE. In order to portray the heterogeneity of various real-world problems, a new definition of the fractional derivative is introduced in the classical Fisher equation [14] and its solution is obtained by utilizing the Sumudu decomposition method. A spectral collocation scheme involving Chebyshev polynomials is presented in [15] to solve the fractional-order Fisher equation incorporating Caputo's interpretation for derivatives. The solution of the temporal-fractional Fisher equation is derived in [16] by utilizing a collocation scheme. The discretization along the temporal domain is done via a finite difference approximation, while spatial derivatives are interpolated with extended cubic B-splines. Complex traveling wave solutions for the temporal-fractional Burger–Fisher equation are presented in [17] by using the subequation approach. Diaz and Gallegos [18] proposed an explicit logarithmic approach for solving the reaction–diffusion equation, in which a spatial derivative is formulated in terms of the Riez sense. Additionally, the monotonic character of the technique is proved together with the study of stability and convergence through Gronwall's inequality.

This study includes the implementation of an improvised spline collocation technique to solve the fractional-order Burger–Fisher model, where the fractional operator is discretized using a finite difference scheme. It has a simple computational procedure for solving a variety of differential equations, as it does not require the calculation of integrals, as in variational methods, to obtain the discretized system of equations. Due to their minimal compact support and smoothness, B-splines are well-suited for the numerical simulation of linear and nonlinear fractional-order differential equations. In the classical approach of cubic B-spline collocation, the unknown function and its derivatives are replaced by an approximation in terms of a spline function and its derivatives, respectively. However, to develop an improvised collocation technique based on cubic B-splines, some posterior corrections are applied to the approximation of the second-order derivative of the function by taking an additional term in the Taylor-series expansion for double derivatives, which results in an optimal order of convergence.

The rest of the chapter is organized as follows: In Section 2, the properties of cubic spline functions and subsequent improvements are discussed. Section 3 is devoted to the implementation of the proposed methodology to the Burger–Fisher model of fractional order. The discussion on stability via Von Neumann criteria is given in Section 4 and Section 5 consists of a convergence analysis of the developed technique. In order to demonstrate the robustness of the collocation approach, three examples based on the proposed model are considered in Section 6 and Section 7 briefly summarizes the main findings of the work.

2 Development of the proposed approach

In this section, the properties of cubic spline functions as well as their subsequent enhancements are discussed. First, define $\xi_l = \xi_0 + lh$, $l = 0, 1, \ldots, n$ as equidistant partitioning of spatial interval $[\xi_L, \xi_R]$ into knots of fixed distance h such that $\xi_L = \xi_0 < \xi_1 < \cdots < \xi_{n-1} < \xi_n = \xi_R$ and n denotes the number of collocation points.

The representation of cubic splines basis is as follows [19]:

$$\zeta_l(\xi) = \frac{1}{h^3}\begin{cases} (\xi - \xi_{l-2})^3, & \xi \in [\xi_{l-2}, \xi_{l-1}] \\ h^3 + 3h^2(\xi - \xi_{l-1}) + 3h(\xi - \xi_{l-1})^2 - 3(\xi - \xi_{l-1})^3, & \xi \in [\xi_{l-1}, \xi_l] \\ h^3 + 3h^2(\xi_{l+1} - \xi) + 3h(\xi_{l+1} - \xi)^2 - 3(\xi_{l+1} - \xi)^3, & \xi \in [\xi_l, \xi_{l+1}] \\ (\xi_{l+2} - \xi)^3, & \xi \in [\xi_{l+1}, \xi_{l+2}] \\ 0, & \text{otherwise.} \end{cases}$$

The approximate solution $v(\xi, \tau)$ with respect to the exact solution $y(\xi, \tau)$ of the considered equation can be expressed as a linear combination of the provided cubic polynomial spline basis as follows:

$$v(\xi, \tau) = \sum_{l=-1}^{n+1} \lambda_l(\tau)\zeta_l(\xi), \tag{7}$$

where $\lambda_{-1}(\tau), \lambda_0(\tau), \lambda_1(\tau), \ldots, \lambda_{n+1}(\tau)$ denote time-dependent unknown parameters that need to be calculated from initial as well as final conditions. The numerical solution $v(\xi, \tau)$ along with its first- and second-order derivatives at the knot points ξ_l are related to the time-dependent parameters

$$\begin{aligned} v(\xi_l, \tau) &= \lambda_{l-1}(\tau) + 4\lambda_l(\tau) + \lambda_{l+1}(\tau), \quad l = 0, 1, \ldots, n, \\ v_\xi(\xi_l, \tau) &= \frac{-3}{h}(\lambda_{l-1}(\tau) - \lambda_{l+1}(\tau)), \\ v_{\xi\xi}(\xi_l, \tau) &= \frac{6}{h^2}(\lambda_{l-1}(\tau) - 2\lambda_l(\tau) + \lambda_{l+1}(\tau)). \end{aligned} \tag{8}$$

2.1 Formation of improvised cubic B-splines

Assume that the cubic interpolating polynomial for the unknown function satisfies the following conditions:

(I) the interpolatory condition, for $l = 0, 1, \ldots, n$

$$v(\xi_l, \tau) = y(\xi_l, \tau); \tag{9}$$

(II) at final knot points, i. e., for $l = 0$ and n

$$v_{\xi\xi}(\xi_l, \tau) = y_{\xi\xi}(\xi_l, \tau) - \frac{h^2}{12} y_{\xi\xi\xi\xi}(\xi_l, \tau). \tag{10}$$

Theorem 1. *The following results are valid for the unique approximating function $v(\xi, \tau)$ corresponding to $y(\xi, \tau)$, with $y(\xi, \tau)$ being a sufficiently smooth function defined on the interval $[\xi_L, \xi_R]$ and satisfying equations* (9) *and* (10)

$$v_{\xi}(\xi_l,\tau) = y_{\xi}(\xi_l,\tau) + O(h^4), \tag{11}$$

$$v_{\xi\xi}(\xi_l,\tau) = y_{\xi\xi}(\xi_l,\tau) - \frac{h^2}{12} y_{\xi\xi\xi\xi}(\xi_l,\tau) + O(h^4). \tag{12}$$

Furthermore

$$\|y^{(m)} - v^{(m)}\|_{\infty} = O(h^{4-m}), \quad m = 0,1,2, \tag{13}$$

where $y^{(m)}$ and $v^{(m)}$ denote the mth-order derivative with respect to the spatial variable ξ.

Proof. Provided in [20, 21]. □

The expansion of Theorem 1 can be obtained by utilizing equations (9) and (10). Additionally, using [20, 21], some relations between $y_{\xi\xi\xi\xi}(\xi,\tau)$ and $v_{\xi\xi}(\xi,\tau)$ are established at different knot points, ensuring an optimal order of convergence throughout the spatial domain.

3 Numerical procedure

3.1 Approximation of the fractional-order derivative

A numerical procedure to deal with the Burger–Fisher model incorporating a fractional-order operator over the temporal interval $[0,T]$ is included in this section. Assume knot points on the time domain $[0,T]$ defined by $\tau_k = \tau_0 + kd\tau$, for $k = 0,1,\dots,M$, where $d\tau$ denotes the step length on the temporal domain for some positive integer M. An approximation to discretize the fractional-order derivative term defined in equation (6) at time instant $\tau = \tau_{k+1}$ is as follows [22, 29]:

$$\begin{aligned}
{}^{C}\mathfrak{D}_{\tau}^{\alpha}(y(\xi,\tau_{k+1})) &= \frac{1}{\Gamma(1-\alpha)} \int_{0}^{\tau_{k+1}} (\tau_{k+1} - y)^{-\alpha} \frac{\partial y(\xi,u)}{\partial u} du, \\
&= \frac{1}{\Gamma(1-\alpha)} \sum_{\phi=0}^{k} \int_{\tau_\phi}^{\tau_{\phi+1}} (\tau_{k+1} - y)^{-\alpha} \frac{\partial y(\xi,u)}{\partial u} du, \\
&= \frac{1}{\Gamma(1-\alpha)} \sum_{\phi=0}^{k} \frac{y(\xi,\tau_{\phi+1}) - y(\xi,\tau_\phi)}{d\tau} \int_{\tau_\phi}^{\tau_{\phi+1}} (\tau_{k+1} - u)^{-\alpha} du + e_{d\tau}^{k+1}, \\
&= \frac{1}{\Gamma(1-\alpha)} \sum_{\phi=0}^{k} \frac{y(\xi,\tau_{\phi+1}) - y(\xi,\tau_\phi)}{d\tau} \int_{\tau_{k-\phi}}^{\tau_{k+1-\phi}} \frac{d\vartheta}{\vartheta^{\alpha}} + e_{d\tau}^{k+1}.
\end{aligned}$$

Further simplification results in [23]

$$
{}^{C}\mathfrak{D}_{\tau}^{\alpha}(y(\xi,\tau_{k+1})) = \begin{cases} \gamma \sum_{\phi=0}^{k} \theta_{\phi}(y^{k-\phi+1} - y^{k-\phi}) + O((d\tau)^{2-\alpha}), & 0 < \alpha < 1 \\ \frac{y(\xi,\tau_{k+1}) - y(\xi,\tau_k)}{d\tau} + O(d\tau), & \alpha = 1, \end{cases} \tag{14}
$$

in which $\gamma = (d\tau)^{-\alpha}/\Gamma(2-\alpha)$, $\theta_{\phi} = (\phi+1)^{1-\alpha} - (\phi)^{1-\alpha}$, and $\vartheta = (\tau_{k+1} - u)$. The properties that the coefficients θ_{ϕ} satisfy are as follows:

$$
\begin{cases} \theta_0 = 1, \\ \theta_0 > \theta_1 > \theta_2 > \cdots > \theta_i,\ \theta_{\phi} \to 0 \quad \text{as}\ \phi \to \infty, \\ \theta_{\phi} > 0 \quad \text{when}\ \phi = 0,1,2,\ldots,k, \\ \sum_{\phi=0}^{k}(\theta_{\phi} - \theta_{\phi+1}) + \theta_{k+1} = (1-\theta_1) + \sum_{\phi=1}^{k-1}(\theta_{\phi} - \theta_{\phi+1}) + \theta_k = 1. \end{cases}
$$

3.2 Implementation of the proposed method

On substituting equation (14) together with the Crank–Nicolson scheme into equation (1), the resulting equation becomes

$$
\begin{aligned} &\gamma \sum_{\phi=0}^{k} \theta_{\phi}(y^{k-\phi+1} - y^{k-\phi}) - \eta\Big[\frac{(y_{\xi\xi})^{k+1} + (y_{\xi\xi})^{k}}{2}\Big] + \beta\Big[\frac{(y^p y_{\xi})^{k+1} + (y^p y_{\xi})^{k}}{2}\Big] \\ &- \frac{\epsilon}{2}[[y^{k+1} + y^k] - [(y^{p+1})^{k+1} + (y^{p+1})^{k}]] + S(\xi,\tau_{k+1}), \end{aligned} \tag{15}
$$

where $k = 0,1,\ldots,M$.

Linearizing the nonlinear terms by the approach mentioned in [24] as

$$
\begin{aligned} (y^{p+1})^{k+1} + (y^{p+1})^{k} &= (1-p)(y^{p+1})^{k} + (1+p)(y^p)^k y^{k+1} + O(d\tau)^2 \\ (y^p y_{\xi})^{k+1} + (y^p y_{\xi})^{k} &= (1-p)(y^p y_{\xi})^{k} + (y^p)^k (y_{\xi})^{k+1} + p(y)^{k+1}(y^{p-1}y_{\xi})^k + O(d\tau)^2. \end{aligned}
$$

After utilizing the above linearization procedure, the expression produced at various temporal stages is

$$
\begin{aligned} &\Big[\Big(\frac{p\beta}{2}(y^{p-1}y_{\xi})^k\Big) - \frac{\epsilon}{2}(1-(1+p)(y^p)^k) + \gamma\Big]w^{k+1} + \Big[\frac{\beta}{2}(y^p)^k\Big]y_{\xi}^{k+1} - \Big[\frac{\eta}{2}\Big]y_{\xi\xi}^{k+1} \\ &= \gamma \sum_{\phi=0}^{k-1}(\theta_{\phi} - \theta_{\phi+1})y^{k-\phi} + \gamma\theta_k y^0 + \frac{\beta}{2}[(p-1)(y^p y_{\xi})^k] + \frac{\epsilon}{2}[(y)^k + (p-1)(y^{p+1})^k] \\ &+ \Big[\frac{\eta}{2}\Big]y_{\xi\xi}^{k} + S(\xi,\tau_{k+1}). \end{aligned} \tag{16}
$$

The resultant equation at any lth knot point can be expressed as

$$
P_l(y)_l^{k+1} + Q_l(y_{\xi})_l^{k+1} + R_l(y_{\xi\xi})_l^{k+1} = T_l, \tag{17}
$$

for which

$$
\begin{aligned}
P_l &= \left[\left(\frac{p\beta}{2}(y_l^{p-1})^k(y_\xi)_l^k\right) - \frac{\epsilon}{2}(1-(1+p)(y_l^p)^k) + \gamma\right],\\
Q_l &= \frac{\beta}{2}(y_l^p)^k, \quad R_l = -\frac{\eta}{2},\\
T_l &= \gamma\sum_{\phi=0}^{k-1}(\theta_\phi - \theta_{\phi+1})y_l^{k-\phi} + \gamma\theta_k y_l^0 + \frac{\beta}{2}[(p-1)(y_l^p)(y_\xi)_l^k]\\
&\quad + \frac{\epsilon}{2}[(y)_l^k + (p-1)(y_l^{p+1})^k] + \left[\frac{v_1}{2}\right](y_{\xi\xi})_l^k + \mathsf{S}(\xi_l, \tau_{k+1}).
\end{aligned}
\tag{18}
$$

Upon substituting the calculated values of the function $y(\xi,\tau)$ and its higher-order derivatives at the specified knots and combining the coefficients of the time-dependent parameters λ_l^{k+1}, the following system of equations is derived:

For $l = 0$:

$$
\begin{aligned}
&\left(P_0 - \frac{3Q_0}{h} + \frac{7R_0}{h^2}\right)\lambda_{-1}^{k+1} + \left(4P_0 - \frac{33R_0}{2h^2}\right)\lambda_0^{k+1} + \left(P_0 + \frac{3Q_0}{h} + \frac{14R_0}{h^2}\right)\lambda_1^{k+1} - \frac{7R_0}{h^2}\lambda_2^{k+1}\\
&\quad + \frac{3R_0}{h^2}\lambda_3^{k+1} - \frac{R_0}{2h^2}\lambda_4^{k+1} = \mathsf{T}_0 + O(h^4 + d\tau^{2-\alpha}),\\
&\mathsf{b}_0\lambda_{-1}^{k+1} + \mathsf{c}_0\lambda_0^{k+1} + \mathsf{d}_0\lambda_1^{k+1} + \mathsf{e}_0\lambda_2^{k+1} + \mathsf{f}_0\lambda_3^{k+1} + \mathsf{g}_0\lambda_4^{k+1} = \mathsf{T}_0 + O(h^4 + d\tau^{2-\alpha}).
\end{aligned}
\tag{19}
$$

For $l = 1, 2, \ldots, n-1$:

$$
\begin{aligned}
&\frac{R_l}{2h^2}\lambda_{l-2}^{k+1} + \left(P_l - \frac{3Q_l}{h} + \frac{4R_l}{h^2}\right)\lambda_{l-1}^{k+1} + \left(4P_l - \frac{9R_l}{h^2}\right)\lambda_l^{k+1} + \left(P_l + \frac{3Q_l}{h} + \frac{4R_l}{h^2}\right)\lambda_{l+1}^{k+1}\\
&\quad + \frac{R_l}{2h^2}\lambda_{l+2}^{k+1} = \mathsf{T}_l + O(h^4 + d\tau^{2-\alpha}),\\
&\kappa\lambda_{l-2}^{k+1} + \mathsf{c}_l\lambda_{l-1}^{k+1} + \mathsf{d}_l\lambda_l^{k+1} + \mathsf{e}_l\lambda_{l+1}^{k+1} + \kappa\lambda_{l+2}^{k+1} = \mathsf{T}_l + O(h^4 + d\tau^{2-\alpha}).
\end{aligned}
\tag{20}
$$

For $l = n$:

$$
\begin{aligned}
&-\frac{R_n}{2h^2}\lambda_{n-4}^{k+1} + \frac{3R_n}{h^2}\lambda_{n-3}^{k+1} - \frac{7R_n}{h^2}\lambda_{n-2}^{k+1} + \left(P_n - \frac{3Q_n}{h} + \frac{14R_n}{h^2}\right)\lambda_{n-1}^{k+1} + \left(4P_n - \frac{33R_n}{2h^2}\right)\lambda_n^{k+1}\\
&\quad + \left(P_n + \frac{3Q_n}{h} + \frac{7R_n}{h^2}\right)\lambda_{n+1}^{k+1} = \mathsf{T}_n + O(h^4 + d\tau^{2-\alpha}),\\
&\mathsf{b}_n\lambda_{n-4}^{k+1} + \mathsf{c}_n\lambda_{n-3}^{k+1} + \mathsf{d}_n\lambda_{n-2}^{k+1} + \mathsf{e}_n\lambda_{n-1}^{k+1} + \mathsf{f}_n\lambda_n^{k+1} + \mathsf{g}_n\lambda_{n+1}^{k+1} = \mathsf{T}_n + O(h^4 + d\tau^{2-\alpha}).
\end{aligned}
\tag{21}
$$

In order to obtain the unique solution, two equations obtained from the boundary conditions are

$$\lambda_{-1}^{k+1} + 4\lambda_0^{k+1} + \lambda_1^{k+1} = \mathfrak{k}_1((k+1)d\tau)$$

and

$$\lambda_{n-1}^{k+1} + 4\lambda_n^{k+1} + \lambda_{n+1}^{k+1} = \mathfrak{k}_2((k+1)d\tau).$$

The equations mentioned above are expressed in matrix form as

$$\mathfrak{K}\mathfrak{Z} = \mathfrak{T}, \tag{22}$$

where $\mathfrak{Z}_{r,1} = [\lambda_{r-2}^{k+1}]_{r=1}^{N+3}$

$$\mathfrak{T}_{i,1} = \begin{cases} \mathfrak{k}_1((k+1)d\tau), & r = 1 \\ \mathfrak{T}_{r-2}, & 2 \le r \le n+2 \\ \mathfrak{k}_2((k+1)d\tau), & r = n+3. \end{cases}$$

The matrix $\mathfrak{K}$, has $(1, 4, 1)$ as the first three entries in the first row and the last three entries in the last row, while the second row starts with $(\mathsf{b}_0, \mathsf{c}_0, \mathsf{d}_0, \mathsf{e}_0, \mathsf{f}_0, \mathsf{g}_0)$ and the second last row ends with $(\mathsf{b}_n, \mathsf{c}_n, \mathsf{d}_n, \mathsf{e}_n, \mathsf{f}_n, \mathsf{g}_n)$. The remaining elements of the matrix are

$$[\mathfrak{K}]_{r=3,s=1}^{n+1,n+3} = \begin{cases} \kappa, & |r-s| = 2 \\ \mathsf{c}_{r-2}, & r-s = 1 \\ \mathsf{d}_{r-2}, & r = s \\ \mathsf{e}_{r-2}, & s-r = 1 \\ 0, & \text{otherwise.} \end{cases}$$

The resulting system of linear equations is employed to solve for $[\lambda_{-1}^{k+1}, \lambda_0^{k+1}, \lambda_1^{k+1}, \dots, \lambda_{n-1}^{k+1}, \lambda_n^{k+1}, \lambda_{n+1}^{k+1}]^T$. Substituting this vector into equation (7) yields the required approximate solution at the $(k+1)$th temporal stage, where $k = 0, 1, 2, \dots, M$. To start the iterative procedure, the initial solution of the presented model (1) is determined by utilizing initial conditions (2) at each knot point and two extra conditions, i. e., $y_\xi(\xi_0, \tau_0) = y_\xi(\xi_0)$ and $y_\xi(\xi_n, \tau_0) = y_\xi(\xi_n)$. In this work, numerical computations were performed utilizing Matlab R2021a software on an 11th Generation Intel Core i5-1155G7 processor (2.50 GHz).

4 Stability analysis

This section includes the Von Neumann approach to examine the stability of the improvised collocation technique. Initially, the nonlinear terms are linearized by setting $q_1 = \max y^p$ and subsequently the Crank–Nicolson approach is applied to equation (1), yielding the following expression:

$$\gamma \sum_{\phi=0}^{k} \theta_\phi (y_l^{k-\phi+1} - y_l^{k-\phi}) = \eta \left[\frac{(y_{\xi\xi})_l^{k+1} + (y_{\xi\xi})_l^{k}}{2} \right] - \beta q_1 \left[\frac{(y_\xi)_l^{k+1} + (y_\xi)_l^{k}}{2} \right] + \epsilon(1-q_1) \left[\frac{y_l^{k+1} + y_l^{k}}{2} \right] + \mathrm{S}(\xi_l, \tau_{k+1}). \tag{23}$$

Upon separating the terms at different temporal levels and supposing $\epsilon(1-q_1) = q_2$, the equation above can be written as

$$\left[\gamma - \frac{q_2}{2}\right](y)_l^{k+1} + \frac{q_1\beta}{2}(y_\xi)_l^{k+1} - \frac{\eta}{2}(y_{\xi\xi})_l^{k+1} = \left[\frac{q_2}{2}\right](y)_l^{k} + \frac{\eta}{2}(y_{\xi\xi})_l^{k} - \frac{q_1\beta}{2}(y_\xi)_l^{k} + \gamma \sum_{\phi=0}^{k-1} (\theta_\phi - \theta_{\phi+1}) y_l^{k-\phi} + \gamma\theta_k y_l^0 + \mathrm{S}(\xi_l, \tau_{k+1}). \tag{24}$$

Further, the above equation assumes the following form:

$$a_1(y)_l^{k+1} + a_2(y_\xi)_l^{k+1} - a_3(y_{\xi\xi})_l^{k+1} = a_4(y)_l^{k} - a_2(y_\xi)_l^{k} + a_3(y_{\xi\xi})_l^{k} + \gamma \sum_{\phi=0}^{k-1} (\theta_\phi - \theta_{\phi+1}) y_l^{k-\phi} + \gamma\theta_k y_l^0 + \mathrm{S}(\xi_l, \tau_{k+1}), \tag{25}$$

where

$$a_1 = \gamma - \frac{q_2}{2}, \quad a_2 = \frac{q_1\beta}{2}, \quad a_3 = \frac{\eta}{2}, \quad a_4 = \frac{q_2}{2}.$$

After substituting the computed solution and corresponding higher derivatives, the following equation is formed:

$$\begin{aligned} z_1\lambda_{l-2}^{k+1} + z_2\lambda_{l-1}^{k+1} + z_3\lambda_{l}^{k+1} + z_4\lambda_{l+1}^{k+1} + z_1\lambda_{l+2}^{k+1} &= \mathrm{S}(\xi_l, \tau^{k+1}) - z_1\lambda_{l-2}^{k} + z_5\lambda_{l-1}^{k} \\ &\quad + z_6\lambda_l^k + z_7\lambda_{l+1}^k - z_1\lambda_{l+2}^k + \gamma \sum_{\phi=0}^{k-1} (\theta_\phi - \theta_{\phi+1})(\lambda_{l-1}^{k-\phi} + 4\lambda_l^{k-\phi} + \lambda_{l+1}^{k-\phi}) \\ &\quad + \gamma\theta_k(\lambda_{l-1}^0 + 4\lambda_l^0 + \lambda_{l+1}^0), \end{aligned} \tag{26}$$

where

$$z_1 = -\frac{a_3}{2h^2}, \quad z_2 = a_1 - \frac{8a_3}{2h^2} - \frac{3a_2}{h}, \quad z_3 = 4a_1 + \frac{18a_3}{2h^2},$$
$$z_4 = a_1 - \frac{8a_3}{2h^2} + \frac{3a_2}{h}, \quad z_5 = a_4 + \frac{8a_3}{2h^2} + \frac{3a_2}{h}, \quad z_6 = 4a_4 - \frac{18a_3}{2h^2}, \quad z_7 = a_4 + \frac{8a_3}{2h^2} - \frac{3a_2}{h}.$$

Inserting the solution $\lambda_l^k = \mathrm{A}\psi^k \exp(\iota l \mathrm{w} h)$ into equation (26), where A represents the size of the amplitude, w is the mode number, h denotes the size of the spatial step, and $\iota = \sqrt{-1}$, leads to the following equation:

$$
\begin{aligned}
&z_1\psi^{k+1}e^{\iota(l-2)wh} + z_2\psi^{k+1}e^{\iota(l-1)wh} + z_3\psi^{k+1}e^{\iota lwh} + z_4\psi^{k+1}e^{\iota(l+1)wh} + z_1\psi^{k+1}e^{\iota(l+2)wh}\\
&= -z_1\psi^{k}e^{\iota(l-2)wh} + z_5\psi^{k}e^{\iota(l-1)wh} + z_6\psi^{k}e^{\iota lwh} + z_7\psi^{k}e^{\iota(l+1)wh} - z_1\psi^{k}e^{\iota(l+2)wh}\\
&\quad + \gamma\sum_{\phi=0}^{k-1}(\theta_\phi - \theta_{\phi+1})(\psi^{k-\phi}e^{\iota(l-1)wh} + 4\psi^{k-\phi}e^{\iota lwh} + \psi^{k-\phi}e^{\iota(l+1)wh})\\
&\quad + \gamma\theta_k(\psi^{0}e^{\iota(l-1)wh} + 4\psi^{0}e^{\iota lwh} + \psi^{0}e^{\iota(l+1)wh}) + S(\xi_l,\tau_{k+1}).
\end{aligned}
\tag{27}
$$

On further simplification and after substituting the values of the z_is, the resulting equation obtained is as follows:

$$
\begin{aligned}
\psi^{k+1} &= \frac{\psi^k[\frac{\eta}{2h^2}\cos(2wh) + (q_2 + \frac{4\eta}{h^2})\cos(wh) + 2q_2 - \frac{18\eta}{4h^2} - \iota\frac{3q_1\beta}{h}\sin(wh)]}{-\frac{\eta}{2h^2}\cos(2wh) + (2\gamma - q_2 - \frac{4\eta}{h^2})\cos(wh) + 4\gamma - 2q_2 + \frac{18\eta}{4h^2} + \iota\frac{3q_1\beta}{h}\sin(wh)}\\
&+ \frac{\gamma(2\cos(wh)+4)[\sum_{\phi=0}^{k-1}(\theta_\phi - \theta_{\phi+1})\psi^{k-\phi} + \theta_k\psi^0] + S(\xi_l,\tau_{k+1})e^{-\iota lwh}}{-\frac{\eta}{2h^2}\cos(2wh) + (2\gamma - q_2 - \frac{4\eta}{h^2})\cos(wh) + 4\gamma - 2q_2 + \frac{18\eta}{4h^2} + \iota\frac{3q_1\beta}{h}\sin(wh)}.
\end{aligned}
\tag{28}
$$

The proposed technique is stable if the magnitude of $\psi^{k+1} < 1, \forall k$. From the above expression, it has been discovered that the value of the numerator becomes lower than that of the denominator for varying choices of γ, α, w, and h. The result mentioned above confirms that the presented approach is unconditionally stable.

5 Convergence analysis

This section discusses convergence analysis using some notations and a lemma reported in [25]. The theorem provided below ensures fourth-order convergence over the spatial domain and convergence of order $(2-\alpha)$ along the temporal direction.

Theorem 2. *Assuming that $y(\xi,\tau)$ and $v(\xi,\tau)$ represent solutions to the fractional-order model* (1) *and equation* (24)*, respectively, and $y(\xi,\tau)$ satisfies the smoothness criterion specified in* [25]*, then the relation provided below holds for all $\tau \in [0,T]$ and sufficiently small $d\tau$ and h*

$$
\|\mathfrak{E}^{k+1}\| \le O(h^4 + (d\tau)^{2-\alpha}),
$$

where $\mathfrak{E}_l^{k+1} = y(\xi_l,\tau_{k+1}) - v(\xi_l,\tau_{k+1})$.

6 Numerical results and discussion

In this section, the numerical solution of the fractional Burger–Fisher equation is obtained by utilizing the developed approach. The accuracy of the method has been evaluated through the use of the Euclidean error norm L_2 and the maximum error norm L_∞ provided in [26].

Example 1. In FBF, equation (1) takes the coefficients $\beta = 1$, $\epsilon = 1$, $\eta = 1$ along with the term $S(\xi, \tau) = 0$ and $p = 2$ over the temporal interval $[0, T]$ and the spatial domain $[0, 1]$ and, in the case of $\alpha = 1$, the expression for the exact solution of the mentioned example is given by

$$y(\xi, \tau) = \sqrt{\frac{1}{2}\left[1 + \tanh\left(\frac{\xi}{3} + \frac{10\tau}{9}\right)\right]}. \tag{29}$$

The boundary as well as initial conditions are derived using the exact solution specified above. For different values of the fractional order α, the comparison in terms of absolute error is demonstrated in Table 1, considering $d\tau = 0.01$ and $n = 50$, respectively. This table clearly indicates that the reduction in the absolute error is observed as the value of the fractional order approaches 1 and it is also evident that the proposed technique performs better than [28] when the same set of parameters are used. For $\alpha = 1$, the minimal absolute error is observed, validating the reliability of the proposed approach. A 3D illustration of the absolute error for distinct choices of the fractional parameter α, with $n = 30$ and $d\tau = 0.001$ is provided in Figures 1(a–c), exhibiting the fact that the increasing values of the fractional parameter α yield a convergent solution. Moreover, Figure 2(a) shows the graphical behavior of the numerical solution for some choices of the fractional order α by taking $d\tau = 0.01$ and $n = 30$ at the temporal stage $T = 0.5$, respectively. Furthermore, it is observed that the obtained numerical findings are compatible with the actual results.

Table 1: Analysis of the absolute error for Example 1 under different values of α.

ξ	T	$\alpha = 0.5$		$\alpha = 0.75$		$\alpha = 1$	
		ICSCM	[28]	ICSCM	[28]	ICSCM	[28]
	0.01	1.1707E−03	3.20E−2	4.8016E−04	7.60E−2	3.8916E−09	0.0000
	0.03	7.7503E−03	5.10E−1	5.8510E−04	1.51E−2	6.2272E−09	1.0E−10
0.01	0.05	8.3124E−03	6.18E−1	6.1244E−04	2.02E−1	8.0668E−09	2.00E−9
	0.07	8.5122E−03	6.91E−1	6.0804E−04	2.42E−1	9.6187E−09	1.07E−8
	0.09	8.6643E−03	7.43E−1	5.8988E−04	2.74E−1	1.0933E−08	3.74E−8
	0.01	4.3425E−03	3.45E−1	1.6953E−03	1.01E−2	1.2750E−08	0.0000
	0.03	4.3254E−03	1.25E−1	2.1394E−03	1.76E−2	2.2217E−08	2.0E−10
0.04	0.05	4.1648E−03	6.41E−1	2.2658E−03	2.27E−2	2.9631E−08	1.90E−9
	0.07	3.9791E−03	7.14E−1	2.2620E−03	2.67E−2	3.5859E−08	1.07E−8
	0.09	3.7966E−03	7.65E−1	2.2012E−03	2.99E−2	4.1122E−08	3.74E−8

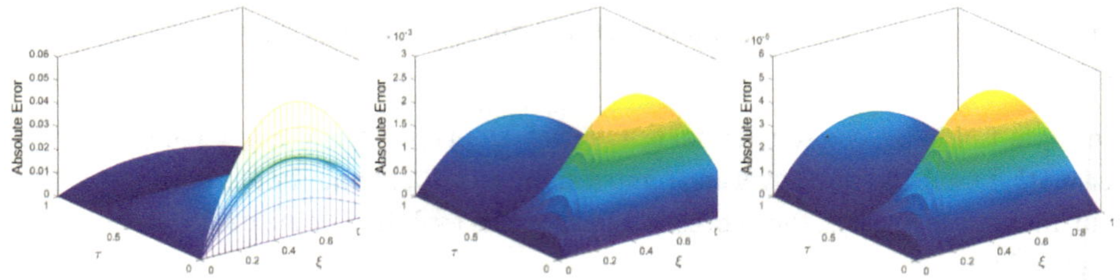

(a) α=0.05 **(b)** α=0.65 **(c)** α=0.85

Figure 1: (a,b,c) 3D behavior of the absolute error for different values of α.

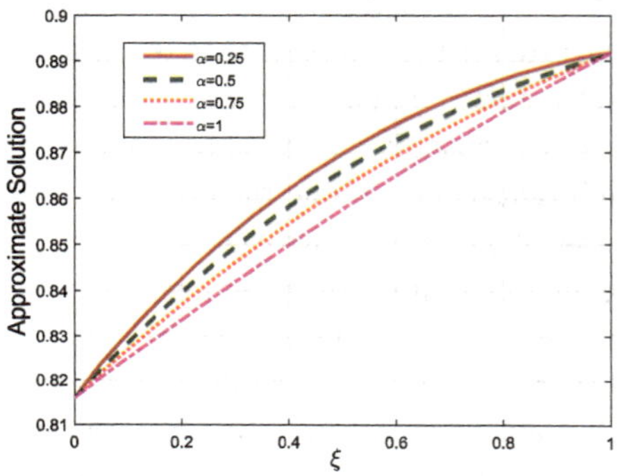

Figure 2: 2D plot of the approximate solution for some values of α.

Example 2. Examine the generalized fractional Fisher equation (5) by assuming parameters $p = 3$, $\eta = 1$, $\epsilon = 1$, satisfying the initial condition

$$y(\xi, 0) = \xi^2 \exp(2\xi), \quad 0 \le \xi \le 1 \tag{30}$$

and the following end conditions:

$$y(0, \tau) = 0, \quad y(1, \tau) = (1 + \tau^2) \exp(2), \quad \tau \ge 0. \tag{31}$$

The forcing term $S(\xi, \tau)$ is

$$\begin{aligned} S(\xi, \tau) = {} & \frac{2\xi^2 \tau^{2-\alpha} \exp(2\xi)}{\Gamma(3 - \alpha)} - 2\eta(1 + \tau^2)[1 + 4\xi + 2\xi^2] \exp(2\xi) \\ & - [(1 + \tau^2)\xi^2 \exp(2\xi)][1 - ((1 + \tau^2)\xi^2 \exp(2\xi))^3]. \end{aligned}$$

The exact solution to the problem is provided below [27]

$$y(\xi, \tau) = (1 + \tau^2)\xi^2 \exp(2\xi). \tag{32}$$

Table 2: Absolute error comparison for Example 2 at different temporal levels T.

ξ	$T = 0.5$ ICSCM ($d\tau = 0.001$)	$T = 0.5$ [27] ($d\tau = 0.0003$)	$T = 1$ ICSCM ($d\tau = 0.001$)	$T = 1$ [27] ($d\tau = 0.0003$)
0.1	7.6117E−07	1.03E−05	1.4670E−06	1.70E−06
0.2	9.3774E−07	1.20E−06	2.9440E−06	1.00E−06
0.3	1.5502E−07	4.00E−07	4.3043E−07	0.0000
0.4	1.5973E−06	3.00E−07	5.2002E−06	1.70E−06
0.5	3.8078E−06	5.70E−06	5.0263E−06	1.00E−07
0.6	5.3638E−06	5.20E−06	3.1773E−06	1.00E−06
0.7	4.9266E−06	2.12E−05	1.1572E−07	8.00E−07
0.8	2.2601E−06	2.00E−07	3.5822E−06	2.50E−06
0.9	9.1683E−07	1.40E−06	6.8845E−06	3.30E−06

Table 3: Error norms obtained for Example 2 across various temporal levels.

T	$\alpha = 0.65$ ICSCM		$\alpha = 0.95$ ICSCM ($d\tau = 0.001$)		$\alpha = 0.95$ [27] ($d\tau = 0.0003$)		CPU Time
	L_2	L_∞	L_2	L_∞	L_2	L_∞	
0.25	3.1059E−04	2.4172E−03	6.7305E−05	1.0930E−04	-	-	0.13957
0.5	5.6625E−05	4.0639E−04	2.8172E−06	5.4901E−06	2.4900E−06	2.1200E−05	0.33128
0.75	1.8054E−05	9.4992E−05	4.3834E−06	6.8062E−06	3.0500E−06	2.1300E−05	0.59932
1	1.0911E−05	3.5704E−05	4.3192E−06	9.0344E−06	5.1050E−07	3.3000E−06	0.92102

The absolute error analysis of Example 2 corresponding to different temporal levels T is demonstrated in Table 2 for $n = 100$ and $\alpha = 0.95$. Table 3 presents a comparison of the error norms between the estimated outcomes and the results mentioned in [27] for distinct time stages T, taking into account $h = 0.01$, $\alpha = 0.95$, $d\tau = 0.001$, and for certain values of the fractional parameter α, the estimation of absolute error at some knot points is provided in Table 4 for different temporal levels, considering $n = 100$ and $d\tau = 0.002$. It is apparent from the tables that the variation from the actual profile becomes increasingly significant, highlighting the sensitivity of the computed solution to variations in the order of the fractional derivative. For distinct time stages T, the impact of the fractional parameter α on the physical behavior of the obtained numerical solution is shown in Figure 3(a), assuming $\alpha = 0.5$, $n = 100$, and $d\tau = 0.005$. The graphical depiction clearly reflects the close alignment between the actual solution and the computational results of the suggested technique. Figure 3(b) displays a 3D surface plot illustrating the physical behavior of the approximate solution in the case of $\alpha = 0.85$, $n = 40$, and $d\tau = 0.001$. From the tabular data and graphical illustration, the consistency between the actual results and the computational outcomes can be easily seen. The effectiveness of the developed technique over [27] is reflected in its straightforward

Table 4: Evaluation of absolute error for Example 2 at distinct temporal levels when $0 \le \xi \le 1$, $n = 100$, and $\alpha = 0.45, 0.75$.

α	ξ	$T = 1$	$T = 3$	$T = 5$
	0.1	5.0183E−06	3.0104E−06	1.3645E−06
	0.3	1.7767E−05	6.0228E−06	2.0181E−07
0.45	0.5	2.5387E−05	5.1172E−06	7.6827E−06
	0.7	3.9949E−06	2.1200E−05	2.2846E−05
	0.9	3.2294E−04	3.7204E−03	6.2252E−02
	0.1	6.3423E−06	1.6703E−06	4.7417E−07
	0.3	1.9787E−05	2.8852E−06	8.7432E−07
0.75	0.5	2.5393E−05	5.9775E−06	7.7570E−06
	0.7	2.9437E−06	2.1300E−05	2.2912E−05
	0.9	2.7004E−05	5.2877E−05	5.6599E−05

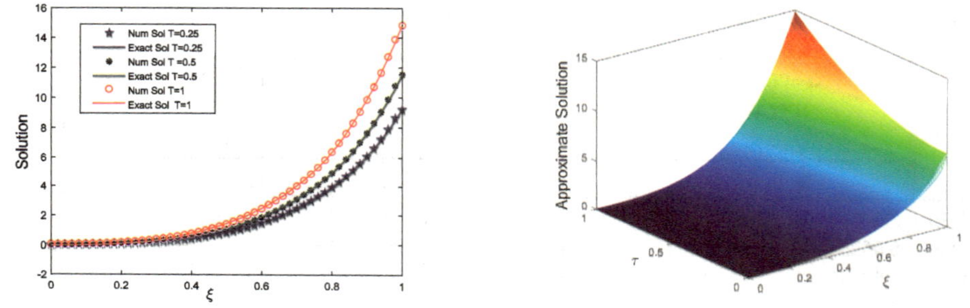

(a) (b)

Figure 3: (a) Solution dynamics at $\alpha = 0.5$ for certain values of T, (b) 3D representation of the solution behavior for $\alpha = 0.85$ with $T \le 1$.

implementation, applicability to fractional-order equations and fourth-order spatial accuracy.

Example 3. Assume the general form of the fractional-order Burgers model (4) by including the coefficients $p = 3$, $\eta = 1$, $\beta = 1$ with respect to the following condition:

$$y(\xi, 0) = 0, \quad 0 \le \xi \le 1, \tag{33}$$

along with boundary conditions

$$y(0, \tau) = 0, \quad y(1, \tau) = 0, \quad \tau \ge 0. \tag{34}$$

The nonlinear term $S(\xi, \tau)$ is

$$S(\xi, \tau) = \frac{2\xi(\xi - 1)\tau^{2-\alpha}}{\Gamma(3 - \alpha)} - 2\eta\tau^2 + (2\xi - 1)\tau^2((\xi^2 - \xi)\tau^2)^3.$$

The following expression provides the exact solution to the considered example above:

$$y(\xi, \tau) = \xi(\xi - 1)\tau^2. \tag{35}$$

For a few values of temporal step size $d\tau$, the computed L_2 and L_∞ error norms are reported in Table 5, considering $n = 200$ and $h = 0.005$. By taking $d\tau = 0.01$ and $h = 0.01$ on the computational domain $[0, 1]$, Table 6 presents the error norms for $\alpha = 0.25, 0.75$. From the tabular analysis, it is clear that as the value of $\alpha \to 1$, the computed solution overlaps with the exact one. For some temporal scales, Figure 4(a) depicts the

Table 5: Computation of error norms for some values of $d\tau$ by taking $h = 0.005$, $T = 2$, and $n = 200$.

$d\tau$	$\alpha = 0.3$		$\alpha = 0.85$		CPU Time
	L_∞	L_2	L_∞	L_2	
1	1.9276E−01	1.3944E−01	1.4826E−01	1.0725E−01	0.00144
0.5	4.2205E−02	2.7875E−02	1.0279E−02	1.8619E−02	0.01146
0.25	1.2788E−02	8.0281E−03	6.0027E−03	3.2812E−03	0.00850
0.125	5.6575E−03	3.7160E−03	2.5093E−03	1.4658E−03	0.01626

Table 6: Numerical evaluation of error norms corresponding to different temporal stages for $n = 100$ and $d\tau = 0.01$.

T	$\alpha = 0.25$		$\alpha = 0.75$	
	L_∞	L_2	L_∞	L_2
0.25	8.5299E−05	6.0555E−05	1.1928E−04	8.4735E−05
0.5	1.4772E−04	1.0490E−04	1.6460E−04	1.1689E−04
0.75	2.0307E−04	1.4420E−04	1.9365E−04	1.3750E−04
1	2.5424E−04	1.8053E−04	2.1558E−04	1.5306E−04

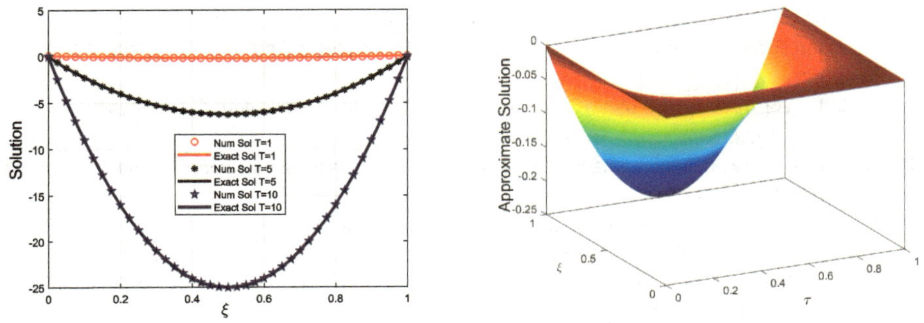

(a) (b)

Figure 4: (a) Behavior of the solution for $\alpha = 0.5$ corresponding to some values of T, (b) 3D surface visualization of the numerical solution at $\alpha = 0.25$ and $T \leq 1$.

close similarity between the numerical outcomes and the exact findings, by choosing $\alpha = 0.5$, $d\tau = 0.01$, and $n = 40$. The space–time graphical illustration of the solution surface considering $d\tau = 0.01$, $\alpha = 0.25$, and $n = 50$ is provided in Figure 4(b). It is worth noting that the numerical results obtained from the developed technique effectively replicate the exact profile throughout the considered domain.

7 Conclusions

In this work, the presented technique is successfully applied to solve the Burger–Fisher model incorporating a fractional derivative with respect to the temporal variable τ. An appropriate finite difference scheme is utilized to interpolate the temporal fractional operator, while an improvised splines basis is employed to carry out the discretization along the spatial direction. The capability of the presented technique is demonstrated by successfully solving three different test examples. Fourier-series analysis is applied to confirm the unconditional stability of the collocation technique. In order to validate the effectiveness of the suggested approach, L_2 and L_∞ error norms are also measured in addition to presenting the computational results graphically and in tabular form. The findings of the study indicate that the presented approach is both simple and effective for dealing with fractional-order differential equations. Future work will focus on extending the current methodology to solve coupled fractional PDEs characterized by variable fractional orders. Additionally, the research will be devoted to studying different formulations of fractional-order derivatives to explore their influence on real-life physical systems governed by fractional PDEs.

Bibliography

[1] Podlubny, I. (1999). *Fractional Differential Equations*. San Diego: Academic Press.

[2] Hermann, R. (2014). *Fractional Calculus: An Introduction for Physicists* (2nd ed.). New Jersey: World Scientific.

[3] Taniuti, T. (1974). Reductive perturbation method and far fields of wave equations. *Progress of Theoretical Physics Supplement*, *55*, 1–35.

[4] Burgers, J. M. (1948). A mathematical model illustrating the theory of turbulence. *Advances in Applied Mechanics*, *1*, 171–199.

[5] Sugimoto, N. (1991). Burgers' equation with a fractional derivative: hereditary effects on nonlinear acoustic waves. *Journal of Fluid Mechanics*, *225*, 631–653.

[6] Liu, C., & Chang, J. (2018). Recovering a source term in the time-fractional Burgers' equation by an energy boundary functional equation. *Applied Mathematics Letters*, *79*, 138–145.

[7] Merdan, M. (2012). Solutions of time-fractional reaction–diffusion equation with modified Riemann–Liouville derivative. *International Journal of Physical Sciences*, *7*, 2317–2326.

[8] Sardar, T., Rana, S., & Chattopadhyay, J. (2015). A mathematical model of dengue transmission with memory. *Communications in Nonlinear Science and Numerical Simulation*, *22*, 511–525.

[9] Caputo, M. (1967). Linear models of dissipation whose Q is almost frequency independent–II. *Geophysical Journal of the Royal Astronomical Society*, *13*, 529–539.

[10] Caputo, M., & Fabrizio, M. (2015). A new definition of fractional derivative without singular kernel. *Progress in Fractional Differentiation and Applications*, *1*, 73–85.

[11] Atangana, A. (2015). *Derivative With a New Parameter: Theory, Methods and Applications*. San Diego: Academic Press.

[12] Hilfer, R. (2000). *Applications of Fractional Calculus in Physics*. Singapore: World Scientific.

[13] Zhang, J., Wei, P., & Wang, M. (2012). Exact solutions of the generalized time-delayed Burgers–Fisher equation with positive fractional power terms. *Applied Mathematical Modelling*, *36*(5), 2192–2196.

[14] Atangana, A. (2016). On the new fractional derivative and application to nonlinear Fisher's reaction–diffusion equation. *Applied Mathematics and Computation*, *273*, 948–956.

[15] Khader, M. M., & Saad, K. M. (2018). A numerical approach for solving the fractional Fisher equation using Chebyshev spectral collocation method. *Chaos, Solitons and Fractals*, *110*, 169–177.

[16] Akram, T., Abbas, M., & Ali, A. (2021). A numerical study on time fractional Fisher equation using an extended cubic B-spline approximation. *Journal of Mathematical and Computational Science*, *22*, 85–96.

[17] Yokus, A., Durur, H., Kaya, D., Ahmad, H., & Nofal, T. A. (2021). Numerical comparison of Caputo and conformable derivatives of time fractional Burgers–Fisher equation. *Results in Physics*, *25*, 104247.

[18] Macias-Diaz, J. E., & Gallegos, A. (2022). Design and numerical analysis of a logarithmic scheme for nonlinear fractional diffusion–reaction equations. *Journal of Computational and Applied Mathematics*, *404*, 113118.

[19] Prenter, P. M. (1975). *Splines and Variational Methods*. New York: Wiley-Interscience.

[20] Daniel, J. W., & Swartz, B. K. (1975). Extrapolated collocation for two-point boundary-value problems using cubic splines. *IMA Journal of Applied Mathematics*, *16*(2), 161–174.

[21] Lucas, T. R. (1974). Error bounds for interpolating cubic splines under various end conditions. *SIAM Journal on Numerical Analysis*, *11*(3), 569–584.

[22] Liu, F., Zhuang, P., Anh, V., Turner, I., & Burrage, K. (2007). Stability and convergence of difference methods for space–time fractional advection–diffusion equation. *Applied Mathematics and Computation*, *191*(1), 12–20.

[23] Lin, Y., & Xu, C. (2007). Finite difference/spectral approximations for the time fractional diffusion equation. *Journal of Computational Physics*, *225*(2), 1533–1552.

[24] Bellman, R. E., & Kalaba, R. E. (1965). *Quasilinearization and Nonlinear Boundary Value Problems*. New York: Elsevier.

[25] Lopez-Marcos, J. C. (1990). A difference scheme for a nonlinear partial integro-differential equation. *SIAM Journal on Numerical Analysis*, *27*(1), 20–31.

[26] Kumari, A., & Kukreja, V. K. (2021). Analysis of RLW and MRLW equations using an improvised collocation technique with SSP-RK43 scheme. *Wave Motion*, *105*, 102761.

[27] Majeed, A., Kamran, M., Abbas, M., & Singh, J. (2020). An efficient numerical technique for solving time-fractional generalized Fisher's equation. *Frontiers in Physics*, *8*, 293.

[28] Tamboli, V. K., & Tandel, P. V. (2022). Solution of the time-fractional generalized Burger–Fisher equation using the fractional reduced differential transform method. *Journal of Ocean Engineering and Science*, *7*, 399–407.

[29] Devi, A., Kumari, A., Parumasur, N., Singh, P., & Kukreja, V. K. (2025). Numerical study of fractional order Burgers'–Huxley equation using modified cubic splines approximation. *Fractal and Fractional*, *9*, 780.

Gagandeep Kaur, Sourav Kumawat, and Bharat Bhushan

Solutions of the time- and spatial-fractional Sine-Gordon equation with the help of a two-variable $(\frac{G'}{G}, \frac{1}{G})$-expansion method

Abstract: In this chapter, we have analyzed the solutions of the Time- and Spatial-Fractional Sine-Gordon equation, produced by the application of the two-variable $(\frac{G'}{G}, \frac{1}{G})$-expansion method, using different orders of fractional derivative on the non-linear evolution equation. This method is obtained by extending the $(\frac{G'}{G})$-expansion method by including an additional term, namely $(\frac{1}{G})$. To find the solutions for the equation concerned, using the above method, we have used the Conformable fractional derivative of order β. Furthermore, we have shown the effect of fractional order on the solutions of the equation by plotting their graphs for different values of β.

Keywords: Solitons, Nonlinear evolution equations, $(\frac{G'}{G}, \frac{1}{G})$-expansion method, Time- and spatial-fractional sine-Gordon equation

MSC 2020: 35R11, 35C07, 37K40, 35L71

1 Introduction

Nonlinear evolution equations (NLEEs) [25] constitute a fundamental framework for modeling and analyzing complex natural phenomena, wave propagation, soliton dynamics, and transfer of energy in nonlinear systems. These equations are defined as partial differential equations that depict how the function quantities evolve over time, where this modification suggests the relationship between these functions and their derivatives is not linear. These equations describe the behavior and dynamics of "solitons" [2, 3], which are solitary waves that retain their shape while moving at a constant speed. These waves are self-reinforcing and remain stable due to the balanced cancelation of nonlinearity with dispersion. These waves were first discovered by the Scottish engineer, John Scott Russell, after his observation of the motion of waves in a channel in the nineteenth century.

Acknowledgement: The second author sincerely acknowledges the financial support provided by Akal University, Talwandi Sabo, which made this research work possible.

Gagandeep Kaur, Sourav Kumawat, Bharat Bhushan, Department of Mathematics, Akal University, Talwandi Sabo-151302, Bathinda, Punjab, India, e-mail: souravkumawat6@gmail.com

https://doi.org/10.1515/9783112234525-018

One such NLEE is the sine-Gordon equation [4], which is written as

$$u_{tt} + u_{xx} + \sin(u) = 0, \tag{1.1}$$

where $u = u(x,t)$ is a function of x and t, denoting space and time, respectively. The $\sin(u)$ introduces the nonlinearity and provides insight into its dynamic behavior. This equation is renowned for its elegant mathematical structure and notable applications throughout various disciplines, such as in physics (in magnetic flux propagation in Josephson junctions, crystal dislocations, and nonlinear optics), field theory, quantum mechanics, and engineering (in transmission lines and mechanical systems).

Here, we are interested in the fractional version of the mentioned NLEE, i. e., the time and spatial fractional sine-Gordon (TSF-SG) equation, given as [21]

$$\frac{\partial^{2\beta} u(x,t)}{\partial t^{2\beta}} - \frac{\partial^{2\beta} u(x,t)}{\partial x^{2\beta}} + u(x,t) - \frac{1}{6}u^3(x,t) = 0; \quad t > 0, 0 < \beta < 1. \tag{1.2}$$

Introducing the fractional derivative of space and time into the NLEE expands our understanding of the behavior of the phenomena explained by the standard form of the equation [6, 18, 19]. There exist various fractional derivatives, as shown in [7, 22]. We give a brief review of the derivatives, such as the Riemann–Liouville derivative, the Caputo derivative, the Weyl derivative, the Marchaud derivative, the Conformable derivative, and the Atangana–Baleanu derivative. In this chapter, $D_t^{\gamma}(u)$ and $D_x^{\gamma}(u)$ denote the conformable derivatives of order γ of u w. r. t. x and t, respectively, which we will study in the next section in detail [17].

To date, mathematicians and physicists have developed numerous techniques to solve the NLEEs. Some of these are Hirota's direct method [8], the sine-cosine method [15], the $(\frac{G'}{G})$-expansion method [9], the $(\frac{G'}{G^2})$-expansion method [1, 10], the new version fractional F-expansion method [20], the modified Kudryashov method [9, 12], the reduced differential transform method [16], the tanh method [23], the $(\frac{G'}{G}, \frac{1}{G})$-expansion method [13, 14, 24], and many more. The selection of the technique depends upon the NLEEs.

In order to find the solution of the TSF-SG equation, given in equation (1.2), we are going to apply a two-variable $(\frac{G'}{G}, \frac{1}{G})$-expansion method, introduced by Li et al. [13], which was derived as the extension of the $(\frac{G'}{G})$-expansion method by adding another term $(\frac{1}{G})$ that allows for more precise and flexible solutions of evolution equations and thereby provides a deeper understanding of the behavior of NLEEs. This method overcomes the limitations of the $(\frac{G'}{G})$-expansion method, being its extension, by offering a wider range of solutions and increased flexibility in capturing the diverse nature of the nonlinear dynamics of differential equations, thus improving the likelihood of obtaining unique and more accurate solutions.

This chapter consists of five sections, starting with the introduction. The second section is the preliminaries. The third section is devoted to the methodology. The fourth section provides the application of the method mentioned above to the equation concerned. The last section contains the results and conclusion of our research.

2 Preliminaries

Conformable fractional derivative (CFD):

CFD is one of the fractional derivatives. For the function $g(t)$, a CFD [5] of order γ is defined as

$$D_t^{\gamma} g(t) = \lim_{\epsilon \to 0} \frac{g(t + \epsilon t^{1-\gamma}) - g(t)}{\epsilon}, \quad \text{where } (0 < \gamma \leq 1),$$

which preserves the various properties of the standard integral derivative and extends the integer order of the derivative to fractional order. CFD was introduced by Khalil et al. [11] and it adheres to the following properties:

1. *Leibniz rule:*
   Let $\gamma \in (0, 1]$ and $p(t)$ and $q(t)$ be γ-differentiable at point $t > 0$, then

$$D_t^{\gamma}(p(t)q(t)) = [D_t^{\gamma} p(t)]q(t) + [D_t^{\gamma} q(t)]p(t).$$

2. *Linearity:*
   Let $c(t)$ and $d(t)$ be γ-differentiable at point $t > 0$, then

$$D_t^{\gamma}(ic(t) + jd(t)) = iD_t^{\gamma}(c(t)) + jD_t^{\gamma}(d(t)), \quad \text{for all } i, j \in \mathbb{R}.$$

3. *Chain rule:*
   Let $k(t)$ and $j(t)$ be γ-differentiable at point $t > 0$, then

$$D_t^{\gamma} k(j(t)) = [D_{j(t)}^{\gamma} k(j(t))][D_t^{\gamma} j(t)]j(t)^{1-\gamma}.$$

3 Methodology

Before discussing the steps of the method specified above, firstly, we consider Riccati's linear ordinary differential equation

$$\frac{d^2 G}{d\xi^2} + \Lambda G = \eta \tag{3.1}$$

and, considering the notations

$$R = \frac{G'}{G} \quad \text{and} \quad T = \frac{1}{G},$$

then, equation (3.1) satisfies

$$R' = -R^2 + \eta T - \Lambda, \quad \text{and} \quad T' = -RT. \tag{3.2}$$

Let S_1 and S_2 be two arbitrary constants. Now, the solution of equation (3.1) depends on the value of Λ and can be classified into three cases, which are as follows:

Case 1: When $\Lambda < 0$, the general solution of equation (3.1), is the hyperbolic-type function, given by

$$G = S_1 \sinh(\sqrt{-\Lambda}\xi) + S_2 \cosh(\sqrt{-\Lambda}\xi) + \frac{\eta}{\Lambda}, \tag{3.3}$$

having

$$T^2 = \frac{-\Lambda(R^2 - 2\eta T + \Lambda)}{\Lambda^2\sigma_1 + \eta^2}, \quad \text{where } \sigma_1 = S_1^2 - S_2^2. \tag{3.4}$$

Case 2: When $\Lambda > 0$, the general solution of equation (3.1) is a trigonometric-type function

$$G = S_1 \sin(\sqrt{\Lambda}\xi) + S_2 \cos(\sqrt{\Lambda}\xi) + \frac{\eta}{\Lambda}, \tag{3.5}$$

such that

$$T^2 = \frac{\Lambda(R^2 - 2\eta T + \Lambda)}{\Lambda^2\sigma_2 - \eta^2}, \quad \text{where } \sigma_2 = S_1^2 + S_2^2. \tag{3.6}$$

Case 3: When $\Lambda = 0$, we obtain a rational-type function as a solution to equation (3.1), which is

$$G = \frac{\eta}{2}\xi^2 + S_1\xi + S_2, \tag{3.7}$$

with

$$T^2 = \frac{R^2 - 2\eta T}{S_1^2 - 2\eta S_2}. \tag{3.8}$$

Now, the steps involved in the method are as follows:

Step 1: Consider the NLEE in two variables x and t with a fractional derivative of order β

$$I(u, D_t^\beta(u), D_x^\beta(u), \dots) = 0 \tag{3.9}$$

and let

$$u(x,t) = v(\xi), \quad \text{where } \xi = k\frac{x^\beta}{\beta!} - c\frac{t^\beta}{\beta!}. \tag{3.10}$$

Step 2: Deploying the equation (3.10) into equation (3.9), we obtain an ordinary differential equation

$$J(v, v', v'', v''', \dots) = 0, \tag{3.11}$$

where J is a polynomial in $v(\xi)$ and its derivatives are denoted by v', v'', v''', and so on.

Step 3: Now, solving equation (3.11), we obtain the three cases. Let

$$v(\xi) = \sum_{i=0}^{N} e_i \left(\frac{G'}{G}\right)^i + \sum_{i=1}^{N} f_i \left(\frac{G'}{G}\right)^{i-1} \left(\frac{1}{G}\right) \tag{3.12}$$

be the solution of equation (3.11), where e_i and f_i are the constants. The positive integer N is calculated by solving the equation formed by equating the degree of the highest-order derivative term to the degree of the nonlinear term in the differential equation. This principle is known as the homogeneous balance principle.

Step 4: We will derive a system of equations for the hyperbolic-type solution (3.3), (3.4), trigonometric-type solution (3.5), (3.6), and rational-type solution (3.7), (3.8) by substituting the value of $v(\xi)$ (3.12) into equation (3.11) and using equation (3.1) for the value of $G'' = \frac{d^2G}{d\xi^2}$ and then equating the constant term and coefficients of $R^i T^j$ to zero. Then, we solve this system of equations for the variables (e_0, e_1, f_1, c, k, Λ, and η) involved in the equation.

Step 5: Now, we substitute the solutions obtained as the outcome of Step 4 in equation (3.12) with equation (3.10) to obtain the solutions of equation (3.9). To understand the solution equations, we plot them graphically using the computational package.

In the following section, we utilized our proposed method to find the soliton solutions of the TSF-SG equation.

4 Application of the method on the TSF-SG equation

In this section, we will find the solutions for the TSF-SG equation (1.2). Applying the transformation (3.10) on equation (1.2) and with the aid of CFD, we obtain

$$(c^2 - k^2)v'' + v(\xi) - \frac{v^3(\xi)}{3!} = 0. \tag{4.1}$$

Using step 3 of the above described method on equation (4.1) we obtain

$$v(\xi) = e_0 + e_1 R(\xi) + f_1 T(\xi). \tag{4.2}$$

Now, as the concerned method suggests, we have the following three types of solutions:

Hyperbolic solutions ($\Lambda < 0$)

Using equation (3.3) and the value of ξ in equation (4.2), we have

$$v(x,t) = e_0 + e_1 \frac{S_1\sqrt{-\Lambda}\cosh(\sqrt{-\Lambda}(\frac{kx^\beta}{\beta!} - \frac{ct^\beta}{\beta!})) + S_2\sqrt{-\Lambda}\sinh(\sqrt{-\Lambda}(\frac{kx^\beta}{\beta!} - \frac{ct^\beta}{\beta!}))}{S_1\sinh(\sqrt{-\Lambda}(\frac{kx^\beta}{\beta!} - \frac{ct^\beta}{\beta!})) + S_2\cosh(\sqrt{-\Lambda}(\frac{kx^\beta}{\beta!} - \frac{ct^\beta}{\beta!})) + \frac{\eta}{\Lambda}}$$
$$+ f_1 \frac{1}{S_1\sinh(\sqrt{-\Lambda}(\frac{kx^\beta}{\beta!} - \frac{ct^\beta}{\beta!})) + S_2\cosh(\sqrt{-\Lambda}(\frac{kx^\beta}{\beta!} - \frac{ct^\beta}{\beta!})) + \frac{\eta}{\Lambda}}, \tag{4.3}$$

where S_1, S_2, Λ, η, and c are arbitrary constants.

Now, following step 4 of the methodology, we have derived the following system of equations:

$$(RT)^0 = e_0 - \frac{e_0^3}{6} + \frac{e_0 f_1^2 \Lambda^2}{2(\Lambda^2(S_1^2 - S_2^2) + \eta^2)} + \frac{f_1^3 \Lambda^3 \eta}{3(\Lambda^2(S_1^2 - S_2^2) + \eta^2)^2}$$
$$+ \frac{f_1 c^2 \Lambda^2 \eta}{\Lambda^2(S_1^2 - S_2^2) + \eta^2} - \frac{f_1 k^2 \Lambda^2 \eta}{\Lambda^2(S_1^2 - S_2^2) + \eta^2},$$
$$R = e_1 + 2e_1 c^2 \Lambda - 2e_1 k^2 \Lambda - \frac{e_0^2 e_1}{2} + \frac{e_1 f_1^2 \Lambda^2}{2(\Lambda^2(S_1^2 - S_2^2) + \eta^2)},$$
$$R^2 = -\frac{e_0 e_1^2}{2} + \frac{e_0 f_1^2 \Lambda}{2(\Lambda^2(S_1^2 - S_2^2) + \eta^2)} + \frac{f_1^3 \Lambda^2 \eta}{3(\Lambda^2(S_1^2 - S_2^2) + \eta^2)^2}$$
$$+ \frac{f_1 c^2 \Lambda \eta}{\Lambda^2(S_1^2 - S_2^2) + \eta^2} - \frac{f_1 k^2 \Lambda \eta}{\Lambda^2(S_1^2 - S_2^2) + \eta^2};$$
$$R^3 = 2e_1 c^2 - 2e_1 k^2 - \frac{e_1^3}{6} + \frac{e_1 f_1^2 \Lambda}{2(\Lambda^2(S_1^2 - S_2^2) + \eta^2)},$$
$$R^0\mathrm{T} = f_1 + f_1 c^2 \Lambda - f_1 k^2 \Lambda - \frac{e_0^2 f_1}{2} - \frac{e_0 f_1^2 \Lambda \eta}{\Lambda^2(S_1^2 - S_2^2) + \eta^2} + \frac{f_1^3 \Lambda^2}{6(\Lambda^2(S_1^2 - S_2^2) + \eta^2)}$$
$$- \frac{2f_1^3 \Lambda^2 \eta^2}{3(\Lambda^2(S_1^2 - S_2^2) + \eta^2)^2} - \frac{2f_1 c^2 \Lambda \eta^2}{\Lambda^2(S_1^2 - S_2^2) + \eta^2} + \frac{2f_1 k^2 \Lambda \eta^2}{\Lambda^2(S_1^2 - S_2^2) + \eta^2},$$
$$RT = -e_0 e_1 f_1 - 3e_1 c^2 \eta + 3e_1 k^2 \eta - \frac{e_1 f_1^2 \Lambda \eta}{\Lambda^2(S_1^2 - S_2^2) + \eta^2},$$
$$R^2 T = 2f_1 c^2 - 2f_1 k^2 - \frac{e_1^2 f_1}{2} + \frac{f_1^3 \Lambda}{6(\Lambda^2(S_1^2 - S_2^2) + \eta^2)}.$$

On solving the above system using a computational package program like Mathematica, we have extracted the solutions provided below as results with corresponding values of $u(x,t)$ and their graphs:

Result 1. We have the first result as follows

$$e_0 = 0, \quad e_1 = 0, \quad f_1 = 2\sqrt{3}\sqrt{{S_2}^2 - {S_1}^2}, \quad k = \pm\frac{\sqrt{c^2\Lambda - 1}}{\sqrt{\Lambda}}. \tag{4.4}$$

On substituting equation (4.4) into $u(x,t)$ above, the solution becomes

$$u_{11}(x,t) = \frac{2\sqrt{3}\sqrt{{S_2}^2 - {S_1}^2}}{S_1 \sinh\left(\sqrt{-\Lambda}\left(\frac{kx^\beta}{\beta!} - \frac{ct^\beta}{\beta!}\right)\right) + S_2 \cosh\left(\sqrt{-\Lambda}\left(\frac{kx^\beta}{\beta!} - \frac{ct^\beta}{\beta!}\right)\right) + \frac{\eta}{\Lambda}}. \tag{4.5}$$

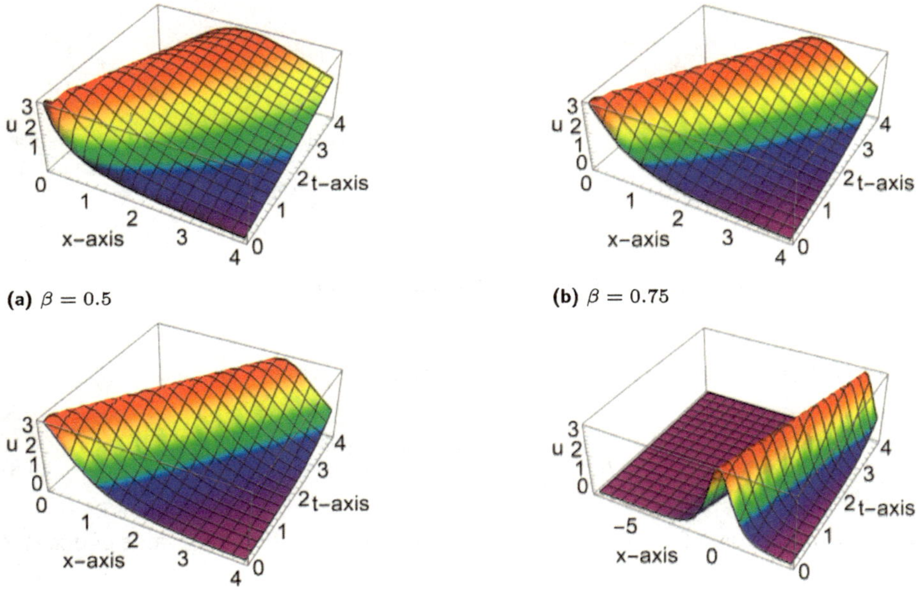

(a) $\beta = 0.5$ **(b)** $\beta = 0.75$ **(c)** $\beta = 0.9$ **(d)** $\beta = 1$

Figure 1: The solution of equation (4.5) when $S_1 = 1$, $S_2 = 5$, $\Lambda = -1.5$, $\eta = 0$, and $c = 1$.

Result 2. We obtained second set as

$$e_0 = 0, \quad e_1 = 0, \quad f_1 = -2\sqrt{3}\sqrt{{S_2}^2 - {S_1}^2}, \quad k = \pm\frac{\sqrt{c^2\Lambda - 1}}{\sqrt{\Lambda}}. \tag{4.6}$$

Using the same procedure as in Result 1 and replacing equation (4.4) with equation (4.6), we obtain

$$u_{12}(x,t) = -\frac{2\sqrt{3}\sqrt{{S_2}^2 - {S_1}^2}}{S_1 \sinh\left(\sqrt{-\Lambda}\left(\frac{kx^\beta}{\beta!} - \frac{ct^\beta}{\beta!}\right)\right) + S_2 \cosh\left(\sqrt{-\Lambda}\left(\frac{kx^\beta}{\beta!} - \frac{ct^\beta}{\beta!}\right)\right) + \frac{\eta}{\Lambda}}. \tag{4.7}$$

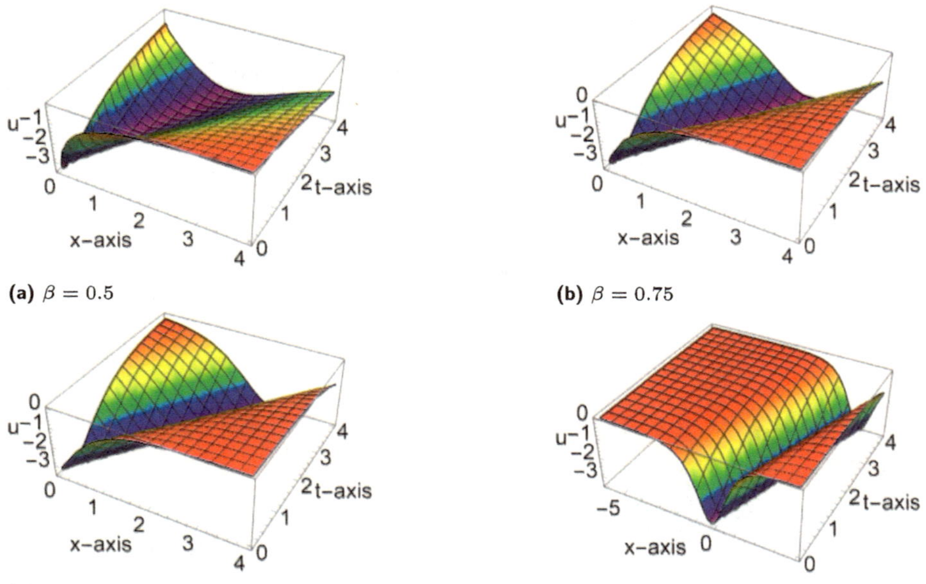

(a) $\beta = 0.5$ (b) $\beta = 0.75$ (c) $\beta = 0.9$ (d) $\beta = 1$

Figure 2: The solution of equation (4.7) when $S_1 = 2$, $S_2 = 5$, $\Lambda = -1.5$, $\eta = 0$, and $c = 1$.

Result 3. In this case, we have the following values

$$e_0 = 0, \quad e_1 = -\frac{i\sqrt{6}}{\sqrt{\Lambda}}, \quad f_1 = -\frac{\sqrt{6}\sqrt{{S_1}^2\Lambda^2 - {S_2}^2\Lambda^2 + \eta^2}}{\Lambda}, \quad k = -\frac{\sqrt{c^2\Lambda + 2}}{\sqrt{\Lambda}}. \tag{4.8}$$

Using the procedure of Result 1 with equation (4.8), the corresponding value of $u(x,t)$ is

$$\begin{aligned} u_{13}(x,t) = {} & -\frac{\sqrt{6}\sqrt{{S_1}^2\Lambda^2 - {S_2}^2\Lambda^2 + \eta^2}}{\Lambda(S_1\sinh(\sqrt{-\Lambda}(\frac{kx^\beta}{\beta!} - \frac{ct^\beta}{\beta!})) + S_2\cosh(\sqrt{-\Lambda}(\frac{kx^\beta}{\beta!} - \frac{ct^\beta}{\beta!})) + \frac{\eta}{\Lambda})} \\ & - \left(\frac{i\sqrt{6}}{\sqrt{\Lambda}}\right)\frac{S_1\sqrt{-\Lambda}\cosh(\sqrt{-\Lambda}(\frac{kx^\beta}{\beta!} - \frac{ct^\beta}{\beta!})) + S_2\sqrt{-\Lambda}\sinh(\sqrt{-\Lambda}(\frac{kx^\beta}{\beta!} - \frac{ct^\beta}{\beta!}))}{S_1\sinh(\sqrt{-\Lambda}(\frac{kx^\beta}{\beta!} - \frac{ct^\beta}{\beta!})) + S_2\cosh(\sqrt{-\Lambda}(\frac{kx^\beta}{\beta!} - \frac{ct^\beta}{\beta!})) + \frac{\eta}{\Lambda}}. \end{aligned} \tag{4.9}$$

Result 4. This result provides the following set of parameter values

$$e_0 = 0, \quad e_1 = \frac{i\sqrt{6}}{\sqrt{\Lambda}}; \quad f_1 = \frac{\sqrt{6}\sqrt{S_1^2\Lambda^2 - S_2^2\Lambda^2 + \eta^2}}{\Lambda}, \quad k = \frac{\sqrt{c^2\Lambda + 2}}{\sqrt{\Lambda}}. \tag{4.10}$$

Similarly, on substituting (4.10) in equations (4.2) and (4.1) with the vision of the above results, the solution equation $u(x,t)$ becomes

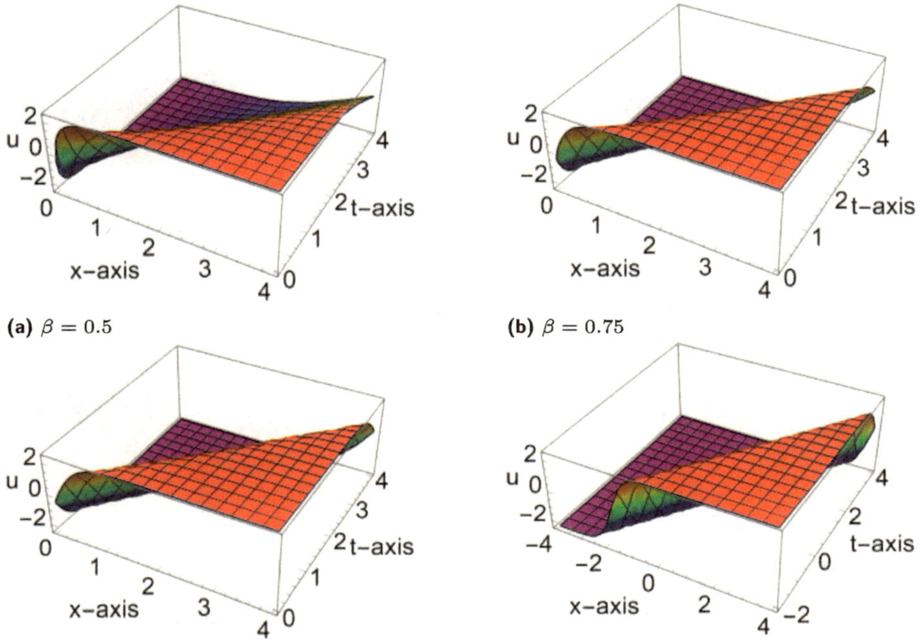

(a) $\beta = 0.5$ (b) $\beta = 0.75$

(c) $\beta = 0.9$ (d) $\beta = 1$

Figure 3: The solution of equation (4.9) when $S_1 = 5$, $S_2 = 2$, $\Lambda = -1.5$, $\eta = 0$, and $c = -3.5$.

$$u_{14}(x,t) = +\frac{\sqrt{6}\sqrt{{S_1}^2\Lambda^2 - {S_2}^2\Lambda^2 + \eta^2}}{\Lambda(S_1\sinh(\sqrt{-\Lambda}(\frac{kx^\beta}{\beta!} - \frac{ct^\beta}{\beta!})) + S_2\cosh(\sqrt{-\Lambda}(\frac{kx^\beta}{\beta!} - \frac{ct^\beta}{\beta!})) + \frac{\eta}{\Lambda})} + \left(\frac{i\sqrt{6}}{\sqrt{\Lambda}}\right)\frac{S_1\sqrt{-\Lambda}\cosh(\sqrt{-\Lambda}(\frac{kx^\beta}{\beta!} - \frac{ct^\beta}{\beta!})) + S_2\sqrt{-\Lambda}\sinh(\sqrt{-\Lambda}(\frac{kx^\beta}{\beta!} - \frac{ct^\beta}{\beta!}))}{S_1\sinh(\sqrt{-\Lambda}(\frac{kx^\beta}{\beta!} - \frac{ct^\beta}{\beta!})) + S_2\cosh(\sqrt{-\Lambda}(\frac{kx^\beta}{\beta!} - \frac{ct^\beta}{\beta!})) + \frac{\eta}{\Lambda}}. \tag{4.11}$$

Result 5. This result yields the following values

$$e_0 = \sqrt{6}, \quad e_1 = 0; \quad f_1 = -\frac{2\sqrt{6}\eta}{\Lambda}, \quad k = -\frac{\sqrt{c^2\Lambda + 2}}{\sqrt{\Lambda}}, \quad S_2 = S_1. \tag{4.12}$$

On substituting equation (4.12) into equations (4.2) and (4.1), the solution equation $u(x,t)$ becomes

$$u_{15}(x,t) = \sqrt{6} - \frac{2\sqrt{6}\eta}{\Lambda(S_1\sinh(\sqrt{-\Lambda}(\frac{kx^\beta}{\beta!} - \frac{ct^\beta}{\beta!})) + S_2\cosh(\sqrt{-\Lambda}(\frac{kx^\beta}{\beta!} - \frac{ct^\beta}{\beta!})) + \frac{\eta}{\Lambda})}. \tag{4.13}$$

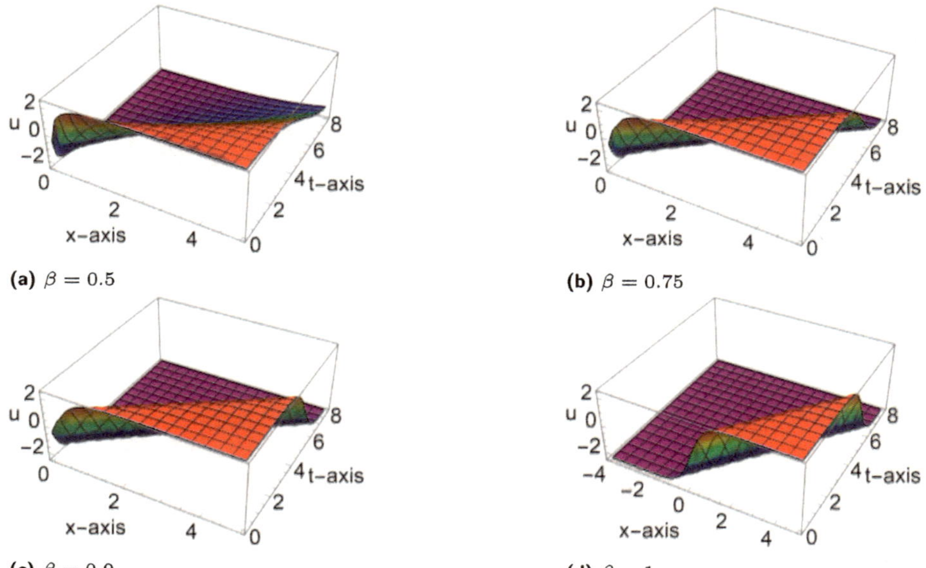

(a) $\beta = 0.5$ (b) $\beta = 0.75$

(c) $\beta = 0.9$ (d) $\beta = 1$

Figure 4: The solution of equation (4.11) when $S_1 = 2.9$, $S_2 = 1.5$, $\Lambda = -4.5$, $\eta = -12.5$, and $c = 1.5$.

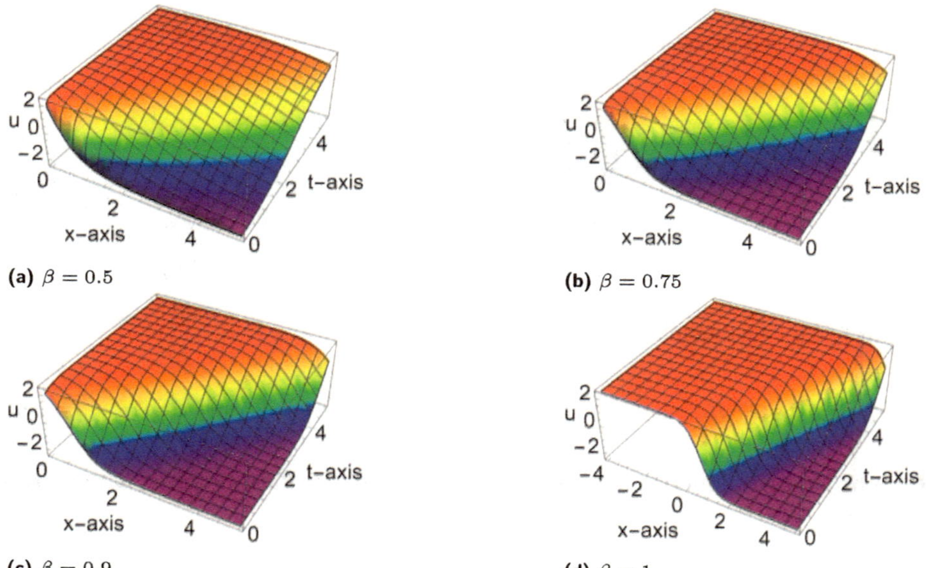

(a) $\beta = 0.5$ (b) $\beta = 0.75$

(c) $\beta = 0.9$ (d) $\beta = 1$

Figure 5: The solution of equation (4.13) when $S_1 = 8.5$, $\Lambda = -1.75$, $\eta = -1$, and $c = -2.75$.

Result 6. This result leads to

$$e_0 = -\sqrt{6}, \quad e_1 = 0, \qquad f_1 = \frac{2\sqrt{6}\eta}{\Lambda}, \quad k = -\frac{\sqrt{c^2\Lambda + 2}}{\sqrt{\Lambda}}, \quad S_2 = -S_1. \tag{4.14}$$

Similarly, from the above results, on substituting equation (4.14) into equations (4.2) and (4.1), the solution equation $u(x,t)$ becomes

$$u_{16}(x,t) = -\sqrt{6} + \frac{2\sqrt{6}\eta}{\Lambda(S_1 \sinh(\sqrt{-\Lambda}(\frac{kx^\beta}{\beta!} - \frac{ct^\beta}{\beta!})) + S_2 \cosh(\sqrt{-\Lambda}(\frac{kx^\beta}{\beta!} - \frac{ct^\beta}{\beta!})) + \frac{\eta}{\Lambda})}. \tag{4.15}$$

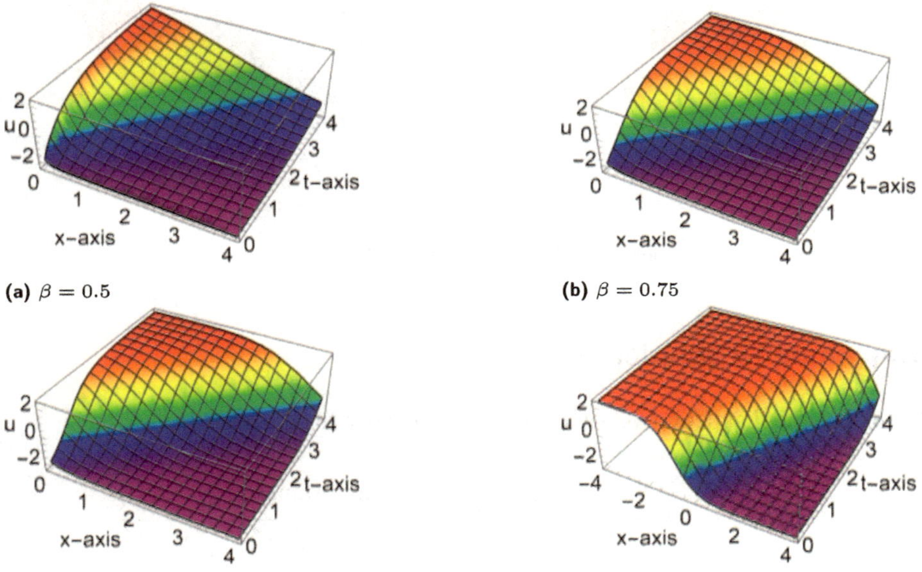

(a) $\beta = 0.5$

(b) $\beta = 0.75$

(c) $\beta = 0.9$

(d) $\beta = 1$

Figure 6: The solution of equation (4.15) when $S_1 = 4$, $\Lambda = -2.5$, $\eta = 1.25$, and $c = -1.75$.

Result 7. This result produces the below values

$$e_0 = -\sqrt{6}, \quad e_1 = 0; \quad f_1 = \frac{2\sqrt{6}\eta}{\Lambda}, \quad k = \frac{\sqrt{c^2\Lambda + 2}}{\sqrt{\Lambda}}, \quad S_2 = -S_1. \tag{4.16}$$

On substituting the values from equation (4.16) into equations (4.2) and (4.1), the solution equation $u(x,t)$ takes the form

$$u_{17}(x,t) = -\sqrt{6} + \frac{2\sqrt{6}\eta}{\Lambda(S_1 \sinh(\sqrt{-\Lambda}(\frac{kx^\beta}{\beta!} - \frac{ct^\beta}{\beta!})) + S_2 \cosh(\sqrt{-\Lambda}(\frac{kx^\beta}{\beta!} - \frac{ct^\beta}{\beta!})) + \frac{\eta}{\Lambda})}. \tag{4.17}$$

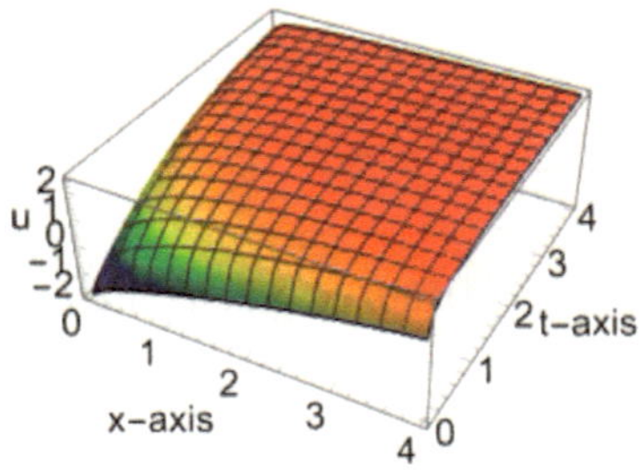

(a) $\beta = 0.5$

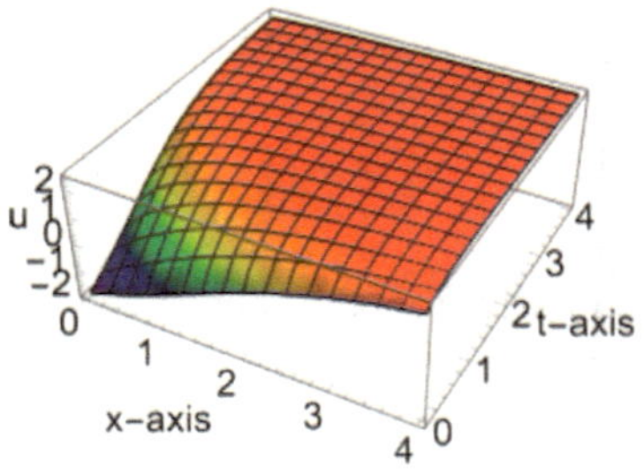

(b) $\beta = 0.75$

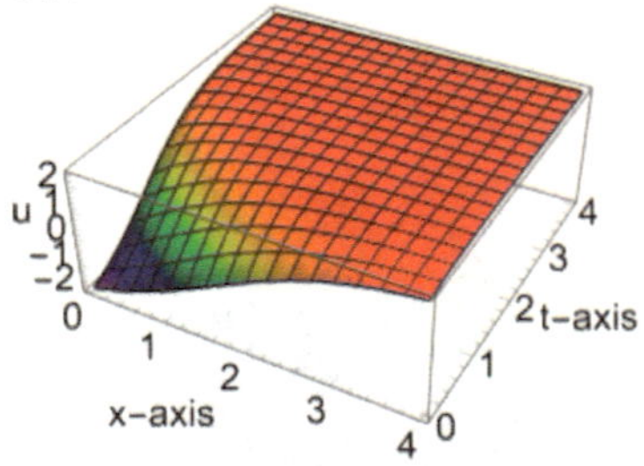

(c) $\beta = 0.9$

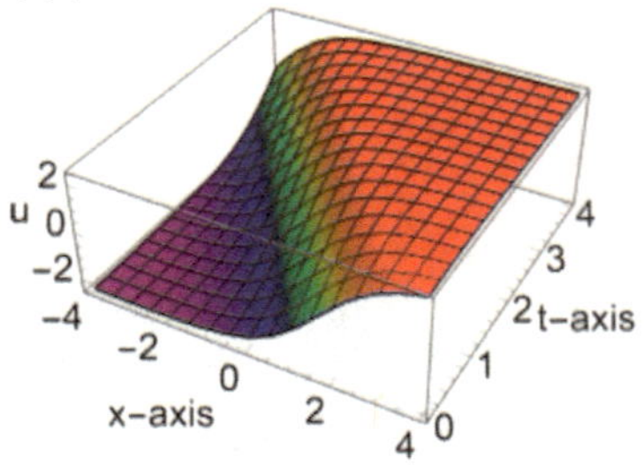

(d) $\beta = 1$

Figure 7: The solution of equation (4.17) when $S_1 = 8.5$, $\Lambda = -1.75$, $\eta = 1.5$, and $c = -1.5$.

Trigonometric solutions ($\Lambda > 0$)

With the help of equation (3.6) and using the value of ξ in equation (4.2), we have

$$
\begin{aligned}
v(x,t) = e_0 + e_1 &\frac{S_1\sqrt{\Lambda}\cos(\sqrt{\Lambda}(\frac{kx^\beta}{\beta!} - \frac{ct^\beta}{\beta!})) - S_2\sqrt{\Lambda}\sin(\sqrt{\Lambda}(\frac{kx^\beta}{\beta!} - \frac{ct^\beta}{\beta!}))}{S_1\sin(\sqrt{\Lambda}(\frac{kx^\beta}{\beta!} - \frac{ct^\beta}{\beta!})) + S_2\cos(\sqrt{\Lambda}(\frac{kx^\beta}{\beta!} - \frac{ct^\beta}{\beta!})) + \frac{\eta}{\Lambda}} \\
&+ f_1\frac{1}{S_1\sin(\sqrt{\Lambda}(\frac{kx^\beta}{\beta!} - \frac{ct^\beta}{\beta!})) + S_2\cos(\sqrt{\Lambda}(\frac{kx^\beta}{\beta!} - \frac{ct^\beta}{\beta!})) + \frac{\eta}{\Lambda}},
\end{aligned}
\tag{4.18}
$$

where S_1, S_2, Λ, η, and c are arbitrary constants.

Following the same procedure as the previous case with equations (3.1), (3.2), and (3.6), we have

$$
\begin{aligned}
(RT)^0 = e_0 - \frac{e_0^3}{6} - \frac{e_0 f_1^2\Lambda^2}{2(\Lambda^2(S_1^2 + S_2^2) - \eta^2)} + \frac{f_1^3\Lambda^3\eta}{3(\Lambda^2(S_1^2 + S_2^2) - \eta^2)^2} \\
- \frac{f_1 c^2\Lambda^2\eta}{\Lambda^2(S_1^2 + S_2^2) - \eta^2} + \frac{f_1 k^2\Lambda^2\eta}{\Lambda^2(S_1^2 + S_2^2) - \eta^2}, \\
R = e_1 + 2e_1c^2\Lambda - 2e_1k^2\Lambda - \frac{e_0^2 e_1}{2} - \frac{e_1 f_1^2\Lambda^2}{2(\Lambda^2(S_1^2 + S_2^2) - \eta^2)},
\end{aligned}
$$

$$R^2 = -\frac{e_0 e_1^2}{2} - \frac{e_0 f_1^2 \Lambda}{2(\Lambda^2(S_1^2+S_2^2)-\eta^2)} + \frac{f_1^3 \Lambda^2 \eta}{3(\Lambda^2(S_1^2+S_2^2)-\eta^2)^2} - \frac{f_1 c^2 \Lambda \eta}{\Lambda^2(S_1^2+S_2^2)-\eta^2} + \frac{f_1 k^2 \Lambda \eta}{\Lambda^2(S_1^2+S_2^2)-\eta^2},$$

$$R^3 = 2e_1 c^2 - 2e_1 k^2 - \frac{e_1^3}{6} - \frac{e_1 f_1^2 \Lambda}{2(\Lambda^2(S_1^2+S_2^2)-\eta^2)},$$

$$R^0 T = f_1 + f_1 c^2 \Lambda - f_1 k^2 \Lambda - \frac{e_0^2 f_1}{2} + \frac{e_0 f_1^2 \Lambda \eta}{\Lambda^2(S_1^2+S_2^2)-\eta^2} - \frac{f_1^3 \Lambda^2}{6(\Lambda^2(S_1^2+S_2^2)-\eta^2)} - \frac{2 f_1^3 \Lambda^2 \eta^2}{3(\Lambda^2(S_1^2+S_2^2)-\eta^2)^2} + \frac{2 f_1 c^2 \Lambda \eta^2}{\Lambda^2(S_1^2+S_2^2)-\eta^2} - \frac{2 f_1 k^2 \Lambda \eta^2}{\Lambda^2(S_1^2+S_2^2)-\eta^2},$$

$$RT = -e_0 e_1 f_1 - 3e_1 c^2 \eta + 3 e_1 k^2 \eta + \frac{e_1 f_1^2 \Lambda \eta}{\Lambda^2(S_1^2+S_2^2)-\eta^2},$$

$$R^2 T = 2 f_1 c^2 - 2 f_1 k^2 - \frac{e_1^2 f_1}{2} - \frac{f_1^3 \Lambda}{6(\Lambda^2(S_1^2+S_2^2)-\eta^2)}.$$

By working out the above algebraic equations through a computational package program such as Mathematica, we have the following result

$$e_0 = 0; \quad e_1 = 0; \quad f_1 = -2\sqrt{3}\sqrt{S_1^2+S_2^2}; \quad \eta = 0; \quad k = \frac{\sqrt{c^2\Lambda - 1}}{\sqrt{\Lambda}}. \tag{4.19}$$

Substituting the above solution in equation (4.18), we obtain

$$u_{21}(x,t) = -\frac{2\sqrt{3}\sqrt{{S_2}^2+{S_1}^2}}{S_1 \sin\left(\sqrt{\Lambda}\left(\frac{kx^\beta}{\beta!} - \frac{ct^\beta}{\beta!}\right)\right) + S_2 \cos\left(\sqrt{\Lambda}\left(\frac{kx^\beta}{\beta!} - \frac{ct^\beta}{\beta!}\right)\right) + \frac{\eta}{\Lambda}}. \tag{4.20}$$

5 Conclusion

In this chapter, we generated traveling wave solutions of the time- and spatial-fractional Sine-Gordon equation (TSFSGE) using a two-variable $(\frac{G'}{G}, \frac{1}{G})$-expansion method. We obtain soliton solutions (a special type of traveling wave solutions) in three forms: a hyperbolic-type solution, which gives us a bright soliton (1), a dark soliton (2), a kink soliton (3), (4), (7), and an antikink soliton (5), (6). In this, we see variation accordingly negative and positive values of constant k in Results 6 and 7. Second, a trigonometric-type solution provides us with a singular solution (8). The variation in the trigonometric-type solution accords with different values of CFD order β. Third is a rational-type solution, which is a trivial solution. The applied method is direct, effective, and reliable, and gives a variety of new solutions to the concerned equation. Our solutions to the mentioned

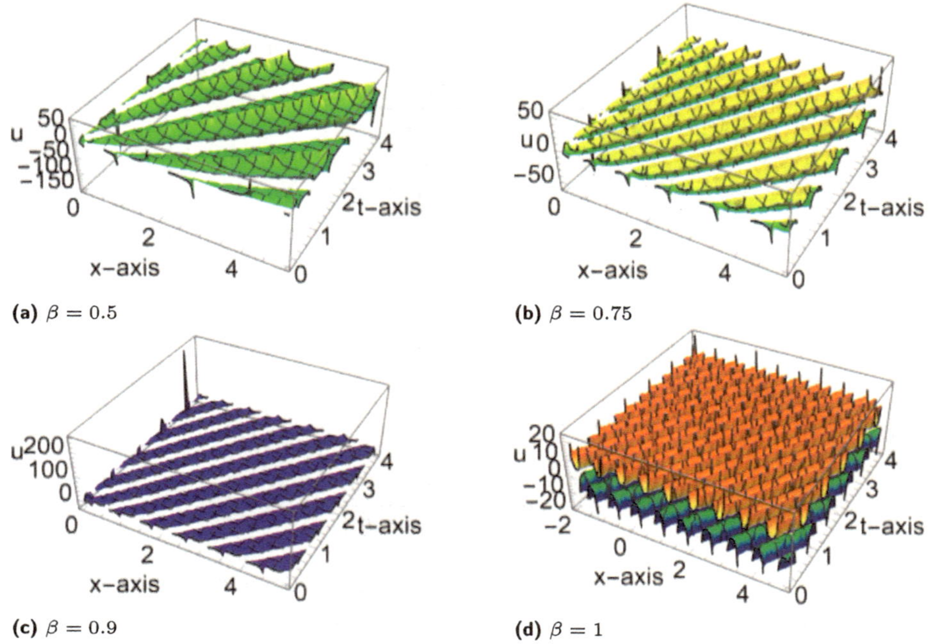

(a) $\beta = 0.5$ (b) $\beta = 0.75$ (c) $\beta = 0.9$ (d) $\beta = 1$

Figure 8: The solution of equation (4.20) when $S_1 = 5$, $S_2 = 2$, $\Lambda = -1.5$, and $\eta = 0$.

equation may be helpful in the fields of engineering physics, solid-state physics, and others.

Bibliography

[1] Aljahdaly, N. H. (2019). Some applications of the modified (G'/G^2)-expansion method in mathematical physics. *Results in Physics*, *13*, 102272.

[2] Allen, J. E. (1998). The early history of solitons (solitary waves). *Physica Scripta*, *57*(3), 436.

[3] Arora, G., Rani, R., & Emadifar, H. (2022). Soliton: a dispersion-less solution with existence and its types. *Heliyon*, *8*(12), e12000.

[4] Barone, A., Esposito, F., Magee, C. J., & Scott, A. C. (1971). Theory and applications of the sine-Gordon equation. *La Rivista del Nuovo Cimento*, *1*(2), 227–267.

[5] Chung, W. S. (2015). Fractional Newton mechanics with conformable fractional derivative. *Journal of Computational and Applied Mathematics*, *290*, 150–158.

[6] Farooq, K., Abdullah, A. R., Hussain, E., & Murad, M. A. S. (2025). Chaos, Lyapunov exponent, and sensitivity demonstration of the coupled nonlinear integrable model with soliton solutions. *AIMS Mathematics*, *10*(10), 22929–22957.

[7] Guswanto, B. H., Andini, L. R., & Triyani (2022). On conformable, Riemann–Liouville, and Caputo fractional derivatives. *Bulletin of Applied Mathematics and Mathematics Education*, *2*(2), 59–64.

[8] Hirota, R. (2004). *The Direct Method in Soliton Theory*. Cambridge University Press.

[9] Ionescu, C., Babalic, C. N., Constantinescu, R., & Efrem, R. (2022). The functional expansion approach for solving NPDEs as a generalization of the Kudryashov and G'/G methods. *Symmetry*, *14*(4), 827.

[10] Kaewta, S., Sirisubtawee, S., Koonprasert, S., & Sungnul, S. (2021). Applications of the (G'/G^2)-expansion method for solving certain nonlinear conformable evolution equations. *Fractal and Fractional*, *5*(3), 88.

[11] Khalil, R., Al Horani, M., Yousef, A., & Sababheh, M. (2014). A new definition of fractional derivative. *Journal of Computational and Applied Mathematics*, *264*, 65–70.

[12] Kumar, D., Seadawy, A. R., & Joardar, A. K. (2018). Modified Kudryashov method via new exact solutions for some conformable fractional differential equations arising in mathematical biology. *Chinese Journal of Physics*, *56*(1), 75–85.

[13] Li, L.-X., Li, E.-Q., & Wang, M.-L. (2010). The $(G'/G, 1/G)$-expansion method and its application to traveling wave solutions of the Zakharov equations. *Applied Mathematics—A Journal of Chinese Universities*, *25*, 454–462.

[14] Miah, M. M., Ali, H. M. S., Akbar, M. A., & Seadawy, A. R. (2019). New applications of the two-variable $(G'/G, 1/G)$-expansion method for closed-form traveling wave solutions of integro-differential equations. *Journal of Ocean Engineering and Science*, *4*(2), 132–143.

[15] Mirzazadeh, M., Eslami, M., Zerrad, E., Mahmood, M. F., Biswas, A., & Belic, M. (2015). Optical solitons in nonlinear directional couplers by sine–cosine function method and Bernoulli's equation approach. *Nonlinear Dynamics*, *81*(4), 1933–1949.

[16] Mohamed, M. S., & Al-Qarshi, T. T. (2017). Approximate solutions for the time–space fractional nonlinear partial differential equations using reduced differential transform method. *Global Journal of Pure and Applied Mathematics*, *13*(6), 2153–2162.

[17] Mahmood, S. S., Boulaaras, S. M., Murad, M. A. S., & Malik, S. (2025). Optical solutions to the time fractional Akbota equation arising in nonlinear optics via two distinct methods. *Scientific Reports*, *15*(1), 28823.

[18] Murad, M. A. S. (2022). Modified integral equation combined with the decomposition method for time fractional differential equations with variable coefficients. *Applied Mathematics—A Journal of Chinese Universities*, *37*(3), 404–414.

[19] Murad, M. A. S., Tedjani, A. H., Li, Z., & Hussain, E. (2025). Optical solutions to time-fractional improved $(2+1)$-dimensional nonlinear Schrödinger equation in optical fibers. *Scientific Reports*, *15*(1), 29208.

[20] Pandir, Y., & Duzgun, H. H. (2017). New exact solutions of time fractional Gardner equation using a new version of the f-expansion method. *Communications in Theoretical Physics*, *67*(1), 9.

[21] Roy, R., & Akbar, M. A. (2020). A new approach to study nonlinear space–time fractional sine-Gordon and Burgers equations. *IOP SciNotes*, *1*(3), 035003.

[22] Teodoro, G. S., Machado, J. A. T., & De Oliveira, E. C. (2019). A review of definitions of fractional derivatives and other operators. *Journal of Computational Physics*, *388*, 195–208.

[23] Wazwaz, A.-M. (2005). The tanh method: exact solutions of the sine-Gordon and the sinh-Gordon equations. *Applied Mathematics and Computation*, *167*(2), 1196–1210.

[24] Zayed, E. M. E., Ibrahim, S. A. H., & Abdelaziz, M. A. M. (2012). Traveling wave solutions of the nonlinear $(3+1)$-dimensional Kadomtsev–Petviashvili equation using the two-variable $(G'/G, 1/G)$-expansion method. *Journal of Applied Mathematics*, *2012*, 1–8.

[25] Zheng, S. (2004). *Nonlinear Evolution Equations*. Chapman & Hall/CRC.

Sudhakar Kumar Chaubey, Kawthar Al-Ajmi, Devendra Kumar, Amani Al-Saidi, Sheikha Al-Shihhi, and Marwa Al-Jabri

Riemannian manifolds equipped with a semisymmetric metric ξ-connection and Ricci–Bourguignon solitons

Abstract: In this chapter, we study almost Ricci–Bourguignon solitons on n-dimensional Riemannian manifolds endowed with a semisymmetric metric ξ-connection. We establish the existence of such solitons when the potential vector field coincides with the Reeb vector field ξ. Furthermore, when the potential vector field is pointwise collinear with ξ, we show that the underlying manifold becomes quasi-Einstein and derive conditions under which the associated scalar function is harmonic, subharmonic, or superharmonic. We also investigate gradient almost Ricci–Bourguignon solitons and discuss several of their geometric properties. These results contribute to the understanding of Ricci–Bourguignon-type geometric structures and their behavior under generalized curvature flows.

Keywords: Riemannian manifolds, semisymmetric metric ξ-connection, almost Ricci–Bourguignon solitons, gradient almost Ricci–Bourguignon solitons, quasi-Einstein manifolds

MSC 2020: 53C21, 53C25, 53C44, 53C15

1 Introduction

Geometric flows play a central role in understanding the evolution of Riemannian metrics on smooth manifolds. Among them, the *Ricci flow*, introduced by Hamilton [1], has been extensively studied due to its profound applications in geometry and topology, including the proof of the Poincaré conjecture. The Ricci flow deforms a Riemannian

Acknowledgement: The authors would like to express their sincere gratitude to the reviewer for his/her careful reading and valuable comments to improve our chapter. By virtue of his/her unmatched efforts, we could make our chapter better than the first draft. Moreover, we want to give our gratitude to the Editor for providing his/her valuable suggestions. This research project was funded by the University of Technology and Applied Sciences through the Internal Research Funding Program-2024, grant number (UTAS-Shinas-cy01-2024-003).

Sudhakar Kumar Chaubey, Kawthar Al-Ajmi, Devendra Kumar, Amani Al-Saidi, Sheikha Al-Shihhi, Marwa Al-Jabri, Unit of Sciences and Mathematics, Department of Supportive Requirements, University of Technology and Applied Sciences-Shinas, P. O. Box 77, Postal Code 324, Oman

https://doi.org/10.1515/9783112234525-019

metric $g(t)$ according to its Ricci curvature, leading to a nonlinear parabolic partial differential equation of the form

$$\frac{\partial}{\partial t}g(t) = -2\,S,$$

where $g_0 = g(0)$ and S is the Ricci tensor of the Riemannian metric g.

A natural generalization of the Ricci flow is the *Ricci–Bourguignon flow*, first proposed by Bourguignon [2], which introduces a real parameter ρ to interpolate between the Ricci flow and other curvature-normalizing flows. Formally, the Ricci–Bourguignon flow is defined by

$$\frac{\partial}{\partial t}g(t) = -2\,S + 2\rho r g,$$

where S is the Ricci tensor, r is the scalar curvature, and $\rho \in \mathbb{R}$ is a constant. This flow reduces to the classical Ricci flow when $\rho = 0$ and exhibits different geometric behaviors for other values of ρ, allowing the study of a wider class of curvature evolution phenomena.

Closely related to geometric flows are *soliton solutions*, which evolve under the flow only by diffeomorphisms and scaling. A Riemannian manifold (M^n, g) is called a *Ricci–Bourguignon soliton* if there exists a vector field V and a constant μ such that

$$S + \frac{1}{2}\mathcal{L}_V g = \mu g + \rho r g,$$

where $\mathcal{L}_V g$ denotes the Lie derivative of the metric along V. Depending on the value of the constant μ, a Ricci–Bourguignon soliton on M is referred to as *steady* ($\mu = 0$), *shrinking* ($\mu > 0$), or *expanding* ($\mu < 0$). The soliton is considered *trivial* if the potential vector field V is Killing, meaning that the metric changes only through isometries. The conformal version of this problem under the assumption of constant scalar curvature has been studied by Fischer [18]. If V is the gradient of a smooth function f, the soliton is said to be a *gradient Ricci–Bourguignon soliton*, satisfying

$$S + \nabla^2 f = \mu g + \rho r g.$$

Ricci–Bourguignon solitons generalize Ricci solitons and serve as fixed points or self-similar solutions of the flow, providing important insights into the long-term behavior and singularity formation of the flow. Understanding these solitons is crucial in the classification of manifolds with special curvature properties and in the study of geometric evolution equations. When a Riemannian metric g satisfies the above equation, it is called a *gradient Ricci–Bourguignon soliton*, which Catino and Mazzieri [14] termed a *gradient ρ-Einstein soliton*. In their work, they showed that every gradient ρ-Einstein soliton is rectifiable. Several important results concerning Ricci–Bourguignon solitons have also been obtained in [15–17, 19]. If we choose μ to be a smooth function in the above equations, then we recover the equations of almost Ricci–Bourguignon solitons and gradient almost Ricci–Bourguignon solitons, respectively.

Throughout the chapter, we denote an n-dimensional Riemannian manifold endowed with a semisymmetric metric ξ-connection by $(M, \widetilde{\nabla})$.

The chapter is organized as follows. After the introduction, Section 2 collects the basic preliminaries of $(M, \widetilde{\nabla})$. In Section 3, we establish the existence of an almost Ricci–Bourguignon soliton (g, V, μ, ρ) on $(M, \widetilde{\nabla})$ when the potential vector field coincides with the Reeb vector field ξ of M. We further consider the case where the potential vector field V of (g, V, μ, ρ) in $(M, \widetilde{\nabla})$ is pointwise collinear with ξ, that is, $V = h\xi$ for the smooth function h in M. In this setting, we show that $(M, \widetilde{\nabla})$ is a quasi-Einstein manifold. We also discuss the conditions under which the function h is harmonic, subharmonic, or superharmonic. In Section 4, we investigate the properties of gradient almost Ricci–Bourguignon solitons on $(M, \widetilde{\nabla})$.

2 Preliminaries

Let ξ be a unit vector field of on n-dimensional Riemannian manifold endowed with a semisymmetric ξ-connection $(M, \widetilde{\nabla})$. ξ is named as the Reeb vector field of $(M, \widetilde{\nabla})$. Suppose η is a 1-form associated with ξ as

$$g(X, \xi) = \eta(X), \quad \eta(\xi) = 1, \tag{1}$$

for arbitrary vector field X on M. Let $\widetilde{\nabla}$ and ∇ denote, respectively, the semisymmetric metric connection and the Levi-Civita connection on M. Then, we have

$$\widetilde{\nabla}_X Y = \nabla_X Y + \eta(Y)X - g(X, Y)\xi \tag{2}$$

for all $X, Y \in \mathfrak{X}(M)$ [8]. Here, $\mathfrak{X}(M)$ is the set of all smooth vector fields of M. A semisymmetric metric connection $\widetilde{\nabla}$ on M is said to be a semisymmetric metric ξ-connection [3] if $\widetilde{\nabla}\xi = 0$, which is equivalent to

$$\nabla\xi = -I + \eta \otimes \xi, \tag{3}$$

where I is the unit transformation and '$\otimes$' denotes the tensor product. This vector field has been studied by several authors; see, for instance, [4, 6, 7, 20–23] and the references therein. Note that the structure $(M, \widetilde{\nabla})$, that is, M satisfies equation (3) is a particular case of the Riemannian concircular structure manifold ($\alpha = -1$) [5, 9]. From (3), we note that

$$(\nabla_X \eta)(Y) = g(\nabla_X \xi, Y) = -g(X, Y) + \eta(X)\eta(Y), \tag{4}$$

for $X, Y \in \mathfrak{X}(M)$. This equation shows that the 1-form η is closed, that is, $d\eta = 0$.

It is observed in [3, 6, 7] that $(M, \tilde{\nabla})$ satisfies the following curvature identities

$$R(X,Y)\xi = \eta(X)Y - \eta(Y)X, \tag{5}$$

$$R(\xi,X)Y = \eta(Y)X - g(X,Y)\xi, \tag{6}$$

$$\eta(R(X,Y)Z) = \eta(Y)g(X,Z) - \eta(X)g(Y,Z) \tag{7}$$

$\forall X, Y, Z \in \mathfrak{X}(M)$, R represents the Riemann curvature tensor of M. The contraction of equation (5) along the vector field X gives

$$S(Y,\xi) = -(n-1)\eta(Y) \Longleftrightarrow Q\xi = -(n-1)\xi, \tag{8}$$

where $S(X,Y) = \sum_{i=1}^{n} g(R(e_i,X)Y,e_i) = g(QX,Y)$ for a set of orthonormal vector fields $e_i, e_2, \ldots, e_n$ on M. Here, S and Q are named as the Ricci tensor and the Ricci operator of g.

From equation (8), we observe that $-(n-1)$ is an eigenvalue of the Ricci operator Q corresponding to the eigenvector ξ. The covariant derivative of equation (8) along an arbitrary vector field X gives

$$(\nabla_X S)(Y,\xi) = S(X,Y) + (n-1)g(X,Y), \tag{9}$$

which is equivalent to

$$(\nabla_X Q)(\xi) = QX + (n-1)X, \tag{10}$$

since equations (1), (3), (4), and (8) are used. From equation (10), it is obvious that $\xi(r) = 2(r + n(n-1))$, which infers that the scalar curvature r is constant if and only if $r = -n(n-1)$.

3 Almost Ricci–Bourguignon soliton on $(M, \tilde{\nabla})$

Let (g, V, μ, ρ) denote an almost Ricci–Bourguignon soliton on $(M, \tilde{\nabla})$ and $V = \xi$. Therefore

$$(\pounds_\xi g)(X,Y) = g(\nabla_X \xi, Y) + g(X, \nabla_Y \xi) = -2[g(X,Y) - \eta(X)\eta(Y)], \tag{11}$$

where equations (1) and (3) are used. The almost Ricci–Bourguignon soliton equation on $(M, \tilde{\nabla})$ for $V = \xi$ is given as

$$(\pounds_\xi g)(X,Y) + 2S(X,Y) = 2(\mu + \rho r)g(X,Y),$$

which reduces to

$$S(X,Y) = -(\mu + \rho r)g(X,Y) - \eta(X)\eta(Y), \tag{12}$$

$\forall X, Y \in \mathfrak{X}(M)$. Equation (12) shows that the manifold $(M, \tilde{\nabla})$ admitting an almost Ricci–Bourguignon soliton (g, ξ, λ) is a special quasi-Einstein manifold with the associated scalars $a = -(\mu + \rho r)$ and $b = -1$. For more details, see [10–13]. Contracting equation (12) over X and Y, we obtain

$$r = (\mu + \rho r + 1)n - 1. \tag{13}$$

Again, take $Y = \xi$ in equation (12) and following equations (1) and (8) we obtain

$$\mu + \rho r = -(n-1). \tag{14}$$

The last two equations give

$$r = -n(n-2) - 1 = \text{constant}.$$

It is well-known that the scalar curvature of $(M, \tilde{\nabla})$ satisfies the identity $\xi(r) = 2(r+n(n-1))$. Since r is constant and therefore after comparing the scalar curvature of $(M, \tilde{\nabla})$, we have $n = 1$, which is inadmissible. Thus, we can write:

Theorem 1. *There does not exist an almost Ricci–Bourguignon soliton $(g, V = \xi, \mu, \rho)$ on $(M, \tilde{\nabla})$.*

Let the soliton vector V of an almost Ricci–Bourguignon soliton (g, V, μ, ρ) be pointwise collinear with the Reeb vector field ξ of $(M, \tilde{\nabla})$. That is, $V = h\xi$. The covariant derivative of this equation along X gives

$$\nabla_X V = X(h)\xi - h(X - \eta(X)\xi). \tag{15}$$

Using equation (15) in

$$\begin{aligned}(\pounds_V g)(X,Y) &= g(\nabla_X V, Y) + g(X, \nabla_Y V)\\ &= X(h)\eta(Y) + Y(h)\eta(X) - 2h\{g(X,Y) - \eta(X)\eta(Y)\}.\end{aligned} \tag{16}$$

The almost Ricci–Bourguignon soliton equation

$$(\pounds_V g)(X,Y) + 2S(X,Y) = 2(\mu + \rho r)g(X,Y),$$

becomes

$$X(h)\eta(Y) + Y(h)\eta(X) - 2h\{g(X,Y) - \eta(X)\eta(Y)\} + 2S(X,Y) = 2(\mu + \rho r)g(X,Y). \tag{17}$$

Let us consider $Y = \xi$ in the above equation. Then, we have

$$X(h) + \xi(h)\eta(X) - 2(n-1)\eta(X) = 2(\mu + \rho r)\eta(X), \tag{18}$$

where equation (8) is used. Again, taking $X = \xi$, we find

$$\xi(h) = \mu + \rho r + (n-1). \tag{19}$$

Using equation (19) in equation (18), we find

$$X(h) = \xi(h)\eta(X). \tag{20}$$

Using equations (19) and (20) in equation (17), we obtain

$$S(X,Y) = (\mu + \rho r + h)g(X,Y) - (h + \xi(h))\eta(X)\eta(Y), \tag{21}$$

which shows that $(M, \tilde{\nabla})$ admits an almost Ricci–Bourguignon soliton (g, V, μ, ρ) with $V = f\xi$ a special type of quasi-Einstein manifold. Thus, we state the following:

Theorem 2. *An $(M, \tilde{\nabla})$ admit an almost Ricci–Bourguignon soliton (g, V, μ, ρ) with $V = h\xi$ is a quasi-Einstein manifold and Ricci tensor S satisfies equation* (21).

If possible, we suppose that $\xi(h) + h = 0$, then $(M, \tilde{\nabla})$ under consideration is an Einstein manifold. Now, we write the following:

Corollary 1. *Let $(M, \tilde{\nabla})$ admit an almost Ricci–Bourguignon soliton (g, V, μ, ρ), where the potential vector field is of the form $V = h\xi$. If the function h satisfies $\xi(h) + h = 0$, then (M, g) is an Einstein manifold.*

From equation (20), we have

$$Dh = \xi(h)\xi.$$

Hence, the gradient of the smooth function h is pointwise collinear with the Reeb vector field ξ. By taking the covariant derivative of the above equation in the direction of X, we obtain

$$\nabla_X Dh = X(\xi(h)\xi + \xi(h)\nabla_X \xi, \tag{22}$$

which after contraction over X gives

$$\triangle h = \xi(\xi(h) - (n-1)\xi(h), \tag{23}$$

since $\operatorname{div} \xi = -(n-1)$. Here, $\triangle$ denotes the Laplacian operator of g. The Laplacian operator associated with a Riemannian metric g, known as the Laplace–Beltrami operator, is defined for a smooth function $\mathfrak{f} \in C^\infty(M)$ by

$$\Delta_g \mathfrak{f} = \operatorname{div}_g(\nabla \mathfrak{f}).$$

It plays a fundamental role in differential geometry and geometric analysis. The Laplacian arises naturally in the study of heat and diffusion processes, spectral geometry, and harmonic functions on Riemannian manifolds. Moreover, it appears in the evolution equations of curvature quantities in geometric flows such as the Ricci flow and the Ricci–Bourguignon flow. The operator is also an essential tool in establishing rigidity and comparison results under suitable curvature conditions and has important applications in mathematical physics.

Let $\xi = \frac{\partial}{\partial t}$. Then, equation (23) becomes

$$\triangle h = \frac{\partial}{\partial t}\left(\frac{\partial h}{\partial t} - (n-1)h\right). \tag{24}$$

- *Spatial diffusion:* The Laplacian $\triangle h$ on the left-hand side represents the diffusion of the quantity h in space. In physical terms, it measures how h spreads out from regions of high concentration to regions of low concentration, as in heat conduction.
- *Temporal evolution with damping:* Expanding the right-hand side gives

$$\frac{\partial^2 h}{\partial t^2} - (n-1)\frac{\partial h}{\partial t},$$

  which corresponds to a second-order time evolution with a damping term proportional to $(n-1)$. The term $\frac{\partial^2 h}{\partial t^2}$ models inertial or wave-like behavior, while $-(n-1)\frac{\partial h}{\partial t}$ introduces dissipation, causing the system to relax over time.
- *Role of dimension:* The factor $(n-1)$ shows that the damping depends on the dimension of the space. Higher dimensions lead to stronger damping, causing faster decay of temporal variations.

In summary, equation (24) describes a quantity h that diffuses spatially, evolves dynamically in time, and experiences dimension-dependent damping. Such behavior arises in models of heat propagation with memory, viscoelastic media, and certain geometric evolution equations where spatial diffusion is coupled with temporal dissipation.

Let $\mathcal{F}$ be a smooth function of an n-dimensional Riemannian manifold M. Then, $\mathcal{F}$ is

(i) harmonic if and only if $\triangle\mathcal{F} = 0$;
(ii) superharmonic if and only if $\triangle\mathcal{F} \leq 0$;
(iii) subharmonic if and only if $\triangle\mathcal{F} \geq 0$.

The above definition together with equation (24) states the following:

Theorem 3. *Let $(M, \tilde{\nabla})$ admit an almost Ricci–Bourguignon soliton (g, V, μ, ρ). If the soliton vector V is pointwise collinear with the Reeb vector ξ of M, that is, $V = h\xi$, for some smooth function h, then h satisfies the partial differential equation $\triangle h = \frac{\partial}{\partial t}(\frac{\partial h}{\partial t} - (n-1)h)$. Also, the smooth h is*

(i) *harmonic if* $\frac{\partial h}{\partial t} - (n-1)h = constant;$
(ii) *superharmonic if* $\frac{\partial}{\partial t}(\frac{\partial h}{\partial t} - (n-1)h) \leq 0;$
(iii) *subharmonic if* $\frac{\partial}{\partial t}(\frac{\partial h}{\partial t} - (n-1)h) \geq 0.$

These functions play a fundamental role in both geometry and analysis. Some important applications include:

1. *Potential theory:* Harmonic functions generalize the notion of potential fields in physics, modeling steady-state heat, gravitational, or electrostatic potentials on curved spaces.
2. *Maximum principle:* Subharmonic and superharmonic functions satisfy comparison principles, which are crucial for uniqueness results in partial differential equations and for proving rigidity theorems in geometry.
3. *Geometric analysis:* Harmonic functions are used to construct harmonic maps, study minimal surfaces, and analyze the geometry of manifolds via Hodge theory.
4. *Spectral geometry:* Eigenfunctions of the Laplacian, which generalize harmonic functions, provide information about the manifold's shape, volume, and curvature properties.
5. *Brownian motion and probability:* On Riemannian manifolds, harmonic functions describe expected values of stochastic processes, while subharmonic and superharmonic functions help estimate exit probabilities and potential theory in stochastic analysis.
6. *Singularity and boundary behavior:* Subharmonic and superharmonic functions are used to study the behavior of solutions near singularities or boundaries, important in both PDEs and geometric flows.

From equation (19), we have

$$\mu = \xi(h) - (n-1) - \rho r. \tag{25}$$

Let $\xi = \frac{\partial}{\partial t}$. Then, the above equation reduces to

$$\mu = \frac{\partial h}{\partial t} - (n-1) - \rho r.$$

Thus, we state:

Corollary 2. *Let the Riemannian metric g of an $(M, \tilde{\nabla})$ be an almost Ricci–Bourguignon soliton (g, V, μ, ρ). If V is pointwise collinear with the Reeb vector ξ, that is, $V = h\xi$ for smooth function h, then (g, V, μ, ρ) is expanding, shrinking or steady if $\frac{\partial h}{\partial t} < \rho r + (n-1)$, $\frac{\partial h}{\partial t} > \rho r + (n-1)$, or $\frac{\partial h}{\partial t} = \rho r + (n-1)$, respectively.*

Contracting equation (21) over the vector fields X and Y, we find

$$r = (n-1)[\xi(h) + h - n] = (n-1)\left[\frac{\partial h}{\partial t} + h - n\right], \tag{26}$$

where r is a scalar quantity (such as scalar curvature), h is a smooth function, ξ is a vector field (the Reeb vector field), and n is the dimension of the manifold.

- *Curvature measure:* The quantity r represents a scalar curvature-type invariant that depends on both the local variation of h along the vector field ξ and the value of h itself.
- *Rate of change term:* The derivative $\xi(h) = \frac{\partial h}{\partial t}$ measures the rate of change of h along the flow generated by ξ. Physically, this represents how the "density" or "potential" h evolves in the direction of the preferred vector field.
- *Source/damping term:* The term $h-n$ acts as a source or damping contribution. When $h > n$, it increases r, corresponding to expansion or increased curvature; when $h < n$, it decreases r, corresponding to contraction or decreased curvature.
- *Dimension scaling:* The factor $(n-1)$ amplifies the contribution of the terms in higher dimensions, reflecting how local changes in h influence the scalar r more strongly as the manifold's dimension increases.

In summary, this equation shows that the scalar quantity r is determined by a combination of the variation of h along a distinguished vector field and its deviation from a baseline value n, with the effect amplified by the dimension of the manifold.

4 Gradient almost Ricci–Bourguignon solution on $(M, \tilde{\nabla})$

Let $(M, \tilde{\nabla})$ denote an n-dimensional Riemannian manifold endowed with a semisymmetric metric ξ-connection. Suppose $(M, \tilde{\nabla})$ admits a gradient almost Ricci–Bourguignon solution. Therefore

$$\nabla_X Df = -QX + (\mu + \rho r)X. \tag{27}$$

Taking the covariant derivative of equation (27) with respect to Y, we obtain

$$\nabla_Y \nabla_X Df = -(\nabla_Y Q)(X) - Q(\nabla_Y X) + Y(\lambda)X + \lambda \nabla_Y X, \tag{28}$$

where $\lambda = \mu + \rho r$. Interchanging X and Y in equation (28), we obtain

$$\nabla_X \nabla_Y Df = -(\nabla_X Q)(Y) - Q(\nabla_X Y) + X(\lambda)Y + \lambda \nabla_X Y. \tag{29}$$

Using equations (27)–(29) in the curvature identity

$$R(X,Y)Df = \nabla_X\nabla_Y Df - \nabla_Y\nabla_X Df - \nabla_{[X,Y]}Df,$$

we find

$$R(X,Y)Df = (\nabla_Y Q)(X) - Q(\nabla_X Q)(Y) - Y(\lambda)X - X(\lambda)Y. \tag{30}$$

Contracting equation (30) over X, we obtain

$$S(Y,Df) = \frac{1}{2}Y(r) + (n-1)Y(\lambda). \tag{31}$$

The inner product of equation (30) with the Reeb vector field ξ gives

$$g(R(X,Y)Df,\xi) = (\nabla_Y S)(X,\xi) - (\nabla_X S)(Y,\xi) + Y(\lambda)\eta(X) - \eta(Y)X(\lambda). \tag{32}$$

Equation (8) can be rewritten as

$$g(R(X,Y)Df,\xi) = \eta(Y)X(f) - \eta(X)Y(f). \tag{33}$$

Equating equation (32) and equation (33), we obtain

$$(\nabla_Y S)(X,\xi) - (\nabla_X S)(Y,\xi) + Y(\lambda)\eta(X) - X(\lambda)(Y) = \eta(Y)X(f) - \eta(X)Y(f). \tag{34}$$

Equations (9) and (10) lead to

$$(\nabla_X S)(Y,\xi) - (\nabla_Y S)(X,\xi) = 0. \tag{35}$$

Using equation (35) in equation (34), we obtain

$$Y(\lambda)\eta(X) - \eta(Y)X(\lambda) = X(f)\eta(Y) - Y(f)\eta(X). \tag{36}$$

Setting $Y = \xi$ in equation (36), we find

$$\xi(\lambda)\eta(X) - X(\lambda) = X(f) - \xi(f)\eta(X). \tag{37}$$

- *Interpretation of terms:*
  - $\xi(\lambda)\,\eta(X)$ measures the rate of change of λ along the preferred direction ξ, projected along X.
  - $-X(\lambda)$ is the variation of λ in the X-direction.
  - $X(f) - \xi(f)\,\eta(X)$ is the analogous combination for the function f.
- *Overall meaning:* The equation expresses that the difference between the directional change along ξ and the gradient along X for λ is exactly balanced by the corresponding variation in f. Geometrically, it implies that

$$d(\lambda - f) = (\xi(\lambda) - \xi(f))\,\eta,$$

meaning that $\lambda - f$ is constant along directions orthogonal to ξ and varies only along the flow of ξ.

- *Intuitive physical picture:* If ξ represents a preferred flow (e. g., vertical flow of a fluid), then the equation states that the variation of λ in horizontal directions is completely determined by its evolution along the flow and by the variation of f. In other words, the system is *uniform in directions orthogonal to the main flow*, and all changes occur along ξ.

Consider the equation

$$\xi(\lambda)\,\eta(X) - X(\lambda) = X(f) - \xi(f)\,\eta(X),$$

where $\lambda, f \in C^{\infty}(M)$, ξ is a distinguished vector field, η is a 1-form with $\eta(\xi) = 1$, and X is any vector field on M.

The above equation can be rewritten as

$$X(f - \lambda) = \eta(X)(\xi(f) - \xi(\lambda)), \quad \forall X \in \mathfrak{X}(M),$$

which gives

$$d(f - \lambda) = (\xi(f) - \xi(\lambda))\,\eta.$$

Geometric interpretation

- For X orthogonal to ξ, $\eta(X) = 0$ implies $X(f - \lambda) = 0$. Thus, $f - \lambda$ is constant along directions orthogonal to ξ.
- Along the flow of ξ, we have $\xi(f - \lambda) = \xi(f) - \xi(\lambda)$, so $f - \lambda$ may vary only along ξ.

General solution

Let $\phi = f - \lambda$. Then

$$\phi = \phi_0 + \psi(s),$$

where ϕ_0 is constant on leaves orthogonal to ξ, s is the parameter along the flow of ξ, and ψ is a smooth function of s. In other words, the solution is constant in directions orthogonal to ξ and varies only along the flow of ξ.

This shows that the original PDE reduces to an ODE along the distinguished vector field ξ, capturing the behavior of f and λ along a preferred direction on the manifold.

Conflicts of interest

The authors declare no conflicts of interest.

Data availability

Data sharing is not applicable to this chapter as no datasets were generated or analyzed during the current study.

Bibliography

[1] Hamilton, R. S. (1982). Three-manifolds with positive Ricci curvature. *Journal of Differential Geometry, 17*, 255–306.
[2] Bourguignon, J.-P. (1981). Ricci curvature and conformal geometry. *Inventiones Mathematicae, 63*, 475–497.
[3] Chaubey, S. K., Lee, J. W., & Yadav, S. K. (2019). Riemannian manifolds with a semi-symmetric metric P-connection. *Journal of the Korean Mathematical Society, 56*, 1113–1129.
[4] Chaubey, S. K., Suh, Y. J., & De, U. C. (2020). Characterizations of the Lorentzian manifolds admitting a type of semi-symmetric metric connection. *Analysis and Mathematical Physics, 10*, 15.
[5] Chaubey, S. K., & Suh, Y. J. (2022). Riemannian concircular structure manifolds. *Filomat, 36*(19), 6699–6711.
[6] Chaubey, S. K., & De, U. C. (2020). Characterization of three-dimensional Riemannian manifolds with a type of semi-symmetric metric connection admitting Yamabe soliton. *Journal of Geometry and Physics, 157*, 103846.
[7] Chaubey, S. K., & De, U. C. (2022). Three-dimensional Riemannian manifolds and Ricci solitons. *Quaestiones Mathematicae, 45*(5), 765–778.
[8] Yano, K. (1970). On semi-symmetric metric connections. *Revue Roumaine de Mathématiques Pures et Appliquées, 15*, 1579–1586.
[9] Chen, B.-Y., Chaubey, S. K., De, U. C., et al. (2025). Almost Ricci solitons on Riemannian concircular structure 3-manifolds. *Ricerche di Matematica.*
[10] Chaubey, S. K. (2019). Certain results on $N(k)$-quasi Einstein manifolds. *Afrika Mathematica, 30*(1–2), 113–127.
[11] Chaubey, S. K., Bhaishya, K. K., & Siddiqi, M. D. (2021). Existence of some classes of $N(k)$-quasi Einstein manifolds. *Boletim da Sociedade Paranaense de Matemática, 39*(5), 145–162.
[12] Deszcz, R., Glogowska, M., Hotloś, M., & Sentürk, Z. (1998). On certain quasi-Einstein semisymmetric hypersurfaces. *Annales Universitatis Scientiarum Budapestinensis, Sectio Mathematica, 41*, 151–164.
[13] Chaki, M. C., & Maithy, R. K. (2000). On quasi Einstein manifolds. *Publicationes Mathematicae Debrecen, 57*(3–4), 297–306.
[14] Catino, G., & Mazzieri, L. (2016). Gradient Einstein solitons. *Nonlinear Analysis, 132*, 66–94.
[15] Catino, G., Cremaschi, L., Djadli, Z., Mantegazza, C., & Mazzieri, L. (2017). The Ricci–Bourguignon flow. *Pacific Journal of Mathematics, 287*(2), 337–370.
[16] Chaubey, S. K., & Suh, Y. J. (2023). Ricci–Bourguignon solitons and Fischer–Marsden conjecture on generalized Sasakian-space-forms with β-Kenmotsu structure. *Journal of the Korean Mathematical Society, 60*(2), 341–358.
[17] Dwivedi, S. (2020). Some results on Ricci–Bourguignon solitons and almost solitons. *Canadian Mathematical Bulletin.*
[18] Fischer, A. E. (2004). An introduction to conformal Ricci flow. *Classical and Quantum Gravity, 21*(3), S171–S218.
[19] Ho, P. T. (2020). On the Ricci–Bourguignon flow. *International Journal of Mathematics, 31*, 2050044.
[20] Haseeb, A., Chaubey, S. K., Mofarreh, F., & Ahmadini, A. A. H. (2023). A solitonic study of Riemannian manifolds equipped with a semi-symmetric metric ξ-connection. *Axioms, 12*(9), 809.

[21] Li, Y., Siddesha, M. S., Kumara, H. A., & Praveena, M. M. (2024). Characterization of Bach and Cotton tensors on a class of Lorentzian manifolds. *Mathematics*, *12*(19), 3130.

[22] Haseeb, A., Chaubey, S. K., & Khan, M. A. (2023). Riemannian 3-manifolds and Ricci–Yamabe solitons. *International Journal of Geometric Methods in Modern Physics*, *20*(1), 2350015.

[23] De, U. C., Sardar, A., & De, K. (2022). Ricci–Yamabe solitons and 3-dimensional Riemannian manifolds. *Turkish Journal of Mathematics*, *46*(3), 1078–1088.

Suraj Roy and Sharmistha Bhattacharya (Halder)

Face recognition using fractal-based IF set oscillation in the SIFT domain

Abstract: There have been continuous developments in a number of recognition domains, with an emphasis on face identification for applications like security and forensics. Variations in emotions, lighting, position, occlusion, aging, and resolution are among the obstacles that face recognition encounters. In order to address uncertainty, the research presents a novel method dubbed IFO-FNDC, which combines fractal theory and intuitionistic fuzzy set oscillation in the SIFT domain. The technique consists of converting color images to gray, finding key feature points using SIFT, applying fractals to the images, and comparing known and unknown images. The algorithm performs better than other cutting-edge algorithms in terms of efficiency, runtime, and accuracy rate, making it suitable for security and forensic analysis applications when tested on the FacePix and FERET databases using MATLAB 2015a.

Keywords: Face recognition, intuitionistic fuzzy set, fractal neighbor distance, IF oscillation, scale-invariant feature transform

MSC 2020: 68T10, 03E72, 68U10, 62H35

1 Introduction

One essential component of human perceptual organisms' capacity is face recognition. By building a similar algorithmic reproduction of facial recognition, it is an unavoidable challenge for humans. A crucial biometric characteristic, the face has many uses in a variety of industries, including marketing, banking, entertainment, health, education, and security and law enforcement. The computational model of any face recognition system aids in both conceptual comprehension and a variety of practical applications, such as automated crowd monitoring, criminal identification, content-based image database management, and human–computer interface (HCI) access control design. The wide range of applications has encouraged academics worldwide to tackle facial recognition issues through interdisciplinary methods. In the past few years, fractal theory and intuitionistic fuzzy set oscillation have been applied to face recognition. Fractals are geometric shapes with intricate structures at arbitrary tiny scales. A logical statistical

Acknowledgement: The authors acknowledge Tripura University for providing the necessary research facilities.

Suraj Roy, Sharmistha Bhattacharya (Halder), Department of Mathematics, Tripura University, Agartala, Tripura, India, e-mails: surajroy.mathematics@gmail.com, s.bhattacharya@tripurauniv.ac.in

https://doi.org/10.1515/9783112234525-020

measure of a pattern's level of detail complexity is provided by the fractal dimension. The mathematician Benoit Mandelbrot came up with the term "fractal," which comes from the Latin word *fractus*, which means "broken" or "fractured". The authors of the paper [1] demonstrated the application of fractal image encoding for image feature extraction. For line segments, cubes, and squares, the Hausdorff–Besicovitch dimension and topological dimension are identical. However, there are objects for which this does not follow; such objects are known as fractals, and here it gives $d_{H_B} > d_T$.

L. Brouwer first proposed the idea of intuitionism at the turn of the twentieth century. Titani interpreted the intuitionistic fuzzy set in a different way. Gaisi Takeuti and Satako proposed an IF set in 1991. Maki first proposed the idea of an m_X-open set [2]. The concept of the fuzzy m_X framework, including both fuzzy minimal structure and fuzzy minimum space, was introduced by Alimohammady and Roohi [3–6]. The notion of minimal structure in a generic topological space was first introduced by Popa and Noiri [7] as exploratory research. It has been demonstrated that, for a fuzzy set to oscillate, the existence of at least two fuzzy open sets and two fuzzy closed sets is necessary.

1.1 Motivation

The motivation of this work arises from the limitations of conventional face recognition techniques in effectively handling intraclass variations and interclass similarities. By integrating intuitionistic fuzzy m_X^*-oscillation with fractal neighbor distance in the SIFT domain, the proposed method aims to achieve robust and discriminative face recognition performance.

In this chapter, the proposed face recognition algorithm exploits the concepts of fractal dimension, lacunarity at different key locations, and intuitionistic fuzzy m_X^*-operator oscillation. The effectiveness of the proposed technique was validated using MATLAB 2015a after testing on several face datasets with different key feature values. The experimental results indicate that, with an increase in the number of feature coordinates, the proposed algorithm consistently produces the desired and anticipated recognition performance.

2 Preliminaries

In this section a short preliminary discussion related to the contributing components of the proposed method is presented.

2.1 Fractal dimension and lacunarity

The Hausdorff–Besicovitch dimension measures the geometric complexity of a bounded set. For an image I, the fractal dimension is computed by embedding the image into a three-dimensional space where the third coordinate corresponds to pixel intensity

$$\mathrm{FD}_I = \frac{\log N(r)}{\log(1/r)}. \tag{1}$$

Fractal dimension is crucial for measuring texture and roughness [10].

2.2 Theoretical concepts related to fuzzy m_x^* oscillation

An intuitionistic fuzzy set, introduced by Atanassov, extends the classical fuzzy set by incorporating both membership and nonmembership degrees [13, 14]. Let E be a nonempty universe of discourse. An intuitionistic fuzzy set (IFS) A in E is defined as

$$A = \{\langle x, \mu_A(x), \nu_A(x)\rangle \mid x \in E\},$$

where the functions $\mu_A : E \to [0,1]$ and $\nu_A : E \to [0,1]$ represent the degrees of membership and nonmembership of the element $x \in E$, respectively, satisfying the condition

$$0 \le \mu_A(x) + \nu_A(x) \le 1, \quad \forall x \in E.$$

For each element $x \in E$, the quantity

$$\pi_A(x) = 1 - \mu_A(x) - \nu_A(x)$$

denotes the degree of indeterminacy or hesitation. It is evident that a classical fuzzy set is a special case of an intuitionistic fuzzy set, obtained when $\nu_A(x) = 1 - \mu_A(x)$ for all $x \in E$. Let I^X denote the collection of intuitionistic fuzzy sets defined on a universe X [8, 9]. An operator $\Omega^o : I^X \to I^X$ is called a *fuzzy m_X^*-open oscillatory operator* if it is defined by

$$\Omega^o_{a_j}(x) = \Lambda_{a_j}(x) - \mathrm{Int}_{a_j}(x),$$

where $\Lambda_{a_j}(x)$ and $\mathrm{Int}_{a_j}(x)$ represent the associated upper and lower operators corresponding to the attribute a_j. Similarly, an operator $\Omega^c : I^X \to I^X$ defined by

$$\Omega^c_{a_j}(x) = \mathrm{Cl}_{a_j}(x) - V_{a_j}(x)$$

is referred to as a *fuzzy m_X^*-closed oscillatory operator.*

For any object y in a fuzzy m_X^*-structure, if

$$\Omega^o_{a_j}(y) = \alpha,$$

then

$$\Omega^o_{a_j}(y^c) = -\alpha,$$

and conversely, where $\alpha \in I^X$.

If either $\tilde{1}$ or $\tilde{0}$ belongs to m_X^* with $\alpha \in m_X^*$, where α is an intuitionistic fuzzy set, then a subfamily of intuitionistic fuzzy sets in X is said to form an *intuitionistic fuzzy m_X^*-structure* associated with intuitionistic fuzzy m_X^*-oscillation.

The operator $\Lambda : I^X \to I^X$ is defined by

$$\Lambda_{a_j}(x) = \sup_{\nu} \inf_{\mu}\{\langle \mu_{a_j}(x_i), \nu_{a_j}(x_i)\rangle : \mu_{a_j}(x_i) \geq \mu_{a_j}(x),\ \nu_{a_j}(x_i) \leq \nu_{a_j}(x),\ x_i \in G\},$$

where G is an open set and $j = 1, 2, \ldots, n$. If no such open set exists, $\Lambda_{a_j}(x) = \hat{I}$.

Similarly, the interior operator is given by

$$\text{Int}_{a_j}(x) = \inf_{\nu} \sup_{\mu}\{\langle \mu_{a_j}(x_i), \nu_{a_j}(x_i)\rangle : \mu_{a_j}(x_i) \leq \mu_{a_j}(x),\ \nu_{a_j}(x_i) \geq \nu_{a_j}(x),\ x_i \in G\},$$

where G is a closed set. If no such closed set exists, $\text{Int}_{a_j}(x) = \varnothing$.

Here, $\mu_{a_j}(x_i)$ and $\nu_{a_j}(x_i)$ denote the membership and nonmembership values, respectively, of the attribute a_j corresponding to the object x_i.

An *intuitionistic fuzzy m_X^*-oscillator operator* is defined by

$$\Omega^o_{a_j}(x) = \Lambda_{a_j}(x) - \text{Int}_{a_j}(x)$$

and can be expressed as

$$\langle \Lambda_{\mu_{a_j}}(x_i) - \text{Int}_{\nu_{a_j}}(x_i),\ \Lambda_{\nu_{a_j}}(x_i) - \text{Int}_{\mu_{a_j}}(x_i)\rangle.$$

In applications, the magnitude of these values is considered rather than their sign

$$\text{FD}_I = \frac{\log N(r)}{\log(1/r)}. \tag{2}$$

Fractal dimension is crucial for measuring texture and roughness [18, 19].

2.3 Scale-invariant feature transform (SIFT)

SIFT, developed by David Lowe [15, 16], is invariant to scale, rotation, and illumination changes. The four major stages are scale-space extrema detection, key-point localization, orientation assignment, and descriptor computation.

3 IF m_x^* oscillation based on fractal neighbor distance on a SIFT domain approach for face recognition

This section presents a face recognition approach based on *intuitionistic fuzzy m_x^*-oscillation* combined with *fractal neighbor distance (FND)* classification. An image database containing M facial images is used for training and testing. For each image, key feature points are extracted from fixed coordinates using the SIFT algorithm (Figure 4). Images belonging to the same individual are grouped into a single class during training.

For each subject, five images are randomly selected to construct the training library, while the remaining images are used for testing. The training set consists of multiple classes such as $A = \{I_{a1}, I_{a2}, I_{a3}, I_{a4}, I_{a5}\}$, $B = \{I_{b1}, I_{b2}, I_{b3}, I_{b4}, I_{b5}\}$, ..., $O = \{I_{o1}, I_{o2}, I_{o3}, I_{o4}, I_{o5}\}$. To reduce the number of candidate classes for comparison, the fractal dimension, weighted fractal dimension, and lacunarity of the key feature points are evaluated, as illustrated in Figure 1.

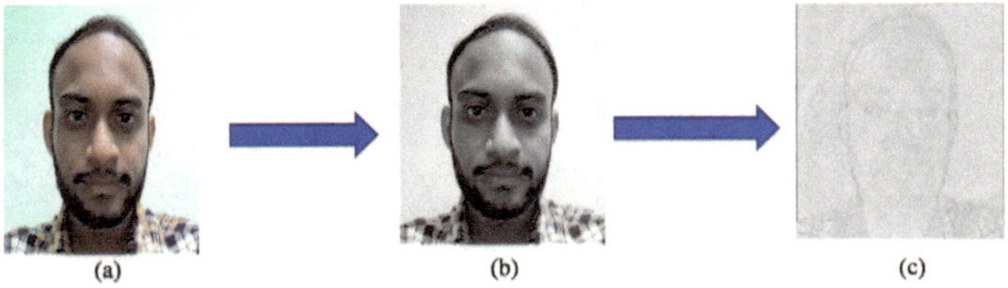

Figure 1: Fractal transformation of original image: (a) Original image, (b) Gray-scale image, (c) Fractally transformed image.

A class-level threshold K_1 is introduced. For convenience, only the first image of each class is initially considered. Let I_P denote the input image. The threshold value is computed as

$$\begin{aligned} \mathrm{T_h} &= \min\{d_{FN}(I_P, I_{a1}), d_{FN}(I_P, I_{b1}), \ldots, d_{FN}(I_P, I_{o1})\} \\ &= \min\{d(f(I_P), f(I_{a1})), d(f(I_P), f(I_{b1})), \ldots, d(f(I_P), f(I_{o1}))\}. \end{aligned} \tag{3}$$

Following [11, 12], the class selection threshold is set as $K_1 = T_h + 1.3$. Only those classes satisfying this condition are retained for further comparison.

To perform within-class matching, the proposed method employs intuitionistic fuzzy m_x^*-oscillation. The overall workflow of the algorithm is shown in Figure 2 and is summarized as follows:

1. *Feature extraction:* Membership and nonmembership pixel values are computed using the extracted feature values. The same coordinates are used for both the input image and the training images.

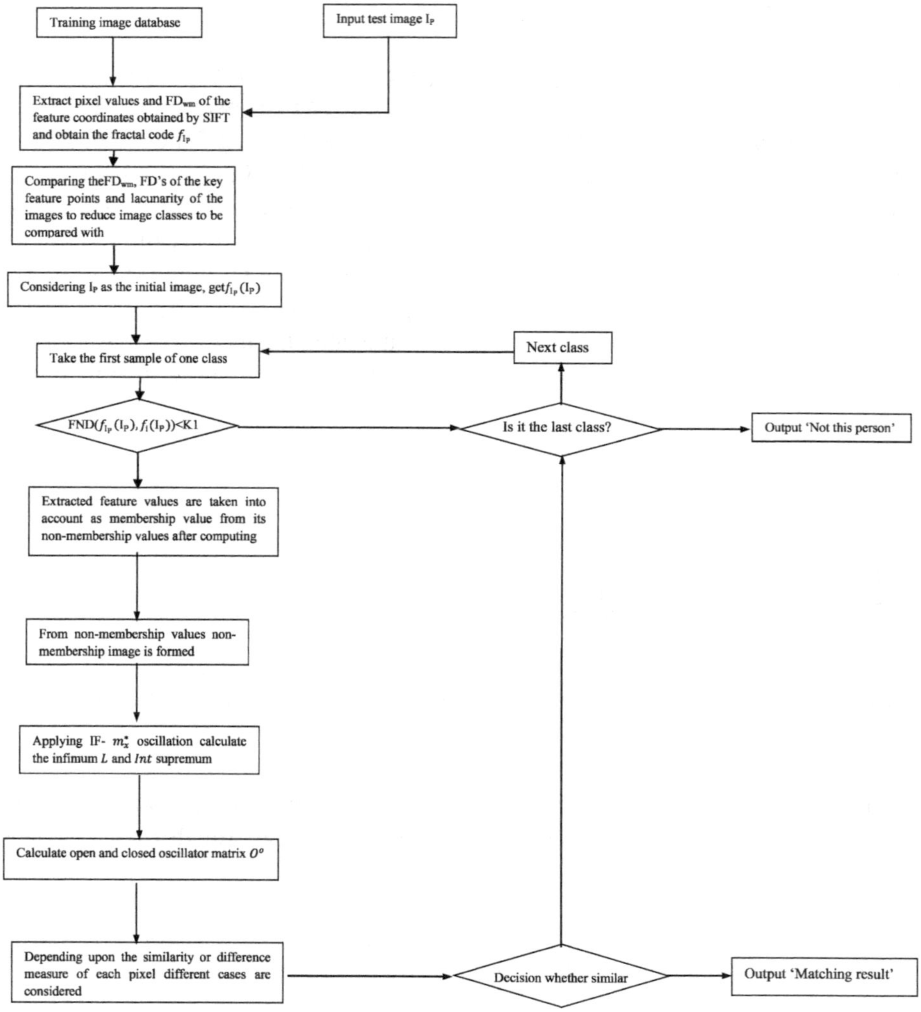

Figure 2: Flow chart of IFO-FNDC-based face recognition.

2. *Operator computation:* The Λ and Int operators are evaluated. The Λ operator is obtained by comparing the feature matrices of the training images and the input image based on lower nonmembership values, while the Int operator is computed using the corresponding supremum values.
3. *Oscillation evaluation:* Open oscillatory values ($\Lambda_{J_\mu} - \mathrm{Int}_{J_\mu}$) from membership images and closed oscillatory values ($\Lambda_{J_\nu} - \mathrm{Int}_{J_\nu}$) from nonmembership images are calculated.
4. *Similarity assessment:* Four possible similarity cases are analyzed using the computed oscillatory values, where pixelwise comparisons determine the final similarity and dissimilarity measures.

4 Experimental performance discussion

A set of facial images was collected for experimental evaluation and used as training and query images (Figure 3). The database contains images of multiple individuals captured under different pose variations. All images are stored in true-color `.jpg` format. The proposed method was implemented and tested using MATLAB 7.9.

Figure 3: Image database used as training and query sets.

Initially, a small number of feature coordinates were manually selected from frontal face images. These coordinates correspond to the centers of the left and right eyes, the nose tip, and the two corners of the mouth, as shown in Figure 4. Pixel values were extracted at these fixed coordinates for each facial expression and used as feature descriptors.

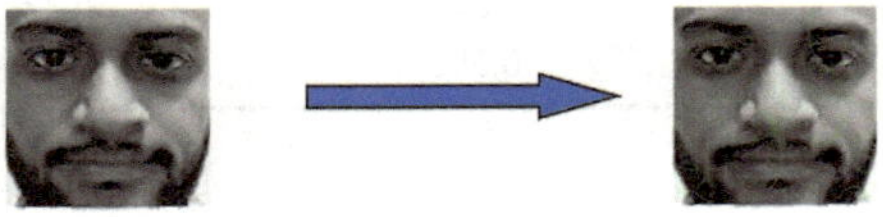

Figure 4: Key points detected after applying SIFT.

The performance of the proposed *intuitionistic fuzzy m_X^*-oscillation with fractal features in the SIFT domain* is evaluated and compared with existing fuzzy oscillation-based face recognition methods, namely Fuzzy m_X^* Oscillation [20], IF m_X^* Oscillation, and Fuzzy

m_x^* Oscillation with SIFT. As observed from Table 1, the recognition accuracy of the proposed method improves with an increase in the number of feature coordinates. Furthermore, Figure 5 shows a strong linear relationship with a coefficient of determination $R^2 = 0.9942$. From Table 2, it is evident that the proposed algorithm outperforms the compared methods.

Table 1: Accuracy table for IF oscillation in the SIFT domain.

Query set	Features	Identical	Different	Accuracy
20	20	7	43	0.896
25	20	11	38	0.910
30	30	4	36	0.931
35	30	12	38	0.945
40	40	3	37	0.959
45	40	2	41	0.972

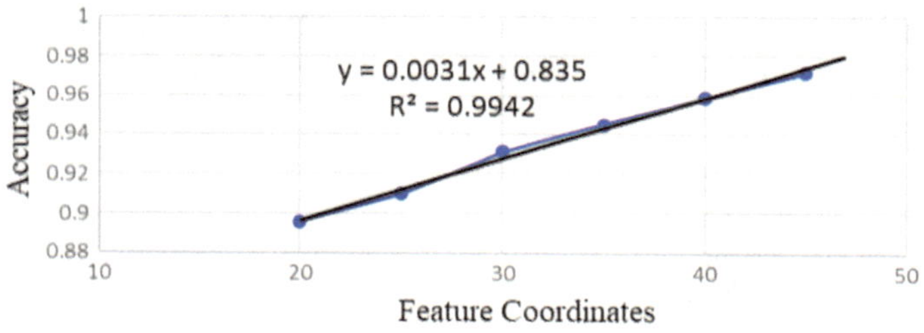

Figure 5: Performance of IF oscillation in SIFT.

Table 2: Various fuzzy oscillation-based algorithms compared for accuracy with the proposed IF m_x^* oscillation in the SIFT domain.

Query set	Feature coordinates	Fuzzy m_x^* oscillation	IF m_x^* oscillation	Fuzzy m_x^* oscillation with SIFT	IF m_x^* oscillation with fractal in SIFT
12	25	0.615	0.751	0.762	0.798
16	35	0.753	0.812	0.847	0.876
19	45	0.865	0.894	0.953	0.976

Finally, the proposed approach is further evaluated against LDA, PCA, FND, and FNDC using the FacePix [21] and FERET [22] databases and the outcomes are shown [17] in Table 3 and Table 4.

Table 3: Recognition rate (%) and recognition time (s) of each method on the FacePix database.

Method	Four training samples		Six training samples		Eight training samples	
	Rate	Time	Rate	Time	Rate	Time
LDA	86.9	0.115	89.5	0.169	92.3	0.193
PCA	86.2	0.128	87.9	0.182	90.7	0.227
FND	88.3	0.467	92.3	0.575	93.7	0.875
FNDC	88.0	0.211	92.1	0.226	94.5	0.268
IFO–FNDC	86.8	0.301	92.2	0.313	95.3	0.426

Table 4: Recognition rate (%) and recognition time (s) of each method on the FERET database.

Method	Three training samples		Four training samples		Five training samples	
	Rate	Time	Rate	Time	Rate	Time
LDA	42.3	2.78	61.9	4.39	66.3	6.87
PCA	41.7	3.56	57.3	9.51	62.8	12.97
FND	41.9	3.69	63.8	4.81	73.8	6.44
FNDC	42.3	2.29	65.5	1.72	73.5	3.09
IFO–FNDC	42.4	2.08	65.3	2.59	74.3	3.35

4.1 Future scope

Future work may extend the proposed IFO–FNDC framework to handle pose-invariant and occlusion-robust face recognition scenarios. The integration of adaptive feature selection and deep learning-based feature representations with intuitionistic fuzzy m_x^*-oscillation can further enhance recognition accuracy. Additionally, real-time implementation and evaluation on large-scale unconstrained face datasets remain promising directions for further research.

5 Conclusion

A novel and efficient face recognition framework based on the mathematical model of *intuitionistic fuzzy* m_x^* *oscillation with fractal neighbor distance* (IFO–FNDC) in the SIFT domain is presented in this study. The proposed approach is motivated by the effectiveness of intuitionistic fuzzy m_x^*-oscillation and fractal theory in capturing discriminative facial features. In the IFO–FNDC model, fractal neighbor distance is employed for between-class comparison, while intuitionistic fuzzy m_x^*-oscillation is used for within-class matching, thereby enhancing the performance of existing fractal-based face recognition methods. Experimental evaluations conducted on the FacePix and FERET databases demonstrate that the recognition accuracy of the proposed IFO–FNDC method consistently improves with an increase in the number of training samples.

Bibliography

[1] Ebrahimpour, H., Chandran, V., & Sridharan, S. (2004). An application of fractal image-set coding in facial recognition. In D. Zhang & A. K. Jain (Eds.), *Biometric Authentication* (pp. 178–186). Springer. https://doi.org/10.1007/978-3-540-25948-0_25.

[2] Maki, H. (1986). Generalized λ-sets and the associated closure operator. *Special issue in commemoration of Prof. Kazusada Ikeda's retirement*, 139–146.

[3] Alimohammady, M., & Roohi, M. (2006). Fuzzy minimal structure and fuzzy minimal vector spaces. *Chaos, Solitons & Fractals*, *27*, 599–605. https://doi.org/10.1016/j.chaos.2005.04.049.

[4] Alimohammady, M., & Roohi, M. (2006). Compactness in fuzzy minimal spaces. *Chaos, Solitons & Fractals*, *28*, 906–912. https://doi.org/10.1016/j.chaos.2005.08.043.

[5] Alimohammady, M., & Roohi, M. (2009). Extreme points in minimal spaces. *Chaos, Solitons & Fractals*, *39*, 1480–1485. https://doi.org/10.1016/j.chaos.2007.06.028.

[6] Alimohammady, M., Roohi, M., & Delavar, M. R. (2009). Transfer closed and transfer open multimaps in minimal spaces. *Chaos, Solitons & Fractals*, *40*, 1162–1168. https://doi.org/10.1016/j.chaos.2007.08.071.

[7] Popa, P. V., & Noiri, N. (2000). On *M*-continuous functions. *Analele Universității Dunărea de Jos Galați, Fasc. II*, *18*, 31–41.

[8] Bhattacharya, S. (Halder), & Roy, S. (2014). On fuzzy oscillation and its application in image processing. *Annals of Fuzzy Mathematics and Informatics*, *7*, 2093–9310.

[9] Bhattacharya, S. (Halder), & Roy, S. B. (2017). Application of IF set oscillation in the field of face recognition. *Pattern Recognition and Image Analysis*, *27*, 625–636. https://doi.org/10.1134/S1054661817030270.

[10] Al-Kadi, O. S., & Watson, D. (2008). Texture analysis of aggressive and nonaggressive lung tumor CE CT images. *IEEE Transactions on Biomedical Engineering*, *55*, 1822–1830. https://doi.org/10.1109/TBME.2008.919735.

[11] Tang, Z., Wu, X., Fu, B., Chen, W., & Feng, H. (2018). Fast face recognition based on fractal theory. *Applied Mathematics and Computation*, *321*, 721–730. https://doi.org/10.1016/j.amc.2017.11.017.

[12] Tang, Z., Wu, X., Leng, X., & Chen, W. (2017). A fast face recognition method based on fractal coding. *Signal, Image and Video Processing*, *11*, 1221–1228. https://doi.org/10.1007/s11760-017-1078-7.

[13] Atanassov, K. T. (1986). Intuitionistic fuzzy sets. *Fuzzy Sets and Systems*, *20*, 87–96. https://doi.org/10.1016/S0165-0114(86)80034-3.

[14] Atanassov, K. T. (1999). Intuitionistic fuzzy sets. In K. T. Atanassov (Ed.), *Intuitionistic Fuzzy Sets: Theory and Applications* (pp. 1–137). Physica-Verlag. https://doi.org/10.1007/978-3-7908-1870-3_1.

[15] Lowe, D. G. (1999). Object recognition from local scale-invariant features. In *Proceedings of the Seventh IEEE International Conference on Computer Vision* (Vol. 2, pp. 1150–1157). https://doi.org/10.1109/ICCV.1999.790410.

[16] Lowe, D. G. (2004). Distinctive image features from scale-invariant keypoints. *International Journal of Computer Vision*, *60*, 91–110. https://doi.org/10.1023/B:VISI.0000029664.99615.94.

[17] Ji, G., Zheng, H., Liu, Z., & Ren, X. (2008). Cell image matching based on fractal neighbor distance. In *International Conference on Natural Computation* (Vol. 4, pp. 609–612). https://doi.org/10.1109/ICNC.2008.399.

[18] Wang, J., & Zheng, N. (2013). A novel fractal image compression scheme with block classification and sorting based on Pearson's correlation coefficient. *IEEE Transactions on Image Processing*, *22*, 3690–3702. https://doi.org/10.1109/TIP.2013.2268977.

[19] Nikonorov, A. V., & Minaev, E. Yu. (2013). Fractal recognition of compact artifacts on color images. *Pattern Recognition and Image Analysis*, *23*, 455–458. https://doi.org/10.1134/S1054661813040111.

[20] Bhattacharya, S., Roy, S. B., & Saha, S. (2015). Application of fuzzy oscillation in the field of face recognition. In *Proceedings of the IEEE International Symposium on Advanced Computing and Communication (ISACC)* (pp. 192–197).

[21] FacePix Database. Centre for Ubiquitous Computing (CUbiC). https://cubic.asu.edu/content/facepix-database (Accessed July 17, 2023).
[22] Face Recognition Technology (FERET). National Institute of Standards and Technology (NIST). https://www.nist.gov/programs-projects/face-recognition-technology-feret (Accessed July 17, 2023).

Antim, Azad Rohilla, Abdallah M. Talafha, Subhasis Panda, and Sandeep Malik

Exploring analytical soliton solutions of the Kadomtsev–Petviashvili equation by a new Kudryashov method

Abstract: This chapter investigates a nonlinear extended Kadomtsev–Petviashvili (KP) equation, a fundamental model widely applied in Bose–Einstein condensates, nonlinear optics, plasma physics, and fluid dynamics. By employing the new Kudryashov technique, we systematically derive exact soliton solutions of the considered KP equation. The obtained results not only extend the applicability of the KP model in various physical contexts but also demonstrate the efficiency of the new Kudryashov method in handling nonlinear evolution equations.

Keywords: Kadomtsev–Petviashvili equation, new Kudryashov method, soliton solutions

MSC 2020: 35Q51, 35C07, 37K40

1 Introduction

Differential equations (DEs) play a central role in describing complex physical processes in fields such as nonlinear optics, atmospheric science, fluid dynamics, and plasma physics. Among these equations, the Kadomtsev–Petviashvili (KP) equation is a well-known, two-dimensional extension of the Korteweg–de Vries (KdV) equation and is widely used to model weakly dispersive and weakly nonlinear wave phenomena [1–3]. Its ability to capture multidimensional wave interactions makes it an important subject of mathematical and physical investigation. Soliton solutions of the KP equation are of particular interest because they describe stable, localized wave structures that retain their shape during propagation and interaction. These solutions provide valuable insights into nonlinear wave behavior and contribute to a deeper understanding of integrability properties, conservation laws, and the long-term evolution of nonlinear

Antim, Azad Rohilla, Department of Mathematics, Sanskaram University, Patauda-124108, Jhajjar, Haryana, India
Abdallah M. Talafha, Department of Mathematics and Natural Sciences, Prince Mohammad Bin Fahd University (PMU), AL Khobar, Saudi Arabia
Subhasis Panda, Department of Mathematics, School of Advanced Sciences, VIT-AP University, Amravati 522241, Andhra Pradesh, India
Sandeep Malik, Department of Mathematics, Akal University, Talwandi Sabo, Bathinda, Punjab 151302, India, e-mail: sndp796@gmail.com

https://doi.org/10.1515/9783112234525-021

systems. Finding exact soliton solutions also aids in validating numerical simulations and improving analytical techniques used in applied mathematics. There are several analytical methodologies to derive such solutions, such as the modified Jacobi elliptic function method [4], Lie symmetry analysis [5, 6], the modified auxiliary equation technique [7], the modified extended tanh function method [8], the Kudryashov auxiliary equation method [9, 10], the Kumar–Malik methodology [11, 12], and the $(G'/G, 1/G)$-expansion method [13]. Along with these methods, the new Kudryashov method is a very effective technique to solve nonlinear PDEs [14–17]. In this study, we will employ this technique to derive analytic solutions of an extended KP equation.

2 The new Kudryashov method

This section describes the algorithm of the new Kudryashov method [14, 15].

Let the given NLPDE be

$$\psi(u, u_t, u_x, u_{tt}, \dots) = 0, \tag{1}$$

where ψ denotes a polynomial function. Now, assume a traveling wave transformation

$$u(x,t) = U(\varpi), \quad \varpi = x - \chi t, \tag{2}$$

where χ is the wave velocity. By using equation (2), equation (1) can be reduced to the following nonlinear ODE:

$$G(U, U_{\varpi}, U_{\varpi\varpi}, U_{\varpi\varpi\varpi}, \dots) = 0. \tag{3}$$

We assume that equation (3) admits a solution of the form

$$U(\varpi) = \sum_{i=0}^{N} c_i Q^i(\varpi), \tag{4}$$

where $c_i, (i = 0, 1, \dots, N)$ are unknown constants of $Q^i(\varpi)$ with $c_N \neq 0$ and $Q(\varpi)$ is defined by

$$Q(\varpi) = \frac{1}{aA^{\Theta\varpi} + bA^{-\Theta\varpi}}, \tag{5}$$

which satisfies the auxiliary equation

$$(Q'(\varpi))^2 = (\Theta(\ln A)Q(\varpi))^2(1 - 4abQ^2(\varpi)). \tag{6}$$

Here, a, b, and A are nonzero constants, such that $A > 0$ and $A \neq 1$. The value of N is determined by the homogeneous balance principle. Substituting the expression (4) into equation (3) and equating the coefficients of same powers of $Q^i(\varpi)$ to zero (noting that $Q(\varpi) \neq 0$), yields a system of algebraic equations. Solving this system provides the unknown constants a, b, and c_is. Once these parameters are obtained, exact solutions can be constructed explicitly. Finally, substituting $\varpi = x - \chi t$ into the resulting expressions completes the solution process for equation (1).

3 The (2 + 1)-dimensional extended Kadomtsev–Petviashvili equation

The KP equation was formulated to model the dynamics of small-amplitude, long waves with a dominant direction of propagation. [18, 19]. While the KdV equation models purely one-dimensional wave propagation, the KP equation incorporates an additional spatial dimension, allowing a more realistic representation of physical systems where wavefronts exhibit lateral modulations. In 2018, Manukure et al. [20] proposed an extended KP equation

$$(u_t + 6uu_x + u_{xxx})_x - u_{yy} + \lambda u_{tt} + \mu u_{ty} = 0, \tag{7}$$

where λ and μ are nonzero constants. Suppose that equation (7) admits a traveling wave solution as

$$u(x,y,t) = U(\varpi), \quad \varpi = x + \psi y - \chi t, \tag{8}$$

where χ denotes the wave speed.

Substituting equation (8) into equation (7) leads to the following nonlinear ODE:

$$U_{\varpi\varpi\varpi\varpi} + 6UU_{\varpi\varpi} + (\lambda\chi^2 - (\mu\psi + 1)\chi - \psi^2)U_{\varpi\varpi} + 6U_{\varpi}^2 = 0. \tag{9}$$

Integrating equation (9) twice with respect to ϖ, yields

$$U_{\varpi\varpi} + 3U^2 + (\lambda\chi^2 - (\mu\psi + 1)\chi - \psi^2)U = 0, \tag{10}$$

where the integration constants are taken to be zero for simplicity. In equation (10), by balancing the terms $U_{\varpi\varpi}$ with U^2, gives $N = 2$. Consequently, substituting $N = 2$ into equation (4) gives

$$U_{\varpi} = c_0 + c_1 Q(\varpi) + c_2 Q(\varpi)^2. \tag{11}$$

Substituting equation (11) into equation (10) and equating the coefficient of $Q(\varpi)$ to zero provides

$$
\begin{aligned}
Q^0 &: 3c_0^2 - c_0(-\lambda\chi^2 + (\mu\psi + 1)\chi + \psi^2) = 0,\\
Q^1 &: c_1\Theta^2 \ln(A)^2 + 6c_0c_1 - c_1(-\lambda\chi^2 + (\mu\psi + 1)\chi + \psi^2) = 0,\\
Q^2 &: 4c_2\Theta^2 \ln(A)^2 + 6c_0c_2 + 3c_1^2 - c_2(-\lambda\chi^2 + (\mu\psi + 1)\chi + \psi^2) = 0,\\
Q^3 &: -8abc_1\Theta^2 \ln(A)^2 + 6c_1c_2 = 0,\\
Q^4 &: -24abc_2\Theta^2 \ln(A)^2 + 3c_2^2 = 0.
\end{aligned} \tag{12}
$$

On solving system (12), the following cases are obtained:

Case 1:

$$
\begin{aligned}
&c_0 = 0, \quad c_1 = 0, \quad c_2 = 8\,\Theta^2(\ln(A))^2 ab,\\
&\chi = \frac{\mu\psi + 1 \pm \sqrt{-16\ln(A)^2\Theta^2\lambda + \mu^2\psi^2 + 4\lambda\,\psi^2 + 2\mu\psi + 1}}{2\lambda}.
\end{aligned} \tag{13}
$$

By inserting equation (13) into equation (8), one obtains the following exact solution

$$
u(x, y, t) = \frac{8\,\Theta^2(\ln(A))^2 ab}{(aA^{\Theta\varpi} + bA^{-\Theta\varpi})^2}, \tag{14}
$$

where $\varpi = x + \psi y - (\frac{\mu\psi+1\pm\sqrt{-16\ln(A)^2\Theta^2\lambda+\mu^2\psi^2+4\lambda\,\psi^2+2\mu\psi+1}}{2\lambda})t.$

Set 2:

$$
\begin{aligned}
&c_0 = -\frac{4}{3}\,\Theta^2(\ln(A))^2, \quad c_1 = 0, \quad c_2 = 8\,\Theta^2(\ln(A))^2 ab,\\
&\chi = \frac{\mu\psi + 1 \pm \sqrt{16\ln(A)^2\Theta^2\lambda + \mu^2\psi^2 + 4\lambda\,\psi^2 + 2\mu\psi + 1}}{2\lambda}.
\end{aligned} \tag{15}
$$

By using equation (15) in equation (8), we obtain the following exact solution:

$$
u(x, y, t) = -\frac{4}{3}\,\Theta^2(\ln(A))^2 + \frac{8\,\Theta^2(\ln(A))^2 ab}{(aA^{\Theta\varpi} + bA^{-\Theta\varpi})^2}, \tag{16}
$$

where $\varpi = x + \psi y - (\frac{\mu\psi+1\pm\sqrt{16\ln(A)^2\Theta^2\lambda+\mu^2\psi^2+4\lambda\,\psi^2+2\mu\psi+1}}{2\lambda})t.$

4 Conclusion

In this chapter, a nonlinear extended KP equation has been successfully analyzed using the new Kudryashov method. As a result, exact soliton solutions were systematically derived, which highlights the capability of this method to efficiently solve nonlinear evolution equations. The obtained solutions enrich the solution structure of the KP-type models and provide useful analytical insight into the underlying nonlinear phenomena.

Owing to the broad relevance of the KP equation in nonlinear optics, plasma physics, and fluid dynamics, the present results further enhance the applicability of the model in these physical phenomena. Overall, the work confirms that the new Kudryashov method is a powerful and reliable tool for constructing exact solutions of nonlinear PDEs, and it may be effectively applied to other nonlinear models arising in mathematical physics.

Bibliography

[1] Ma, W. X. (2015). Lump solutions to the Kadomtsev–Petviashvili equation. *Physics Letters A*, *379*(36), 1975–1978.

[2] Kumar, M., Tiwari, A. K., & Kumar, R. (2017). Some more solutions of Kadomtsev–Petviashvili equation. *Computers & Mathematics with Applications*, *74*(10), 2599–2607.

[3] Sawada, K., & Kotera, T. (1974). A method for finding N-soliton solutions of the KdV equation and KdV-like equation. *Progress of Theoretical Physics*, *51*(5), 1355–1367.

[4] Kopçasız, B., & Sağlam, F. N. K. (2025). The effect of standard Wiener process on the stochastic Davey–Stewartson model via the Jacobi elliptic function expansion approach. *Mathematical Methods in the Applied Sciences*. *49*(5), 4031–4055.

[5] Yadav, R., Kumar, S., Elsherbeny, A. M., Yildirim, Y., Jweeg, M. J., Al-Dulaimi, A. M. K., Moraru, L., & Biswas, A. (2025). Optical solitons for the dispersive concatenation model with Kerr law of self-phase modulation by Lie symmetry. *Ukrainian Journal of Physical Optics*, *26*(4), 4066–4082.

[6] Kumar, M., Srivastava, S., & Tanwar, D. V. (2024). Lie symmetries, soliton dynamics, conservation laws and stability analysis of Bogoyavlensky–Konoplechenko system. *Optical and Quantum Electronics*, *56*(7), 1087.

[7] Tariq, H., Ashraf, H., Rezazadeh, H., & Demirbilek, U. (2024). Travelling wave solutions of nonlinear conformable Bogoyavlenskii equations via two powerful analytical approaches. *Applied Mathematics-A Journal of Chinese Universities*, *39*(3), 502–518.

[8] Demirbilek, U., Tedjani, A. H., & Seadawy, A. R. (2025). Analytical solutions of the combined Kairat-II-X equation: a dynamical perspective on bifurcation, chaos, energy, and sensitivity. *AIMS Mathematics*, *10*(6), 13664–13691.

[9] Murad, M. A. S. (2025). Formation of optical soliton wave profiles of nonlinear conformable Schrödinger equation in weakly non-local media: Kudryashov auxiliary equation method. *Journal of Optics*, *54*(5), 3177–3190.

[10] Murad, M. A. S., Faridi, W. A., Jhangeer, A., Iqbal, M., & Garayev, M. (2025). Optical solutions with Kudryashov's arbitrary type of generalized non-local nonlinearity and refractive index via Kudryashov auxiliary equation method. *Fractals*, *33*(04), 1–15.

[11] Kumar, S., & Malik, S. (2024). A new analytic approach and its application to new generalized Korteweg–de Vries and modified Korteweg–de Vries equations. *Mathematical Methods in the Applied Sciences*, *47*(14), 11709–11726.

[12] Malik, S., Boulaaras, S. M., Talafha, A. M., & Sağlam, F. N. K. (2025). Soliton dynamics of the fourth-order nonlinear Boussinesq equation arising in shallow-water via advanced analytical techniques. *Scientific Reports*, *15*, 45421.

[13] Kumawat, S., Bhushan, B., & Kaur, S. (2025). On the application of the $(G'/G, 1/G)$-expansion method to the time fractional Chaffee–Infante equation. *Palestine Journal of Mathematics*, *14*(Special Issue IV), 127–138.

[14] Malik, S., Hashemi, M. S., Kumar, S., Rezazadeh, H., Mahmoud, W., & Osman, M. S. (2023). Application of new Kudryashov method to various nonlinear partial differential equations. *Optical and Quantum Electronics*, *55*(1), 8.

[15] Bhan, C., Karwasra, R., Malik, S., & Kumar, S. (2024). Bifurcation, chaotic behavior and soliton solutions to the KP-BBM equation through new Kudryashov and generalized Arnous methods. *AIMS Mathematics*, *9*(4), 8749–8767.

[16] Hussain, E., Shah, S. A. A., Bariq, A., Li, Z., Ahmad, M. R., Ragab, A. E., & Az-Zo'bi, E. A. (2024). Solitonic solutions and stability analysis of Benjamin Bona Mahony Burger equation using two versatile techniques. *Scientific Reports*, *14*(1), 13520.

[17] Farooq, K., Hussain, E., Younas, U., Mukalazi, H., Khalaf, T. M., Mutlib, A., & Shah, S. A. A. (2025). Exploring the wave's structures to the nonlinear coupled system arising in surface geometry. *Scientific Reports*, *15*(1), 11624.

[18] Ma, Y. L., Wazwaz, A. M., & Li, B. Q. (2021). New extended Kadomtsev–Petviashvili equation: multiple soliton solutions, breather, lump and interaction solutions. *Nonlinear Dynamics*, *104*(2), 1581–1594.

[19] Kopçasız, B., Sağlam, F. N. K., Bulut, H., & Radwan, T. (2025). Exploration of the soliton solutions of the (n+1) dimensional generalized Kadomstev Petviashvili equation using an innovative approach. *Scientific Reports*, *15*(1), 14542.

[20] Manukure, S., Zhou, Y., & Ma, W. X. (2018). Lump solutions to a (2 + 1)-dimensional extended KP equation. *Computers & Mathematics with Applications*, *75*(7), 2414–2419.

Erhan Güler, Yusuf Yaylı, and Hasan Hilmi Hacısalihoğlu

On hypersurfaces of revolution with the $\mathbb{IV}$-Laplacian condition in four-dimensional Euclidean space

Abstract: A family of hypersurfaces of revolution, denoted by $\mathbf{x} = \mathbf{x}(u, v, w)$, is investigated, which is generated via double rotation in the four-dimensional Euclidean space $\mathbb{E}^4$. Based on the curvature components of $\mathbf{x}$, the fourth Laplace–Beltrami operator is computed. It is shown that this operator satisfies the relation $\Delta^{\mathbb{IV}}\mathbf{x} = \mathcal{A}\mathbf{x}$, where $\mathcal{A}$ is a 4×4 matrix.

Keywords: 4-Space, hypersurfaces of revolution, curvature, Gauss map, fourth fundamental form, fourth Laplace–Beltrami operator

MSC 2020: 53A07, 53A10, 53C42, 58J35

1 Introduction

The study of submanifolds whose immersions into $\mathbb{E}^m$ or $\mathbb{E}^m_v$ can be represented as finite linear combinations of eigenfunctions of the Laplace operator has evolved extensively over the last fifty years. This area of research, initially shaped by the foundational work of Chen [12–15], has since become a significant topic in differential geometry.

According to Takahashi [45], a connected submanifold in the Euclidean space $\mathbb{E}^m$ can be characterized as being of 1-type precisely when it is minimal in $\mathbb{E}^m$ or when it is contained as a minimal submanifold within a hypersphere of $\mathbb{E}^m$. Subsequent studies concerning 2-type closed spherical submanifolds were carried out in [8, 9, 13], with additional geometric developments reported in [8, 9, 13]. An extension of Takahashi's characterization in the same ambient space was provided by Garay [26]. Additional insights were contributed by Cheng and Yau through their study of hypersurfaces possessing constant scalar curvature. Investigations on submanifolds with finite-type Gauss maps in $\mathbb{E}^m$ were carried out by Chen and Piccinni [17], and the construction of hypersurfaces having pointwise-1 Gauss maps in $\mathbb{E}^{n+1}$ was introduced by Dursun [22].

Erhan Güler, Texas Tech University, College of Arts and Sciences, Department of Mathematics and Statistics, 79409 Lubbock, TX, USA, e-mail: eguler@ttu.edu
Yusuf Yaylı, Ankara University, Faculty of Sciences, Department of Mathematics, 06100 Ankara, Türkiye, e-mail: yayli@science.ankara.edu.tr
Hasan Hilmi Hacısalihoğlu, Bilecik Şeyh Edebali University, Faculty of Sciences, Department of Mathematics, 11230 Bilecik, Türkiye, e-mail: hhacisalihogluu@gmail.com

https://doi.org/10.1515/9783112234525-022

Within the three-dimensional Euclidean space $\mathbb{E}^3$, it was shown by Takahashi [45] that the relation $\Delta r = \lambda r$, for some constant $\lambda \in \mathbb{R}$, holds only for minimal surfaces and spheres. The condition $\Delta H = AH$, where $A \in \mathrm{Mat}(3,3)$, was proven by Ferrandez et al. [23] to characterize surfaces that are minimal, spherical, or cylindrical. A classification of minimal helicoids with Gauss maps of pointwise-1 (first kind) was developed by Choi and Kim [19]. A particular subclass of finite-type surfaces of revolution was studied by Garay [25]. In [20], it was demonstrated by Dillen et al. that surfaces satisfying $\Delta r = Ar + B$, with $A \in \mathrm{Mat}(3,3)$ and $B \in \mathrm{Mat}(3,1)$, must be minimal surfaces, spheres, or circular cylinders. The condition $\Delta^{\mathrm{III}} x = Ax$ was examined in the context of surfaces of revolution by Stamatakis and Zoubi [44]. Finite-type helicoidal surfaces M^2 w. r. t. the 1st, 2nd, 3rd fundamental forms – that is, those satisfying $\Delta^J r = Ar$ for $J = \mathrm{I}, \mathrm{II}, \mathrm{III}$ – were constructed by Senoussi and Bekkar [43]. Lastly, investigations concerning the operator of Cheng–Yau with a Gauss map on revolutions were carried out by Kim et al. [36].

In four-dimensional Euclidean space $\mathbb{E}^4$, foundational research on general rotational surfaces was carried out by Moore [40, 41]. The concept of hypersurfaces having vector fields satisfying harmonic H was introduced by Hasanis and Vlachos [33], while Cheng and Wan [18] focused on complete hypersurfaces characterized by constant mean curvature. Investigations into L_1-pointwise 1-type Gauss maps were conducted by Kim and Turgay [37]. Arslan et al. [3] constructed a surface of Vranceanu exhibiting such a Gauss map, and further discussion on generalized rotational surfaces was provided in [4]. Surfaces of tensor product including a pointwise-1 Gauss map were developed by Arslan et al. [5], while rotational surfaces with similar properties were examined by Kahraman Aksoyak and Yaylı [34]. Helicoidal hypersurfaces were addressed in [31], and the Gauss map along with the 3rd Laplace–Beltrami operator for rotational hypersurfaces was computed in [30]. The Cheng–Yau operator for such hypersurfaces was introduced by Güler and Turgay [32]. Additionally, Güler [28] analyzed rotational hypersurfaces satisfying the relation $\Delta^{\mathrm{I}} R = AR$, where $A \in \mathrm{Mat}(4,4)$, and in subsequent work [27] investigated curvature characteristics and the fourth fundamental form of hyperspheres. Further contributions include the classification of rotational λ-hypersurfaces in Euclidean spaces [6], as well as the study of the fourth fundamental form in [29].

In Minkowski space $\mathbb{E}_1^4$, analogs of Moore-type surfaces were investigated by Ganchev and Milousheva [24]. A result by Arvanitoyeorgos et al. [7] established that a Lorentzian submanifold M_1^3 with a mean curvature satisfying $\Delta H = \alpha H$, where α is a constant, must have constant mean curvature. Meridian surfaces of both elliptic and hyperbolic types exhibiting pointwise-1 Gauss maps were introduced by Arslan and Milousheva. Turgay contributed to the classification of surfaces having finite-type Gauss maps, while space-like surfaces with pointwise-1 Gauss maps were analyzed by Dursun and Turgay. A family of right conoid hypersurfaces with lightlike axes was presented by Li, Güler, and Toda [39]. General rotational surfaces in the semi-Euclidean space $\mathbb{E}_2^4$ possessing pointwise-1 Gauss maps were examined by Kahraman Aksoyak and Yaylı [35]. Moreover, surfaces admitting 2-type pseudospherical Gauss maps in $\mathbb{E}_2^5$ were described by Bektaş et al. [10], followed by a classification of 1-type pseudospherical Gauss maps on pseudospherical submanifolds in [11].

A detailed investigation of the differential geometry of 1-type submanifolds and submanifolds with 1-type Gauss map can be found in [16].

This chapter is devoted to the investigation of birotational hypersurfaces embedded in the four-dimensional Euclidean space $\mathbb{E}^4$, with particular attention given to their properties under the action of the fourth-order Laplace–Beltrami operator. The aim is to explore the geometric and analytical behavior of such hypersurfaces through curvature analysis and higher-order differential identities.

In Section 2, a review of fundamental geometric notions in the setting of $\mathbb{E}^4$ is presented. The necessary background concerning hypersurface theory, including definitions of curvature and differential operators, is outlined.

The derivation of curvature formulas for hypersurfaces in four-dimensional space is carried out in Section 3, using the classical tools of differential geometry such as the first and second fundamental forms and the Gauss map.

The construction of birotational hypersurfaces and their structural properties are introduced in Section 4. The parametrization process and the computation of relevant geometric quantities are described based on the rotational symmetry of the profile curve.

In Section 5, the condition $\Delta^{\mathrm{IV}}\mathbf{x} = \mathcal{A}\mathbf{x}$ involving a fourth-order Laplace–Beltrami operator and a diagonal matrix $\mathcal{A} \in \mathrm{Mat}(4,4)$ is analyzed in detail. Conditions under which birotational hypersurfaces satisfy this relation are discussed.

The chapter concludes with a summary of the main results, and potential directions for future research involving higher-order Laplacians and finite-type submanifolds are briefly indicated.

2 Preliminaries

In this section, several essential notions and definitions are provided, and the notational conventions adopted throughout the manuscript are specified. The Euclidean m-space, denoted by $\mathbb{E}^m$, is equipped with the standard metric tensor expressed as $\widetilde{g} = \langle\, ,\, \rangle = \sum_{i=1}^{m} dx_i^2$, where the coordinates $(x_1, x_2, \ldots, x_m)$ are assumed to form a Cartesian frame in $\mathbb{E}^m$.

Let $\pi : M \longrightarrow \widetilde{M}$ be an isometric immersion of an n-dimensional Riemannian submanifold into a Riemannian ambient space. The Levi-Civita connections associated with M and $\widetilde{M}$ are denoted by ∇ and $\widetilde{\nabla}$, respectively.

Throughout this chapter, the symbols X, Y, Z, W (respectively, ξ, η) will be used to represent vector fields tangent (respectively, normal) to the submanifold M. The formulas of Gauss and Weingarten are expressed as follows:

$$\widetilde{\nabla}_X Y = \nabla_X Y + h(X, Y), \tag{1}$$

$$\widetilde{\nabla}_X \xi = -A_\xi(X) + D_X \xi, \tag{2}$$

where h, D, and A, respectively, denote the second fundamental form, the normal connection, and the shape operator associated with the submanifold M.

Given any normal vector $\xi \in T_p^\perp M$, the shape operator A_ξ is understood to act as a symmetric endomorphism on the tangent space T_pM at the point $p \in M$. The relationship between the shape operator and the second fundamental form is formulated by the identity

$$\langle h(X, Y), \xi \rangle = \langle A_\xi X, Y \rangle.$$

The Gauss and Codazzi equations are formulated, respectively, as

$$\langle R(X, Y)Z, W \rangle = \langle h(Y, Z), h(X, W) \rangle - \langle h(X, Z), h(Y, W) \rangle, \tag{3}$$

$$(\bar{\nabla}_X h)(Y, Z) = (\bar{\nabla}_Y h)(X, Z), \tag{4}$$

where R and R^D denote the curvature tensors associated with the Levi-Civita connection ∇ and the normal connection D, respectively. The covariant derivative of the second fundamental form is defined by

$$(\bar{\nabla}_X h)(Y, Z) = D_X h(Y, Z) - h(\nabla_X Y, Z) - h(Y, \nabla_X Z).$$

Consider a hypersurface M that is oriented and smoothly situated in the Euclidean space $\mathbb{E}^{n+1}$. The linear transformation known as the shape operator (or Weingarten map) is denoted by $\mathbf{S}$, and the position vector corresponding to M is represented by $\mathbf{x}$.

A local orthonormal frame field $\{e_1, e_2, \ldots, e_n\}$ is taken such that each vector e_i corresponds to a principal direction of the hypersurface M, and the associated principal curvature is denoted by k_i, for $i = 1, 2, \ldots, n$. The dual basis of this frame field is denoted by $\{\theta_1, \theta_2, \ldots, \theta_n\}$. In this context, Cartan's first structural equation is written as

$$d\theta_i = \sum_{j=1}^{n} \theta_j \wedge \omega_{ij}, \quad i, j = 1, 2, \ldots, n, \tag{5}$$

where ω_{ij} represents the connection 1-forms related to the selected orthonormal frame. From the Codazzi equation (3), the following relations are obtained

$$e_i(k_j) = \omega_{ij}(e_j)(k_i - k_j), \tag{6}$$

$$\omega_{ij}(e_l)(k_i - k_j) = \omega_{il}(e_j)(k_i - k_l), \tag{7}$$

for distinct indices $i, j, l = 1, 2, \ldots, n$.

The quantities s_j are defined as the elementary symmetric functions $\sigma_j(k_1, k_2, \ldots, k_n)$, where σ_j denotes the jth elementary symmetric polynomial given by

$$\sigma_j(a_1, a_2, \ldots, a_n) = \sum_{1 \le i_1 < i_2 < \cdots < i_j \le n} a_{i_1} a_{i_2} \cdots a_{i_j}.$$

For convenience, the notation

$$r_i^j = \sigma_j(k_1, k_2, \ldots, k_{i-1}, k_{i+1}, k_{i+2}, \ldots, k_n)$$

is adopted. By definition, it is assumed that $r_i^0 = 1$, and the relations $s_{n+1} = s_{n+2} = \cdots = 0$ are satisfied. The function s_k is referred to as the kth mean curvature of the hypersurface M.

The quantities $H = \frac{1}{n}s_1$ and $K = s_n$ are well-known curvatures, i. e., mean curvatures with Gauss–Kronecker the hypersurface M, respectively. A hypersurface M is characterized as j-minimal when the condition $s_j \equiv 0$ is satisfied throughout M.

In the Euclidean space $\mathbb{E}^{n+1}$, the ith curvature formulas, denoted by $\mathfrak{C}_i$, are introduced as a unified terminology in this work, although they may be referred to as the ith mean curvature H_i or Gaussian curvature K_i by various authors (e. g., [1], [38]). For each $i = 0, 1, \ldots, n$, the computation of $\mathfrak{C}_i$ begins with the characteristic polynomial of the shape operator $\mathbf{S}$

$$P_{\mathbf{S}}(\lambda) = \det(\mathbf{S} - \lambda \mathcal{I}_n) = \sum_{k=0}^{n} (-1)^k s_k \lambda^{n-k} = 0, \tag{8}$$

where $\mathcal{I}_n$ denotes the identity matrix of order n. From this relation, the curvatures are obtained via the formula $\binom{n}{i}\mathfrak{C}_i = s_i$. In particular, it follows that $\binom{n}{0}\mathfrak{C}_0 = s_0 = 1$ (by definition), $\binom{n}{1}\mathfrak{C}_1 = s_1, \ldots, \binom{n}{n}\mathfrak{C}_n = s_n = K$.

An immersion $x : M \longrightarrow \mathbb{E}^m$ is referred to as being of *finite type* if it can be expressed as a finite linear combination of eigenfunctions of the Laplace operator Δ corresponding to the pair (M, x)

$$x = x_0 + \sum_{i=1}^{k} x_i$$

is satisfied, where x_0 is a constant map and $x_1, \ldots, x_k$ are nonconstant maps satisfying $\Delta x_i = \lambda_i x_i$ for some real constants $\lambda_i \in \mathbb{R}$, $i = 1, \ldots, k$. If all the eigenvalues λ_i are distinct, the submanifold M is referred to as being of *k-type*. Further details can be found in [13].

Let $\mathbf{x} = \mathbf{x}(u, v, w)$ denote an isometric immersion of a three-dimensional submanifold $M^3 \subset \mathbb{E}^3$ into the Euclidean four-space $\mathbb{E}^4$. The triple vector product of vectors $\vec{a} = (a_1, a_2, a_3, a_4)$, $\vec{b} = (b_1, b_2, b_3, b_4)$, and $\vec{c} = (c_1, c_2, c_3, c_4)$ in $\mathbb{E}^4$ is defined by the following expression

$$\vec{a} \times \vec{b} \times \vec{c} = \det \begin{pmatrix} e_1 & e_2 & e_3 & e_4 \\ a_1 & a_2 & a_3 & a_4 \\ b_1 & b_2 & b_3 & b_4 \\ c_1 & c_2 & c_3 & c_4 \end{pmatrix}.$$

For the hypersurface $\mathbf{x}$ in $\mathbb{E}^4$, the symmetric matrices $(g_{ij})_{3\times3}$, $(h_{ij})_{3\times3}$, $(t_{ij})_{3\times3}$, and $(f_{ij})_{3\times3}$ are considered, corresponding, respectively, to the first, second, third, and fourth fundamental forms (i. e., $\mathbb{I}$, $\mathbb{II}$, $\mathbb{III}$, and $\mathbb{IV}$).

The components of the symmetric fundamental form matrices are calculated as follows:

$$g_{11} = \mathbf{x}_u \cdot \mathbf{x}_u, \quad g_{12} = \mathbf{x}_u \cdot \mathbf{x}_v, \quad g_{22} = \mathbf{x}_v \cdot \mathbf{x}_v,$$
$$g_{13} = \mathbf{x}_u \cdot \mathbf{x}_w, \quad g_{23} = \mathbf{x}_v \cdot \mathbf{x}_w, \quad g_{33} = \mathbf{x}_w \cdot \mathbf{x}_w.$$

The entries of the second fundamental form are determined through the following inner products with the unit normal vector e:

$$h_{11} = \mathbf{x}_{uu} \cdot e, \quad h_{12} = \mathbf{x}_{uv} \cdot e, \quad h_{22} = \mathbf{x}_{vv} \cdot e,$$
$$h_{13} = \mathbf{x}_{uw} \cdot e, \quad h_{23} = \mathbf{x}_{vw} \cdot e, \quad h_{33} = \mathbf{x}_{ww} \cdot e.$$

The coefficients of the third fundamental form are expressed as

$$t_{11} = e_u \cdot e_u, \quad t_{12} = e_u \cdot e_v, \quad t_{13} = e_u \cdot e_w,$$
$$t_{22} = e_v \cdot e_v, \quad t_{23} = e_v \cdot e_w, \quad t_{33} = e_w \cdot e_w.$$

In the above formulations, the Euclidean inner product is represented by the dot symbol "$\cdot$". The unit normal vector e associated with the hypersurface $\mathbf{x}$ is defined by

$$e = \frac{\mathbf{x}_u \times \mathbf{x}_v \times \mathbf{x}_w}{\|\mathbf{x}_u \times \mathbf{x}_v \times \mathbf{x}_w\|}, \tag{9}$$

where the numerator involves the triple vector product of the partial derivatives of $\mathbf{x}$, and the denominator ensures normalization.

The shape operator matrix $\mathbf{S}$ associated with the hypersurface $\mathbf{x}$ is expressed as the product

$$(g_{ij})^{-1}(h_{ij}).$$

Similarly, the third fundamental form matrix (t_{ij}) is obtained through the composition

$$(h_{ij})(g_{ij})^{-1}(h_{ij}).$$

Moreover, the matrix denoted by (f_{ij}) is defined by the expression

$$(f_{ij}) = (t_{ij})(g_{ij})^{-1}(h_{ij}).$$

See [30–32] for more details.

3 Curvatures

In the ambient space $\mathbb{E}^4$, the curvature functions $\mathfrak{C}_i$ for $i = 0, 1, 2, 3$ are computed via the characteristic polynomial associated with the shape operator **S**. The polynomial is given in the form

$$P_{\mathbf{S}}(\lambda) = a\lambda^3 + b\lambda^2 + c\lambda + d,$$

where the identity $\det(\mathbf{S} - \lambda\mathcal{I}_3) = 0$ is used to define it.

From this expression, the curvature invariants are obtained as follows:

$$\mathfrak{C}_0 = 1, \quad \binom{3}{1}\mathfrak{C}_1 = -\frac{b}{a}, \quad \binom{3}{2}\mathfrak{C}_2 = \frac{c}{a}, \quad \binom{3}{3}\mathfrak{C}_3 = -\frac{d}{a}.$$

Consequently, the expressions for the curvature functions in $\mathbb{E}^4$ are obtained as follows:

Theorem 1. *Let $\mathbf{x}(u, v, w)$ be a hypersurface immersed in $\mathbb{E}^4$. Then, the associated curvature quantities are computed using the following formulas:*

$$\mathfrak{C}_1 = \frac{\left\{\begin{array}{c} (g_{11}h_{22} + g_{22}h_{11} - 2g_{12}h_{12})g_{33} \\ +(g_{11}g_{22} - g_{12}^2)h_{33} - g_{23}^2h_{11} - g_{13}^2h_{22} \\ -2(g_{13}h_{13}g_{22} - g_{23}h_{13}g_{12} - g_{13}h_{23}g_{12} \\ +g_{11}g_{23}h_{23} - g_{13}g_{23}h_{12}) \end{array}\right\}}{3[(g_{11}g_{22} - g_{12}^2)g_{33} - g_{11}g_{23}^2 + 2g_{12}g_{13}g_{23} - g_{22}g_{13}^2]}, \tag{10}$$

$$\mathfrak{C}_2 = \frac{\left\{\begin{array}{c} (g_{11}h_{22} + g_{22}h_{11} - 2g_{12}h_{12})h_{33} \\ +(h_{11}h_{22} - g_{12}^2)g_{33} - g_{11}h_{23}^2 - g_{22}h_{13}^2 \\ -2(g_{13}h_{13}h_{22} - g_{23}h_{13}h_{12} - g_{13}h_{23}h_{12} \\ +g_{23}h_{23}h_{11} - h_{13}h_{23}g_{12}) \end{array}\right\}}{3[(g_{11}g_{22} - g_{12}^2)g_{33} - g_{11}g_{23}^2 + 2g_{12}g_{13}g_{23} - g_{22}g_{13}^2]}, \tag{11}$$

$$\mathfrak{C}_3 = \frac{(h_{11}h_{22} - h_{12}^2)h_{33} - h_{11}h_{23}^2 + 2h_{12}h_{13}h_{23} - h_{22}h_{13}^2}{(g_{11}g_{22} - g_{12}^2)g_{33} - g_{11}g_{23}^2 + 2g_{12}g_{13}g_{23} - g_{22}g_{13}^2}. \tag{12}$$

Further explanation and derivation of these formulas can be found in [27].

4 Birotational hypersurface

In this section, rotational hypersurfaces in the Euclidean 4-space $\mathbb{E}^4$ are introduced, and their curvature expressions are subsequently derived. It is worth noting that a general formulation of rotational hypersurfaces within Riemannian manifolds was introduced by [21].

A hypersurface of rotation $M \subset \mathbb{E}^{n+1}$ is constructed by revolving a curve of profile γ around a fixed axis $\mathfrak{r}$, which is assumed not to intersect the curve itself. This construction is realized by considering the set of points obtained through the orbit of γ under the action of orthogonal transformations in $\mathbb{E}^{n+1}$ that leave the axis invariant pointwise (see [21, Remark 2.3]).

A profile curve of the form $\gamma(u) = (\mathbf{f}(u), 0, \mathbf{g}(u), 0)$ is considered in $\mathbb{E}^4$. The corresponding rotational hypersurface is generated by applying the following orthogonal transformation:

$$\begin{pmatrix} \cos v & -\sin v & 0 & 0 \\ \sin v & \cos v & 0 & 0 \\ 0 & 0 & \cos w & -\sin w \\ 0 & 0 & \sin w & \cos w \end{pmatrix} \begin{pmatrix} \mathbf{f}(u) \\ 0 \\ \mathbf{g}(u) \\ 0 \end{pmatrix}.$$

Based on this rotational process, the formal definition of the hypersurface is established in the subsequent discussion.

Definition 1. A hypersurface exhibiting birotational symmetry in $\mathbb{E}^4$ can be described by the parametrization

$$\mathbf{x}(u, v, w) = (\mathbf{f}(u) \cos v, \mathbf{f}(u) \sin v, \mathbf{g}(u) \cos w, \mathbf{g}(u) \sin w), \tag{13}$$

where f and g are assumed to be differentiable functions, and the parameters satisfy $0 \leq v, w \leq 2\pi$.

Remark 1. If $\mathbf{f}(u) = 1$ and $\mathbf{g}(u) = 1$ in expression (13), the resulting hypersurface corresponds to the well-known Clifford torus in $\mathbb{E}^4$. Further details on this particular case can be found in [2, 46]. Moreover, when the condition $v = w$ is applied in equation (13), a tensor product surface in $\mathbb{E}^4$ is obtained.

Additional information on this topic can be found in [5, 42].

The first-order partial derivatives of the parametrization given in equation (13) with respect to u, v, and w are computed as

$$\mathbf{x}_u = \begin{pmatrix} \mathbf{f}' \cos v \\ \mathbf{f}' \sin v \\ \mathbf{g}' \cos w \\ \mathbf{g}' \sin w \end{pmatrix}, \quad \mathbf{x}_v = \begin{pmatrix} -\mathbf{f} \sin v \\ \mathbf{f} \cos v \\ 0 \\ 0 \end{pmatrix}, \quad \mathbf{x}_w = \begin{pmatrix} 0 \\ 0 \\ -\mathbf{g} \sin w \\ \mathbf{g} \cos w \end{pmatrix}.$$

From these, the components of the 1st fundamental form are obtained as

$$(g_{ij}) = \operatorname{diag}(\mathbf{f}'^2 + \mathbf{g}'^2, \mathbf{f}^2, \mathbf{g}^2). \tag{14}$$

Accordingly, the determinant of the 1st fundamental form matrix becomes

$$g = \det(g_{ij}) = \mathbf{f}^2\mathbf{g}^2(\mathbf{f}'^2 + \mathbf{g}'^2).$$

By applying formula (9), the unit normal vector of equation (13) is derived as

$$e = \frac{1}{(f'^2 + g'^2)^{1/2}} \begin{pmatrix} -g' \cos v \\ -g' \sin v \\ f' \cos w \\ f' \sin w \end{pmatrix}. \tag{15}$$

The entries of the 2nd fundamental form corresponding to the parametrization in equation (13) are determined through the use of second-order derivatives and the unit normal vector provided in equation (15)

$$(h_{ij}) = \operatorname{diag}\left(\frac{\mathbf{f}'\mathbf{g}'' - \mathbf{g}'\mathbf{f}''}{(\mathbf{f}'^2 + \mathbf{g}'^2)^{1/2}}, \frac{\mathbf{f}\mathbf{g}'}{(\mathbf{f}'^2 + \mathbf{g}'^2)^{1/2}}, -\frac{\mathbf{g}\mathbf{f}'}{(\mathbf{f}'^2 + \mathbf{g}'^2)^{1/2}} \right). \tag{16}$$

The determinant of the 2nd fundamental form matrix is found to be

$$h = \det(h_{ij}) = -\frac{\mathbf{f}\mathbf{g}\mathbf{f}'\mathbf{g}'(\mathbf{f}'\mathbf{g}'' - \mathbf{f}''\mathbf{g}')}{(\mathbf{f}'^2 + \mathbf{g}'^2)^{3/2}}.$$

By utilizing the expressions in equations (14) and (16), the shape operator matrix $\mathbf{S}$ corresponding to the birotational hypersurface (13) is computed as

$$\mathbf{S} = \operatorname{diag}\left(\frac{\mathbf{f}'\mathbf{g}'' - \mathbf{g}'\mathbf{f}''}{(\mathbf{f}'^2 + \mathbf{g}'^2)^{3/2}}, \frac{\mathbf{g}'}{\mathbf{f}(\mathbf{f}'^2 + \mathbf{g}'^2)^{1/2}}, -\frac{\mathbf{f}'}{\mathbf{g}(\mathbf{f}'^2 + \mathbf{g}'^2)^{1/2}} \right). \tag{17}$$

Finally, by substituting equations (14) and (16) into the curvature formulas (10), (11), and (12), the explicit expressions for the curvature functions of the birotational hypersurface (13) are obtained.

Corollary 1. *Suppose that* $\mathbf{x} : M^3 \to \mathbb{E}^4$ *is an immersion described by equation* (13). *Then, the corresponding curvature functions of the hypersurface are given by*

$$\mathfrak{C}_1 = \frac{(\mathbf{f}'\mathbf{g}'' - \mathbf{g}'\mathbf{f}'')\mathbf{f}\mathbf{g} - (\mathbf{f}'^2 + \mathbf{g}'^2)(\mathbf{f}\mathbf{f}' - \mathbf{g}\mathbf{g}')}{3\mathbf{f}\mathbf{g}(\mathbf{f}'^2 + \mathbf{g}'^2)^{3/2}},$$
$$\mathfrak{C}_2 = \frac{(\mathbf{f}\mathbf{f}' - \mathbf{g}\mathbf{g}')(\mathbf{g}'\mathbf{f}'' - \mathbf{f}'\mathbf{g}'') - (\mathbf{f}'^2 + \mathbf{g}'^2)\mathbf{f}'\mathbf{g}'}{3\mathbf{f}\mathbf{g}(\mathbf{f}'^2 + \mathbf{g}'^2)^2},$$
$$\mathfrak{C}_3 = \frac{\mathbf{f}'\mathbf{g}'(\mathbf{g}'\mathbf{f}'' - \mathbf{f}'\mathbf{g}'')}{\mathbf{f}\mathbf{g}(\mathbf{f}'^2 + \mathbf{g}'^2)^{5/2}}.$$

Corollary 2. *Assume that* $\mathbf{x} : M^3 \to \mathbb{E}^4$ *is an immersion defined by the parametrization in equation* (13). *If the profile curve* γ *of* $\mathbf{x}$ *is arc-length parametrized, and the functions are chosen as* $\mathbf{f}(u) = \cos u$, $\mathbf{g}(u) = \sin u$, *then the corresponding birotational hypersurface possesses constant curvature values* $\mathfrak{C}_i = 1$, *for each* $i = 1, 2, 3$.

Corollary 3. *Suppose that* $\mathbf{x} : M^3 \to \mathbb{E}^4$ *is an immersion mentioned by equation* (13). *Let the profile curve* γ *be parametrized by arc length, and the functions are selected as* $\mathbf{f}(u) = \mathbf{g}(u) = \frac{u}{\sqrt{2}}$. *Then, the associated birotational hypersurface admits the following curvature values:*

$$\mathfrak{C}_1 = 0, \text{ i. e., } \mathbf{x} \text{ is 1-minimal;}$$
$$\mathfrak{C}_2 = -\frac{1}{3u^2}, \text{ i. e., } \mathbf{x} \text{ has negative 2-curvature;}$$
$$\mathfrak{C}_3 = 0, \text{ i. e., } \mathbf{x} \text{ is 3-minimal.}$$

By employing the shape operator matrix given in (17) and the definition of the third fundamental form matrix $\mathbb{III} = (t_{ij})$, the components of the third fundamental form corresponding to the hypersurface parametrized in equation (13) are computed as

$$(t_{ij}) = \operatorname{diag}\left(\frac{(\mathbf{f}'\mathbf{g}'' - \mathbf{g}'\mathbf{f}'')^2}{(\mathbf{f}'^2 + \mathbf{g}'^2)^2}, \frac{\mathbf{g}'^2}{\mathbf{f}'^2 + \mathbf{g}'^2}, \frac{\mathbf{f}'^2}{\mathbf{f}'^2 + \mathbf{g}'^2} \right). \tag{18}$$

By utilizing the Gauss map expression provided in equation (15) and the third fundamental form matrix computed in equation (18), the components of the fourth fundamental form matrix $\mathbb{IV} = (f_{ij})$ associated with the hypersurface (13) are determined as follows:

$$(f_{ij}) = \operatorname{diag}\left(\frac{(\mathbf{f}'\mathbf{g}'' - \mathbf{g}'\mathbf{f}'')^3}{(\mathbf{f}'^2 + \mathbf{g}'^2)^{7/2}}, \frac{\mathbf{g}'^3}{\mathbf{f}(\mathbf{f}'^2 + \mathbf{g}'^2)^{3/2}}, -\frac{\mathbf{f}'^3}{\mathbf{g}(\mathbf{f}'^2 + \mathbf{g}'^2)^{3/2}} \right). \tag{19}$$

As a result, the corresponding determinant is derived as follows:

$$f = \det(f_{ij}) = -\frac{\mathbf{f}'^3\mathbf{g}'^3(\mathbf{f}'\mathbf{g}'' - \mathbf{f}''\mathbf{g}')^3}{\mathbf{fg}(\mathbf{f}'^2 + \mathbf{g}'^2)^{13/2}}.$$

5 Birotational hypersurface satisfying $\Delta^{\mathbb{IV}}\mathbf{x} = \mathcal{A}\mathbf{x}$

The Laplace–Beltrami operator w. r. t. the fourth fundamental form for a smooth function is presented in this section, followed by its application to the case of equation (13).

The inverse matrix of the fourth fundamental form $\mathbb{IV}$, that is, (f_{ij}), is expressed as follows:

$$\frac{1}{f}\begin{pmatrix} f_{22}f_{33} - f_{23}f_{32} & -(f_{12}f_{33} - f_{13}f_{32}) & f_{12}f_{23} - f_{13}f_{22} \\ -(f_{21}f_{33} - f_{31}f_{23}) & f_{11}f_{33} - f_{13}f_{31} & -(f_{11}f_{23} - f_{21}f_{13}) \\ f_{21}t_{32} - f_{22}f_{31} & -(f_{11}f_{32} - f_{12}f_{31}) & f_{11}f_{22} - f_{12}f_{21} \end{pmatrix},$$

where

$$\begin{aligned} f &= \det(f_{ij}) \\ &= f_{11}f_{22}f_{33} - f_{11}f_{23}f_{32} + f_{12}f_{31}f_{23} - f_{12}f_{21}f_{33} + f_{21}f_{13}f_{32} - f_{13}f_{22}f_{31}. \end{aligned}$$

Definition 2. The Laplace–Beltrami operator w. r. t. the fourth fundamental form associated with a smooth function $\phi = \phi(x^1, x^2, x^3)|_{\mathbf{D}}$, where $\mathbf{D} \subset \mathbb{R}^3$ and ϕ is of class C^2, is defined as follows:

$$\Delta^{\mathrm{IV}}\phi = \frac{1}{f^{1/2}} \sum_{i,j=1}^{3} \frac{\partial}{\partial x^i}\left(f^{1/2} f^{ij} \frac{\partial \phi}{\partial x^j}\right), \tag{20}$$

where $(f^{ij}) = (f_{ij})^{-1}$ denotes the inverse matrix, and $f = \det(f_{ij})$ is the determinant of the fourth fundamental form matrix.

Equation (20) can also be reformulated in the following form:

$$\Delta^{\mathrm{IV}}\phi = \frac{1}{|f|^{1/2}} \left\{ \begin{array}{l} \frac{\partial}{\partial x^1}(|f|^{1/2} f^{11} \frac{\partial \phi}{\partial x^1}) - \frac{\partial}{\partial x^1}(|f|^{1/2} f^{12} \frac{\partial \phi}{\partial x^2}) + \frac{\partial}{\partial x^1}(|f|^{1/2} f^{13} \frac{\partial \phi}{\partial x^3}) \\ -\frac{\partial}{\partial x^2}(|f|^{1/2} f^{21} \frac{\partial \phi}{\partial x^1}) + \frac{\partial}{\partial x^2}(|f|^{1/2} f^{22} \frac{\partial \phi}{\partial x^2}) - \frac{\partial}{\partial x^2}(|f|^{1/2} f^{23} \frac{\partial \phi}{\partial x^3}) \\ +\frac{\partial}{\partial x^3}(|f|^{1/2} f^{31} \frac{\partial \phi}{\partial x^1}) - \frac{\partial}{\partial x^3}(|f|^{1/2} f^{32} \frac{\partial \phi}{\partial x^2}) + \frac{\partial}{\partial x^3}(|f|^{1/2} f^{33} \frac{\partial \phi}{\partial x^3}) \end{array} \right\}. \tag{21}$$

For any rotational hypersurface, the offdiagonal entries of the fourth fundamental form matrix vanish, that is, $f_{ij} = 0$ whenever $i \neq j$. Accordingly, the fourth Laplace–Beltrami operator can be reformulated, and it takes the following form:

$$\Delta^{\mathrm{IV}}\phi = \frac{1}{|f|^{1/2}} \left\{ \frac{\partial}{\partial x^1}\left(|f|^{1/2} f^{11} \frac{\partial \phi}{\partial x^1}\right) + \frac{\partial}{\partial x^2}\left(|f|^{1/2} f^{22} \frac{\partial \phi}{\partial x^2}\right) + \frac{\partial}{\partial x^3}\left(|f|^{1/2} f^{33} \frac{\partial \phi}{\partial x^3}\right) \right\}. \tag{22}$$

Therefore, the fourth Laplace–Beltrami operator for a rotational hypersurface $\mathbf{x}(u, v, w)$ can be written in a more explicit form as

$$\Delta^{\mathrm{IV}}\mathbf{x} = \frac{1}{|f|^{1/2}} \left\{ \frac{\partial}{\partial u}\left(\frac{f_{22}f_{33}}{|f|^{1/2}} \mathbf{x}_u\right) + \frac{\partial}{\partial v}\left(\frac{f_{11}f_{33}}{|f|^{1/2}} \mathbf{x}_v\right) + \frac{\partial}{\partial w}\left(\frac{f_{11}f_{22}}{|f|^{1/2}} \mathbf{x}_w\right) \right\}. \tag{23}$$

The expressions $\frac{f_{22}f_{33}}{|f|^{1/2}}\mathbf{x}_u$, $\frac{f_{11}f_{33}}{|f|^{1/2}}\mathbf{x}_v$, and $\frac{f_{11}f_{22}}{|f|^{1/2}}\mathbf{x}_w$ are differentiated with respect to u, v, and w, respectively. After performing these differentiations, the resulting terms are substituted into equation (23), from which the following expression is obtained:

$$\Delta^{\mathrm{IV}}\mathbf{x} = \begin{pmatrix} \Delta^{\mathrm{IV}}\mathbf{x}_1 \\ \Delta^{\mathrm{IV}}\mathbf{x}_2 \\ \Delta^{\mathrm{IV}}\mathbf{x}_3 \\ \Delta^{\mathrm{IV}}\mathbf{x}_4 \end{pmatrix} = \begin{pmatrix} \mathcal{F}(u)\cos v \\ \mathcal{F}(u)\sin v \\ \mathcal{G}(u)\cos w \\ \mathcal{G}(u)\sin w \end{pmatrix},$$

where

$$\mathcal{F}(u)=\frac{\left\{\begin{array}{c} -3/2\,\mathbf{fgf'g'^{3}(f'^{2}+g'^{2})^{2}} \\ +\left\{\begin{array}{c} -\,\mathbf{fgg'}\left\{\begin{array}{c} \mathbf{3f^{2}f'^{2}g''^{2}-3f'^{4}g'^{2}} \\ -11/2\,\mathbf{f'^{2}g'^{4}}-5/2\,\mathbf{g'^{6}} \end{array}\right\}\mathbf{f''} \\ \mathbf{-f^{3}gg'^{3}f''^{3}+3f^{3}gf'g'^{2}f''^{2}g''} \\ +\left\{\begin{array}{c} \mathbf{f^{3}gf'^{3}g''^{3}} \\ +\mathbf{fg}(3/2\,\mathbf{f'^{2}+2g'^{2})f'g'^{2}(f'^{2}+g'^{2})g''} \\ -1/2\,\mathbf{f'g'^{3}(fg'+gf')(f'^{2}+g'^{2})^{2}} \end{array}\right\} \end{array}\right\} \end{array}\right\}}{\mathbf{fgg'^{3}(f'^{2}+g'^{2})^{-3/2}(f'g''-g'f'')^{3}}},$$

$$\mathcal{G}(u)=\frac{\left\{\begin{array}{c} -3/2\,\mathbf{fgf'^{3}g'(f'^{2}+g'^{2})^{2}} \\ +\left\{\begin{array}{c} \mathbf{-fg^{3}g'^{3}f''^{3}+3fg^{3}f'g'^{2}f''^{2}g''} \\ -\,\mathbf{fgf'^{2}g'}\left\{\begin{array}{c} \mathbf{3g^{2}g''^{2}-2f'^{4}} \\ -7/2\,\mathbf{f'^{2}g'^{2}}-3/2\,\mathbf{g'^{4}} \end{array}\right\}\mathbf{f''} \\ +\left\{\begin{array}{c} \mathbf{fg^{3}f'^{3}g''^{3}} \\ +\mathbf{fg}(5/2\,\mathbf{f'^{2}+3g'^{2})(f'^{2}+g'^{2})f'^{3}g''} \\ -1/2\,\mathbf{f'^{3}g'(fg'+gf')(f'^{2}+g'^{2})^{2}} \end{array}\right\} \end{array}\right\} \end{array}\right\}}{\mathbf{fgf'^{3}(f'^{2}+g'^{2})^{-3/2}(f'g''-g'f'')^{3}}}.$$

6 Conclusion

Based on the results established in the preceding section, the following conclusions are derived.

Corollary 4. *Assume that* $\mathbf{x} : M^3 \rightarrow \mathbb{E}^4$ *is an immersion defined by equation* (13). *Then, the corresponding hypersurface* $\mathbf{x}$ *supplies the eigenvalue-type relation*

$$\Delta^{\mathrm{IV}}\mathbf{x} = \mathcal{A}\mathbf{x},$$

where the matrix $\mathcal{A} \in \mathrm{Mat}(4,4)$ *is given by*

$$\mathcal{A} = \mathrm{diag}\left(\frac{\mathcal{F}}{\mathbf{f}}\mathcal{I}_2, \frac{\mathcal{G}}{\mathbf{g}}\mathcal{I}_2\right)$$

and $\mathcal{I}_2$ *denotes the* 2×2 *identity matrix.*

Corollary 5. *Suppose that* $\mathbf{x} : M^3 \rightarrow \mathbb{E}^4$ *is an immersion specified by the parametrization in equation* (13). *When the profile curve* γ *corresponding to* $\mathbf{x}$ *is determined by arc length, the hypersurface fulfills the identity*

$$\Delta^{\mathrm{IV}}\mathbf{x} = \mathcal{B}\mathbf{x},$$

where $\mathcal{B} = \mathrm{diag}(\mathfrak{p}\mathcal{I}_2, \mathfrak{q}\mathcal{I}_2)$ *and*

$$\mathfrak{p}(u) = \frac{\begin{Bmatrix} -3/2\,\mathbf{f}'\mathbf{g}'^3 + \mathbf{f}^2\mathbf{f}'^3\mathbf{g}''^3 + 3\mathbf{f}^2\mathbf{f}'\mathbf{g}'^2\mathbf{f}''^2\mathbf{g}'' \\ -\mathbf{g}'\begin{Bmatrix} 3\mathbf{f}^2\mathbf{f}'^2\mathbf{g}''^2 - 3\mathbf{f}'^4\mathbf{g}'^2 \\ -11/2\,\mathbf{f}'^2\mathbf{g}'^4 - 5/2\,\mathbf{g}'^6 \end{Bmatrix}\mathbf{f}'' \\ -\mathbf{f}^2\mathbf{g}'^3\mathbf{f}''^3 + (3/2\,\mathbf{f}'^2 + 2\mathbf{g}'^2)\mathbf{f}'\mathbf{g}'^2\mathbf{g}'' \end{Bmatrix}}{\mathbf{g}'^3(\mathbf{f}'\mathbf{g}'' - \mathbf{g}'\mathbf{f}'')^3} - \frac{\mathbf{f}'(\mathbf{f}\mathbf{g}' + \mathbf{g}\mathbf{f}')}{2\mathbf{f}\mathbf{g}(\mathbf{f}'\mathbf{g}'' - \mathbf{g}'\mathbf{f}'')^3},$$

$$\mathfrak{q}(u) = \frac{\begin{Bmatrix} -3/2\,\mathbf{f}'^3\mathbf{g}' - \mathbf{g}^2\mathbf{g}'^3\mathbf{f}''^3 + 3\mathbf{g}^2\mathbf{f}'\mathbf{g}'^2\mathbf{f}''^2\mathbf{g}'' \\ -\mathbf{f}'^2\mathbf{g}'\begin{Bmatrix} 3\mathbf{g}^2\mathbf{g}''^2 - 2\mathbf{f}'^4 \\ -7/2\,\mathbf{f}'^2\mathbf{g}'^2 - 3/2\,\mathbf{g}'^4 \end{Bmatrix}\mathbf{f}'' \\ +\mathbf{g}^2\mathbf{f}'^3\mathbf{g}''^3 + (5/2\,\mathbf{f}'^2 + 3\mathbf{g}'^2)\mathbf{f}'^3\mathbf{g}'' \end{Bmatrix}}{\mathbf{f}'^3(\mathbf{f}'\mathbf{g}'' - \mathbf{g}'\mathbf{f}'')^3} - \frac{\mathbf{g}'(\mathbf{f}\mathbf{g}' + \mathbf{g}\mathbf{f}')}{2\mathbf{f}\mathbf{g}(\mathbf{f}'\mathbf{g}'' - \mathbf{g}'\mathbf{f}'')^3}.$$

Here, $\mathcal{B}$ represents the 2×2 matrix, and $\mathcal{I}_2$ denotes the 2×2 identity.

As a final demonstration, the next example is included.

Example 1. Consider the birotational hypersurface defined by

$$\xi(u, v, w) = \left(\frac{1-u^2}{1+u^2}\cos v, \frac{1-u^2}{1+u^2}\sin v, \frac{2u}{1+u^2}\cos w, \frac{2u}{1+u^2}\sin w\right). \tag{24}$$

This hypersurface satisfies the identity $\xi = e$, and its associated shape operator matrix is given by $\mathbf{S} = -\mathcal{I}_3$, where $\mathcal{I}_3$ denotes the 3×3 identity matrix. The corresponding curvature values are found to be $\mathfrak{C}_1 = -1$, $\mathfrak{C}_2 = 1$, and $\mathfrak{C}_3 = -1$. As a result, the hypersurface ξ satisfies the following relation:

$$\Delta^{\mathrm{IV}}\xi = -3\xi.$$

Moreover, the curvatures satisfy the identities $\mathfrak{C}_1\mathfrak{C}_2 = \mathfrak{C}_3$, $\mathfrak{C}_1^2 = \mathfrak{C}_2$, $\mathfrak{C}_1^3 = \mathfrak{C}_3$. Therefore, the hypersurface given in equation (24) is a hypersphere having an umbilical property.

Bibliography

[1] Alias, L. J., & Gürbüz, N. (2006). An extension of Takahashi theorem for the linearized operators of the highest order mean curvatures. *Geometriae Dedicata, 121*, 113–127.

[2] Aminov, Y. (2001). *The Geometry of Submanifolds*. Amsterdam: Gordon and Breach Science Publishers.

[3] Arslan, K., Bayram, B. K., Bulca, B., Kim, Y. H., Murathan, C., & Öztürk, G. (2011). Vranceanu surface in $\mathbb{E}^4$ with pointwise 1-type Gauss map. *Indian Journal of Pure and Applied Mathematics, 42*(1), 41–51.

[4] Arslan, K., Bayram, B. K., Bulca, B., & Öztürk, G. (2012). Generalized rotation surfaces in $\mathbb{E}^4$. *Results in Mathematics, 61*(3), 315–327.

[5] Arslan, K., Bulca, B., Kılıç, B., Kim, Y. H., Murathan, C., & Öztürk, G. (2011). Tensor product surfaces with pointwise 1-type Gauss map. *Bulletin of the Korean Mathematical Society, 48*(3), 601–609.

[6] Arslan, K., Sütveren, A., & Bulca, B. (2021). Rotational λ-hypersurfaces in Euclidean spaces. *Creative Mathematics and Informatics, 30*(1), 29–40.

[7] Arvanitoyeorgos, A., Kaimakamis, G., & Magid, M. (2009). Lorentz hypersurfaces in $\mathbb{E}_1^4$ satisfying $\Delta H = aH$. *Illinois Journal of Mathematics, 53*(2), 581–590.

[8] Barros, M., & Chen, B. Y. (1987). Stationary 2-type surfaces in a hypersphere. *Journal of the Mathematical Society of Japan*, *39*(4), 627–648.

[9] Barros, M., & Garay, O. J. (1987). 2-type surfaces in S^3. *Geometriae Dedicata*, *24*(3), 29–336.

[10] Bektaş, B., Canfes, E. Ö., & Dursun, U. (2017). Classification of surfaces in a pseudo-sphere with 2-type pseudo-spherical Gauss map. *Mathematische Nachrichten*, *290*(16), 2512–2523.

[11] Bektaş, B., Canfes, E. Ö., & Dursun, U. (2017). Pseudo-spherical submanifolds with 1-type pseudospherical Gauss map. *Results in Mathematics*, *71*(3), 867–887.

[12] Chen, B. Y. (1983). On submanifolds of finite type. *Dōngwú Shùlĭxué Bào*, *9*, 65–81.

[13] Chen, B. Y. (1984). *Total Mean Curvature and Submanifolds of Finite Type*. Singapore: World Scientific.

[14] Chen, B. Y. (1985). Finite type submanifolds and generalizations. University of Rome.

[15] Chen, B. Y. (1985). Finite type submanifolds in pseudo-Euclidean spaces and applications. *Kodai Mathematical Journal*, *8*(3), 358–374.

[16] Chen, B. Y., Güler, E., Yaylı, Y., & Hacısalihoglu, H. H. (2023). Differential geometry of 1-type submanifolds and submanifolds with 1-type Gauss map. *International Electronic Journal of Geometry*, *16*(1), 4–47.

[17] Chen, B. Y., & Piccinni, P. (1987). Submanifolds with finite type Gauss map. *Bulletin of the Australian Mathematical Society*, *35*, 161–186.

[18] Cheng, Q. M., & Wan, Q. R. (1994). Complete hypersurfaces of $\mathbb{R}^4$ with constant mean curvature. *Monatshefte für Mathematik*, *118*, 171–204.

[19] Choi, M., & Kim, Y. H. (2001). Characterization of the helicoid as ruled surfaces with pointwise 1-type Gauss map. *Bulletin of the Korean Mathematical Society*, *38*, 753–761.

[20] Dillen, F., Pas, J., & Verstraelen, L. (1990). On surfaces of finite type in Euclidean 3-space. *Kodai Mathematical Journal*, *13*, 10–21.

[21] Do Carmo, M. P., & Dajczer, M. (1983). Rotation hypersurfaces in spaces of constant curvature. *Transactions of the American Mathematical Society*, *277*, 685–709.

[22] Dursun, U. (2007). Hypersurfaces with pointwise 1-type Gauss map. *Taiwanese Journal of Mathematics*, *11*(5), 1407–1416.

[23] Ferrandez, A., Garay, O. J., & Lucas, P. (1990). On a certain class of conformally at Euclidean hypersurfaces. In *Global Analysis and Global Differential Geometry* (pp. 48–54). Berlin: Springer.

[24] Ganchev, G., & Milousheva, V. (2014). General rotational surfaces in the 4-dimensional Minkowski space. *Turkish Journal of Mathematics*, *38*, 883–895.

[25] Garay, O. J. (1988). On a certain class of finite type surfaces of revolution. *Kodai Mathematical Journal*, *11*, 25–31.

[26] Garay, O. J. (1990). An extension of Takahashi's theorem. *Geometriae Dedicata*, *34*, 105–112.

[27] Güler, E. (2020). Fundamental form IV and curvature formulas of the hypersphere. *Malaya Journal of Matematik*, *8*(4), 2008–2011.

[28] Güler, E. (2020). Rotational hypersurfaces satisfying $\Delta^I R = AR$ in the four-dimensional Euclidean space. *Journal of Polytechnic*, *24*(2), 517–520.

[29] Güler, E. (2024). Fourth fundamental form and i-th curvature formulas in $\mathbb{E}^4$. *São Paulo Journal of Mathematical Sciences*, *18*, 1779–1792.

[30] Güler, E., Hacısalihoglu, H. H., & Kim, Y. H. (2018). The Gauss map and the third Laplace–Beltrami operator of the rotational hypersurface in 4-space. *Symmetry*, *10*(9), 1–12.

[31] Güler, E., Magid, M., & Yaylı, Y. (2016). Laplace–Beltrami operator of a helicoidal hypersurface in four-space. *Journal of Geometry and Symmetry in Physics*, *41*, 77–95.

[32] Güler, E., & Turgay, N. C. (2019). Cheng–Yau operator and Gauss map of rotational hypersurfaces in 4-space. *Mediterranean Journal of Mathematics*, *16*(3), 1–16.

[33] Hasanis, Th., & Vlachos, Th. (1995). Hypersurfaces in $\mathbb{E}^4$ with harmonic mean curvature vector field. *Mathematische Nachrichten*, *172*, 145–169.

[34] Kahraman Aksoyak, F., & Yaylı, Y. (2016). Flat rotational surfaces with pointwise 1-type Gauss map in $\mathbb{E}^4$. *Honam Mathematical Journal*, *38*(2), 305–316.

[35] Kahraman Aksoyak, F., & Yaylı, Y. (2015). General rotational surfaces with pointwise 1-type Gauss map in pseudo-Euclidean space $\mathbb{E}_2^4$. *Indian Journal of Pure and Applied Mathematics*, *46*(1), 107–118.

[36] Kim, D. S., Kim, J. R., & Kim, Y. H. (2016). Cheng–Yau operator and Gauss map of surfaces of revolution. *Bulletin of the Malaysian Mathematical Sciences Society*, *39*(4), 1319–1327.

[37] Kim, Y. H., & Turgay, N. C. (2013). Surfaces in $\mathbb{E}^4$ with L_1-pointwise 1-type Gauss map. *Bulletin of the Korean Mathematical Society*, *50*(3), 935–949.

[38] Kühnel, W. (2015). *Differential Geometry: Curves–Surfaces–Manifolds* (3rd ed.). Providence, RI: AMS. Translated from German 2013.

[39] Li, Y., Güler, E., & Toda, M. (2024). Family of right conoid hypersurfaces with light-like axis in Minkowski four-space. *AIMS Mathematics*, *9*(7), 18732–18745.

[40] Moore, C. L. E. (1919). Surfaces of rotation in a space of four dimensions. *Annals of Mathematics*, *21*, 81–93.

[41] Moore, C. L. E. (1920). Rotation surfaces of constant curvature in space of four dimensions. *Bulletin of the American Mathematical Society*, *26*, 454–460.

[42] Özkaldı, S., & Yaylı, Y. (2010). Tensor product surfaces in $\mathbb{R}^4$ and Lie groups. *Bulletin of the Malaysian Mathematical Sciences Society*, *33*(1), 69–77.

[43] Senoussi, B., & Bekkar, M. (2015). Helicoidal surfaces with $\Delta^J r = Ar$ in 3-dimensional Euclidean space. *Studia Universitatis Babeş-Bolyai Mathematica*, *60*(3), 437–448.

[44] Stamatakis, S., & Zoubi, H. (2010). Surfaces of revolution satisfying $\Delta^{III} x = Ax$. *Journal for Geometry and Graphics*, *14*(2), 181–186.

[45] Takahashi, T. (1966). Minimal immersions of Riemannian manifolds. *Journal of the Mathematical Society of Japan*, *18*, 380–385.

[46] Yoon, D. W. (2003). Some properties of the Clifford torus as rotation surfaces. *Indian Journal of Pure and Applied Mathematics*, *34*(6), 907–915.